# NATURE IN DANGER

# NATURE IN DANGER

## Threatened Habitats
## and Species

by

Noel Simon

in association with

The World Conservation Monitoring
Centre, Cambridge, England

New York
Oxford University Press
1995

*To Claire, Vivienne and Nicholas*

Published in Great Britain by
Guinness Publishing Ltd.,
33 London Road, Enfield, Middlesex EN2 6DJ

''Guinness'' is a registered trademark of
Guinness Publishing Ltd.

Published in the United States of America 1995 by
Oxford University Press, Inc.,
200 Madison Avenue, New York,
New York 10016 USA

"Oxford" is a registered trademark of
Oxford University Press

Library of Congress Cataloging-in-Publication Data

Simon, Noel.
    (Guinness guide to nature in danger)
    Nature in danger: threatened habitats and species/Noel Simon in association with the
    World Conservation Monitoring Centre.
        p.   cm.
    Originally published: The Guinness guide to nature in danger.
    Enfield, Middlesex: Guinness Pub., 1993.
    Includes bibliographical references and index.
    ISBN 0-19-521152-9
    1. Habitat (Ecology)   2. Endangered species.   3. Nature conservation.
    I. World Conservation Monitoring Centre.   II. Title.
OH541.S5345   1995
333.95137—dc20                                      94-43391
                                                      CIP

ISBN 0-19-521152-9

1   3   5   7   9   8   6   4   2

Designed by Amanda Sedge
Maps by Lovell Johns Ltd.
Picture Research by Image Select
Typeset by Ace Filmsetting Ltd., Frome, Somerset
Printed and bound in Italy by New Interlitho Italia SpA, Milan

Illustration on page ii: Yosemite Falls, Yosemite National Park, California (*Explorer*)
Illustration on page vi: Andes, Bolivia (*WWF/Hartmut Jungias*)
Illustration on page x: Moorland near Waldheim, Tasmania (*Auscape International/Dennis Harding*)

# PREFACE

A wide-ranging book of this kind must inevitably be rooted in the work of others. I have drawn heavily on published material, particularly on the work of Dr Edward Maltby and Patrick Dugan for the Wetlands chapter, but as the nature of the book makes it impossible to credit specific sources within the text or in footnotes in the conventional manner, I gratefully acknowledge my indebtedness to the authors and publishers cited in the Bibliography.

This book is not intended to be comprehensive: its purpose is to draw attention to some of the world's principal natural regions and the wildlife characteristically associated with them, and to give examples showing why they are important in the global context, the problems affecting them, and the measures that have been taken or still need to be taken to conserve them.

Scientific names have wherever possible been relegated to a glossary, except where their appearance in the text is necessary to avoid confusion, an arrangement that applies chiefly to the plants, many of which have no common name.

Particular thanks are due to the authors who have contributed chapters to this book within their specialized fields:

**Grasslands:** Dr Hugh Lamprey, who worked in eastern Africa as a biologist from 1953 to 1990 conducting and directing research on wildlife and habitat ecology, particularly in savanna and arid regions. He participated in the establishment and management of national parks and wildlife reserves and was the first Director of the Serengeti Research Institute, Tanzania.

**Forests:** Dr Mark Collins, Programme Director at the World Conservation Monitoring Centre, established and developed the Centre's Habitats and Geographic Information System Data Unit. A specialist on tropical insects and decomposition ecology, who worked for a decade in South East Asia and Africa, he edited *The Last Rain Forests*, is co-editor of the *Conservation Atlas of Tropical Forests* and co-author of *The IUCN Invertebrate Red Data Book* and *Threatened Swallowtail Butterflies of the World*.

Dr Caroline Harcourt, a specialist in the ecology and behaviour of primates, author of the IUCN Red Data Book on the *Lemurs of Madagascar and the Comoros*, and co-editor of *The Conservation Atlas of Tropical Forests*.

**Endangered Birds:** Dr Michael Rands, Programme Director at the International Council for Bird Preservation, and his wife, Dr Gillian Rands, Assistant Editor of *Trends in Ecology and Evolution*.

**Endangered Plants:** Hugh Synge, co-author of the *IUCN Plant Red Data Book* and *Plants in Danger: What do we know?* Between 1973 and 1987 he worked for IUCN and the World Wide Fund for Nature (WWF), where he created their Plant Conservation Programmes.

I am above all profoundly indebted to Dr Robin Pellew, Director of the World Conservation Monitoring Centre (WCMC), Cambridge, who has consistently encouraged and supported the project since its inception. Under his direction, WCMC has become a major force in international conservation.

Others on the staff of WCMC who have provided valuable data and help in other ways include Dr Timothy Johnson, Head of Species Unit; Dr Chris Magin, Animals Conservation Monitoring Group; Graham Drucker, responsible for the collection, management and dissemination of information on protected areas, in particular the Palaearctic Region; Colin Watkins, Development Coordinator; and Mary Cordiner, Librarian and Information Scientist.

My thanks are also due to Dr Jim Thorsell, Senior Adviser, Natural Heritage, IUCN; to Dr Lawrence S. Hamilton, Research Associate, East-West Center, Environment and Policy Institute, Honolulu, Hawaii; to Dr George B. Schaller, Director Wildlife Conservation International for generously providing hitherto unpublished material on Tibet; to Peter Stone of Mountains Agenda, who also allowed me access to unpublished data; and to Donald M. Gordon for contributing information on Ghana to the Forests chapter.

Throughout the whole of the book's protracted gestation – and delayed parturition – my daughter, Vivienne, has given unstintingly of her time and energy in initiating me into the intricacies of operating a word processor and in setting much of the text on disk. And I have been constantly sustained by my wife, Vanessa, whose perceptive criticism and astute editorial comment have immeasurably improved the text.

Noel Simon

# CONTENTS

# INTRODUCTION

*Man is a part of, not apart from, nature.*
F. Fraser Darling

*H*omo sapiens (for so we describe ourselves) is an exploiter; indeed, apart from the element of chance which must have played a part in the evolutionary development of man as it did in other forms of life, one of the reasons for our success as a species is our ability to turn opportunity to advantage. However necessary, even commendable, that characteristic may have been in the past, today's altered circumstances require a corresponding change of attitude - and in our own self-interest. The world's natural resources are being ground between the upper millstone of the Western market economy, with its philosophy of exploiting for growth, and the lower millstone of the Third World's need to exploit simply to survive. While effectively stripping the earth, many millions of people take for granted that it will continue to provide the natural resources that are basic to life - food to eat, water to drink, fuel for warmth, air to breathe. But our demands can be met only as long as they do not exceed our planet's potential for renewal.

The challenge facing conservationists is to bridge the chasm between these very different worlds by devising policies that are acceptable on the one hand to the affluent West, imbued with the profit motive, and, on the other, to meet the legitimate requirements and reasonable aspirations of people living on the fringes of existence.

This is as much a social as an ecological problem. It is easy to be high-minded on a full stomach, easy for those in the West accustomed to relative affluence to pontificate. But for Third World peasants, hard-pressed to scratch a living from the soil, it is an altogether different matter. For them it is enough simply to survive the immediate present: tomorrow is too remote.

What ecology has to teach us is the interdependence of all forms of life. We make a great mistake in singling out one species - inevitably our own - and holding it sacrosanct while dismissing others as of no account. The philosophy that countenances the extermination of all life but our own is not only morally but ecologically misguided since it flouts the immutable laws of nature that govern all life on earth, including our own.

Sapient man we may be - too clever, perhaps, for our own good - but wise we are not. Maybe we lack the humility to heed what is happening all around us. Motivated by self-interest, we seem incapable of appreciating that we are fouling our own nest to an extent that is rapidly making it uninhabitable.

It is not only change but the accelerating speed at which it is taking place that gives cause for concern. The human lifespan does not register so much as a blip on the geological timescale. Yet the changes that have taken place within the last three score years and ten are phenomenal. Little more than a century has elapsed since the discovery of the source of the Nile. In that short time the population of Africa has doubled and redoubled its numbers every 20 years or so. All suitable land there - and a good deal that is unsuitable - has been brought under cultivation. Much of the forest has been felled,

and the deserts continue their inexorable advance. Starvation stalks the continent; but neither famine nor war can curb human fecundity. Emigration has long been the accepted way of absorbing excessive human populations, with large numbers of people moving from overcrowded Europe into the sparsely inhabited Americas, Australia and Africa. But those outlets are no longer open: it is all these countries can do to accommodate and support their own expanding populations. Hence the phenomena of the boat people of Vietnam, the 'wetbacks' of Mexico, and the hordes of destitute refugees clamouring to enter Europe.

The message of this book is that humanity's own wellbeing, and ultimately survival, is irretrievably bound up with ecological imperatives. All forms of life have a part to play in maintaining the planet in a habitable state. We may not always be certain of the precise role of each, or the exact interdependence of one upon another. But the fact that we do not know is clearly no reason for compounding ignorance with folly by assuming that a particular organism is of no value and its extinction a matter of no consequence.

We hear much about human rights, but little about human responsibilities. There can be few higher responsibilities than to our own children; yet looking back over the last half century, it is difficult to avoid wondering what kind of world our children will inherit. Must progress be equated with squalor and degradation? Can advancement be achieved only at the expense of congenial surroundings, beautiful landscapes? Will wildlife be limited to relict species confined to a few degraded national parks and zoos or phials stored in the vaults of sperm banks?

When we do try to act responsibly our attempts, ironically, often have unforeseen consequences. Time and again, solutions which resolve one particular problem give rise to another - as, for example, the dedicated work of the veterinary services in eradicating many of Africa's cattle diseases. Freed from the scourge of disease, livestock numbers increase, overstocking impoverishes the land, and starvation supplants disease. These are among the factors that have brought Africa to a state of acute crisis. Ecological destabilization on a massive scale is the legacy not only of years of conflict but of long-standing misuse of land. Somalia epitomizes the social and ecological breakdown that is fast spreading to other parts of the continent.

How is it possible to talk of enhancing the quality of life while at the same time condoning the massive destruction of the very resources - land, water, forest, grassland - that govern the quality of life, and that can be maintained only as long as we remain within the limitations imposed by our ecological bounds?

One of the most encouraging developments of the postwar period has been the remarkable way in which the notion of conservation has gripped the imagination of the developed world. Fifty years ago the subject lacked political respectability. Public awareness today has become a potent force and in it lies the greatest hope for our planet's survival.

# RAINFORESTS

*Framed in the prodigality of nature*
Shakespeare, *King Richard III*

Forests are a prime example of a sustainable natural resource: their produce can be used, allowed to regenerate, and used again and again, a process that can be repeated indefinitely as long as the basic stock remains intact. Either consciously or fortuitously, that has been the principle generally followed by people living at subsistence level. Our own civilization has been characterized by increasing profligacy, with a great deal of effort and technological brilliance expended on exploiting natural resources but relatively little given to their perpetuation. While human needs continually expand in line with heightened aspirations and increasing numbers of people, the attitude of our exploitative society to natural resources remains largely motivated by self-interest.

The industrial base from which the West's prosperity grew was founded on exploitation of its forests. The Wealden forests of Kent and Sussex, for example, were exhausted by burning charcoal for smelting the ore to which Saxon and medieval ironmasters owed their prosperity. It is scarcely surprising that the Third World similarly looks to its forests as a ready source of hard currency. Faced with the need to boost their economies – and actively encouraged by foreign entrepreneurs only too anxious to help, but who do not have to live with the after effects – many Third World countries take to exploiting their most readily available natural wealth, the forest. The pattern of deforestation that has long been followed in Europe and North America is now being adopted in the Third World – but with one important difference: modern technology enables the destruction of forests to take place on an unprecedented scale and at a pace greater than ever before.

Extraction of timber is not in itself wrong – sustainable exploitation in accordance with the precepts of conservation is perfectly legitimate, indeed commendable. But non-sustainable exploitation that ravages the forest is calamitous, for forests not only provide such irreplaceable products as timber and fuelwood, they also have other less tangible but even more important attributes and functions. Their root systems keep the soil in place, preventing it from being washed away, and thereby helping to retain fertility; they purify water and regulate its flow by releasing a steady supply; and they influence global climate and weather patterns which have, of course, a bearing on agricultural production. Through evaporation and transpiration, forests enable three-quarters of the world's rainfall to return to the atmosphere. Water rising as vapour falls again as rain in due course. In the interim it remains suspended over the forest in the form of mist or cloud which, by deflecting some of the sun's potentially damaging heat, has a moderating effect that is vital to the wellbeing of the tropical forest. Cloud and forest are interdependent; indeed, their relationship is almost symbiotic, the forest generating the cloud and the cloud protecting the forest. Trees also have a crucial role in the oxygen cycle: by absorbing carbon dioxide and releasing oxygen, they constitute an integral part of one of the world's life support systems. Forests, in short, are of critical importance to the earth. Viewed against such a background, the status and condition of the world's rainforests can be seen as a matter of such global importance that their decline must clearly be arrested in the interests of the world as a whole. This does not mean that all exploitation should cease – far from it – but it does mean that exploitation should be limited to the amount that can be sustained – drawing on the interest but leaving the capital intact – for only in that way can this pivotal natural resource be perpetuated.

## The Extent of the Forest

Over the last half century, numerous attempts have been made to establish the size of the world's forests. Many of these estimates were based on conven-

● The log yard of a forest concession in Sarawak. The Malaysian state of Sarawak increased its timber production threefold between 1977 and 1990, and if deforestation continues at the same rate, virtually all of its primary forest will have gone by the end of the century. (WWF / Chris Elliott)

tional ground-based sources. Twenty years ago, exponents of satellites and remote-sensing technology promised to tell us everything we needed to know, and more, but today there is still no global map of forest resources, and no authoritative source of statistics. Hopes are being pinned on the new spectrometers in the Earth Observing System (EOS) satellite.

The most recent estimates give the total as 35.1 million km$^2$/13.5 million sq mls, of which the tropical forests cover roughly 17.1 million km$^2$/6.6 million sq mls, temperate forests 8 million km$^2$/3 million sq mls, and boreal (northern) forests 10 million km$^2$/3.8 million sq mls. The extent of the forest has fluctuated enormously in the geologically recent past. The last Ice Age, at its climax a mere 18,000 years ago, wiped out huge tracts of temperate forest, reducing the global total to less than remains today. Tropical areas became much drier and the forests more fragmented and deciduous. But, as the ice retreated, the process went into reverse and, by about 10,000 years ago, shortly before forest began to be cleared for agriculture, the world's forest cover expanded to about 52% of the earth's total land area. In the present century, and more particularly since the end of the Second World War, the rate of deforestation has accelerated. While the overall extent of temperate and boreal forests has stabilized, even extended slightly, tropical forests have diminished. In the single decade 1980–1990, the overall size of tropical forests and woodlands has halved.

During this period, tropical forests were disappearing at an annual rate of 169,000 km$^2$/65,250 sq mls – almost half as much again as in the previous decade, and representing an annual reduction of almost 1% – equivalent to losing an area the size of England and Wales every year or, to use another yardstick, 32 football pitches every minute!

## Agriculture
The human population in the northern hemisphere has now more or less stabilized, but numbers in the tropics continue to rise exponentially. Agricultural expansion into forest land is the primary cause of tropical deforestation. This is mostly unplanned, being carried out by subsistence farmers intent on scratching a living from the soil. Shifting agricul-

ture, which is heavily dependent upon nutrients obtained from burning the forest, accounted in 1980 for more than one-third of the deforestation in Latin America, for over two-thirds in Africa – West Africa and Madagascar being particularly badly affected – and for half the deforestation in South and Southeast Asia.

A 1992 study sponsored by UNEP, *Global Assessment of Soil Degradation*, reveals that 19.6 million km$^2$/7.5 million sq mls of land, an area twice the size of the US, has suffered soil degradation since 1945, of which 15% cannot recover without substantial international investment, and 5% is beyond hope of restoration.

The slash-and-burn method of subsistence farming is sustainable when populations are low, but degradation becomes permanent when pressure on the land does not permit it to lie fallow for sufficiently long periods. Such problems have to be seen against a Third World background of distorted distribution of wealth and inequable systems of land tenure. They need to be addressed through major land reforms, but few governments are willing to take such measures, not least because the landowners themselves often wield enormous political influence. Some governments are attempting to overcome the difficulty by resettling people in under-populated areas. Indonesia's transmigration programme, engaged in uprooting large numbers of people and resettling them in the outer islands, and Brazil's Rondônia scheme have become notorious for causing permanent deforestation and degradation of soils that are too infertile for permanent agriculture.

## Grazing and Ranching
Much damage is caused by grazing domestic livestock in tropical woodlands and forests. Grazing, browsing and trampling inhibit regeneration. The most extreme example is in India, where the spread of permanent agriculture has forced pastoralists into the forest margins. Of India's 400 million cattle, 90 million are now believed to live on forest land – three times as many as the government itself estimates the land can carry. Livestock ranching has contributed substantially to the deforestation of Central America also. In Brazil, tax incentives were at one time used to encourage landowners to run their cattle on forested land, a practice that resulted in such

widespread degradation that it has now been stopped. But burning remains a glaring issue. In recent years there has been a growing number of instances of fires started by agriculturalists getting out of hand and destroying huge tracts of natural forest. The worst example was in Kalimantan and Sabah (Borneo) in 1982–83, when more than 40,000 km$^2$/15,500 sq mls of forest was lost – an area larger than the Netherlands.

## Fuelwood
Many millions of people living in the tropics depend on the forests to provide their basic needs – shelter, fuel, food and water. In the southern hemisphere wood – either collected and used directly or converted into charcoal – is the common source of fuel for cooking and heating: one-third of the world's population depends on it. In 1980, fuelwood provided more than half (58%) the energy consumed in Africa, 17% in Asia and 8% in Latin America. Charcoal burned in the forests of northern Thailand, for example, finds a ready market in the deforested regions of Bangladesh. Charcoal is also used for industrial purposes – steel production in Brazil, for example – on a scale that can be a major cause of deforestation.

But the sheer weight of human numbers and consequent pressure on the land show that this cannot continue. In many parts of the world demand for fuelwood already exceeds supply. In 1980, 1.3 billion people were believed to be having difficulty in obtaining fuelwood; FAO estimates the figure will have risen to 2.7 billion by the year 2000. The impact of this increased demand can best be seen in tropical woodlands, for example in the savannalands of West and East Africa, where huge areas of woodland are suffering visible degradation. Whilst difficult to evaluate accurately, it has been estimated that 20,000–25,000 km$^2$/7,700–9,650 sq mls of forest and woodland are cleared annually for fuelwood, and the figure is rising: between 1977 and 1988, consumption of fuelwood in the Third World rose by almost 30%.

## Forestry and the Timber Trade
Both the industrial nations and the Third World produce rather similar quantities of useful wood. But the relative proportions of fuelwood and industrial timber

are quite different: 79% of Third World forestry production consists of fuelwood and charcoal, compared with only 19% in the Western World.

These figures serve to emphasize two important points: firstly, that in the tropics more wood is used for fuel than is extracted by the timber trade and, secondly, that only about one-quarter of the world's industrial timber comes from the tropics. Why then, one may ask, is so much fuss made about the effects of the trade in tropical timber?

The reason is that timber extraction in the tropics takes place mainly in primary forests that have not previously been logged or colonized and are thus valuable reservoirs of biological diversity. The effect of opening these lands to logging is to make them accessible to hunters and agriculturalists, who move in along the logging roads. Unless either the logging company or the government controls these incursions (and they seldom do), the cleared forest comes under permanent cultivation. Most production forests in the industrialized nations are managed on a sustained basis with extraction followed by reafforestation. But less than 1% of tropical forests are managed on sustainable lines. Timber production in tropical countries generally follows a boom and bust pattern. Tropical nations that have boomed and are now bust (as far as timber goes) include the Philippines and Côte d'Ivoire; while Malaysia and Indonesia are eating into their timber reserves very quickly. A decline in their productivity is inevitable.

Sarawak, on the island of Borneo, typifies the dilemma. Only 70% of its original forest cover remains. The peat swamp forests were the first to be logged on a commercial scale and were for many years Sarawak's main source of timber. By 1978, lowland evergreen forest rich in valuable dipterocarps – the natural vegetation of the interior – had become the mainstay of the country's timber industry.

Malaysia is the world's largest supplier of tropical logs, but as the export of logs from Peninsular Malaysia is currently banned, the States of Sabah and Sarawak are supplying the huge quantities of wood required for export. Between 1977 and 1990, timber production increased threefold. Most of it is exported to Japan, with Taiwan and Korea also taking large amounts. The timber industry is of immense economic importance to Sarawak. As well as bringing in about half the State's revenue, it employs about one-tenth of Sarawak's entire workforce.

There is, however, concern that the current level of forest exploitation in the State is unsustainable in the long term and that it will cause widespread degradation of the forest. It is also thought likely to create serious difficulties for the rural population, both those settled in long houses and those living as nomadic hunter-gatherers.

A 1990 report on the management of Sarawak's forests found that if harvesting of the hill forests continues at the 1990 rate of around 13 million m³/17 million cu yds/year, all the country's primary forests would be exhausted within about 11 years. Only immature logged forests would remain. The report concluded that, with present practices, which are damaging both to the environment and to residual stands of trees, timber production could never be sustained. Even so, the level of forest management in Sarawak is better than in most other tropical timber producing countries.

## Atmospheric Pollution

The industrialized world's use of forest resources has slowed as timber for fuel and building has been replaced by coal and steel. Ironically, these industries in their turn have produced widespread atmospheric pollution, killing forests through acid rain, often falling hundreds of kilometres away from source.

Acid rain, caused by solutions of sulphur dioxide, nitrogen oxides and ammonia in rainwater, is now recognized as the main cause of forest die-back in western and eastern Europe, North America, Australia, Japan, China, Mexico and other industrial countries. Toxic gases, released into the atmosphere as by-products of heavy industry, have seriously affected European forests since the mid-1970s.

Statistics published in 1990 by the International Institute for Applied Systems Analysis (IIASA) show that damaging levels of sulphur are falling on 75% of Europe's forests, while 60% are affected by excessive nitrogen deposition, and conclude that timber losses will amount to a staggering 118 million m³/154 million cu yds every year for the next century, at an annual cost of US$30 million! An international treaty on air pollutants has now been adopted in Europe, but is probably too late to save vast tracts of forest in eastern Germany and Czechoslovakia.

## Carbon Balance and Global Warming

Because of the massive quantity of fixed carbon contained within them, the world's forests are inextricably bound up with the issue of global warming. Deforestation, particularly when caused by burning, releases carbon in the form of carbon dioxide, the main 'greenhouse' gas. Although the precise amount is uncertain, deforestation is believed to account for about 20% of the total increase in carbon dioxide arising from human activities. Deforestation can thus be seen as a significant factor in global warming, albeit much less important than emissions from the burning of fossil fuels.

The Intergovernmental Panel on Climate Change (IPPC) judges that the world has warmed by 0.3–0.6 °C over the past century. This may not sound much, but is evidently enough to cause glaciers to retreat. Over the same period sea level has risen by 10–20 cm/4–8 in: coral reefs and mangroves (in Bermuda, for example) have already been affected.

The threat to ecosystems lies less in climate change itself than in the speed of that change. Atmospheric pollution is believed to be causing warming to occur ten times more quickly than in the past 10,000 years, a rate many times faster than that at which ecosystems and species can adapt.

One way of correcting carbon imbalance in the atmosphere is to fix carbon dioxide by growing trees. Young, growing forests absorb carbon dioxide, and the view is widely held that tree planting on a large scale would help combat global warming. An American power company is setting a practical example by planting 500 km²/200 sq mls of forest in Guatemala to offset carbon emissions from a new fossil-fuelled power station in Connecticut.

But tree planting alone will not resolve the issue. A reduction in the excessive use of fossil fuels is essential – particularly by northern hemisphere nations – if global warming is to be brought under control.

## Biodiversity Loss

As yet, scientists have no clear idea of how many species there are in the world. Estimates vary between 5 and 30 million,

● Tropical rainforest burning in Amazonia. Such fires not only destroy irreplaceable ecosystems, they also contribute significantly to global warming. (WWF / Mark Edwards)

with about 10 million the generally accepted figure. This includes not only mammals, birds, fish and other vertebrates but insects and other invertebrates and, of course, plants. The lack of a more precise figure is due to the inadequacy of data relating to tropical forests, where 50–90% of all species are believed to reside. The World Resources Institute estimates that at current rates of deforestation, 4–8% of rainforest species could be extinct by 2015, and 17–35% by 2040.

Using the global figure of 10 million species, this would mean the extinction of 20–75 species a day, the bulk of them unknown insects and other invertebrates. The precise consequences of such massive loss of potentially useful species are impossible to predict, but there can be little doubt that ecological patterns would be severely disrupted.

## International Initiatives

The publication in 1980 of the *World Conservation Strategy* focused attention on deforestation and was instrumental in helping politicians to realize that the massive increase in human numbers cannot be sustained unless environmental constraints are recognized and respected,

with the forests serving as a barometer for evaluating the health of life on earth.

There have been several calls to reduce the rate of deforestation, to manage the forests on a sustainable basis, to break the mould of Western-style development, and to redress the pattern of global economics by means of which money continues to flow from south to north, largely in debt repayments. In 1985, FAO and the World Bank, in association with the international development agencies, established the Tropical Forests Action Programme, designed to invest more resources and capital in developing sustainable forestry. Now, seven years on, some 70 tropical nations have requested a review of their forestry programmes and follow-up investment strategies. But despite several billion US dollars changing hands, there is really no evidence that a single tree has been saved. Some non-government agencies believe the money has made matters worse, through being misdirected at industrial rather than grass-roots development and forest conservation. The programme is currently being 're-vamped', but whether the pupa will turn into a butterfly or simply remain stillborn in its cocoon has yet to be seen.

To further this aim, the International Tropical Timber Organization (ITTO) was formed in 1983. The timber-producing and timber-consuming nations clearly share a common interest in conserving and managing timber stocks in ways that will enable the trade to be maintained in perpetuity. ITTO has produced a number of important guidelines and pledged to place the trade on a sustainable footing by the year 2000. But many observers are depressed by the slow progress and lack of commitment shown by both producers and consumers. There is certainly no indication that the forests are benefiting in any way.

Despite the high level of interest in tropical forests, international discussions and negotiations have not yet reached an understanding or concluded any agreement, let alone agreed the course of action needed to save the forests. Behind the façade of action plans and strategies the business of 'mining' the forests continues. Deforestation and land degradation seem unstoppable. To some extent the problem is one of perceived priorities. Southern nations look upon their forests as assets that have to be realized in order to finance development, while northern countries see those same forests as international resources, important primarily for protection against global warming and as reservoirs of biological diversity.

The 1992 Earth Summit in Rio de Janeiro did little to resolve these difficulties. Discussions on a proposed Forests Convention came nowhere near finding common ground before running out of time. Instead, a non-mandatory 'Statement of Principles on Forests' was concluded shortly before the Summit began. The intensity of negotiations underscored the width of the chasm separating conservation and sound land-use from economic development. Both the 'Statement of Principles' and the summit report 'Agenda 21' contain proposals for combating deforestation and placing forest management on a sustainable footing, but whether these weighty words will improve the state of the world's forests remains to be seen.

# TROPICAL RAINFOREST

LOCATION: Tropical forests are confined to a narrow belt encircling the equator between 4° N and 4° S where the correct ecological conditions are met.

AREA: Rainforest covers an area of about 9 million km² / 3.5 million sq mls, an area roughly the size of the USA. About three-fifths of the world's rainforest is in Central and South America, with the remainder divided between West Africa, Southeast Asia and the Pacific islands. Zaïre and Indonesia each possess about one-tenth of the world's total.

ALTITUDE: Rainforest is generally classified as either lowland or montane. Lowland rainforest is more extensive than montane and occurs up to a height of about 900 m / 3,000 ft (higher in western Amazonia). Above that level, montane rainforest reaches to the treeline.

CLIMATE: A basic requirement of tropical evergreen rainforest is a more or less constant temperature within the parameters of 18–30 °C / 64–86 °F and an evenly distributed rainfall of more than 2,000 mm / 80 in a year.

Characterized by evergreen trees – i.e., trees that do not shed their leaves seasonally – rainforest is the most prolific of plant communities, as can be seen from the profusion of lianas and creepers with which the trees are festooned. The roof of the forest, studded with clusters of epiphytes, is formed by the canopy, which can be 45 m / 150 ft or more above the ground, with emergents protruding still higher. The canopy has a greater profusion of animals than any other part of the forest – marsupials and monkeys, countless birds, frogs, snakes and, of course, prodigious

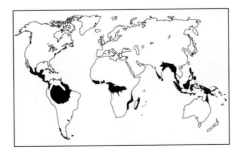

numbers of insects – a community of plants and animals living high above the ground. By enclosing the forest, the canopy prevents sunlight from reaching the forest floor; hence there is little undergrowth. Wherever a gap arises in the canopy, exposing a patch to sunlight, grasses and undergrowth appear. These clearings attract numbers of forest-dwelling animals – in Africa such species as buffalo, bushbuck, bongo and giant forest hog.

Above 900 m / 3,000 ft tropical rainforest rises in a series of altitudinal belts, each with its own specialized assemblage of plants and animals, culminating in bamboo and subalpine shrubs. With increasing height the trees become stunted, gnarled and twisted into weird shapes, with the upper branches – often the boles as well – draped with lichens. But while the forest trees become smaller with altitude, the plants above the treeline become, paradoxically, larger – giant heathers and rhododendrons grace the mountain tops of Asia, giant groundsels and lobelias the alpine moorlands of Africa.

Tropical rainforests are ecologically fragile. The soils on which they grow are impoverished, lacking in humus and deficient in nutrients, incessant rain having leached them out. It may seem strange that impoverished, nutrient-deficient and unstable soils should give rise to such lush vegetation. This is because rainforest is a self-sustaining entity. Unlike temperate forest, where nutrients are held in the soil, tropical rainforest is almost independent of the soil on which it stands, and is based on a system of rapid recycling in which the decomposers – bacteria fungi, termites, ants, and others – have an important function. Many trees have nodules on their roots, containing bacteria capable of converting atmospheric nitrogen into a form the tree can assimilate. Leaves and other organic litter falling to the ground attract the decomposers, which break it down, in the process releasing nutrients which are drawn into the tree roots. Root fungi form a mat of tangled rootlets on the surface of the soil which intercepts nutrients before they can be washed away. They send up vertical rootlets into the forest litter, attaching themselves to leaves and transmitting nutrients directly to the tree. The inter-dependence of the various forms of rainforest life is extremely complex; disruption of one affects the others, and the loss of several may unbalance the whole system.

Remove the trees, and all that remains is an impoverished soil generally unsuited to agriculture. Yet settlers have been encouraged by the government of Brazil to take up land along the trans-Amazonian highway. After a few seasons, productivity inevitably declines, leaving the smallholder with little alternative but to cut his losses and either clear a fresh patch of forest and start the process all over again or join the swelling multitudes in the shanty towns that cluster around the outskirts of the big cities.

## A PENDULOUS-NOSED MONKEY

The most conspicuous feature of the proboscis monkey is its bulbous nose (lacking in the female), up to 10 cm/4 in long. In adult males it droops down over the mouth to the extent that it has to be pushed aside when feeding, but blows up like a balloon when the animal honks its characteristic alarm call. This large and powerfully built monkey swings through the trees by its arms, and has the distinction of being a competent swimmer, frequently plunging into the water from a considerable height.

Endemic to the island of Borneo, it is confined to the coastal lowlands, where it lives in association with riverine rainforest, mangrove and peat swamp forest. Even within its restricted range this monkey's distribution is extremely patchy, and its numbers difficult to determine. There are believed to be about 6,400 in Sarawak and 300 in Sabah. No figures are available either for Kalimantan or Brunei.

The proboscis monkey's habit of sleeping in trees close to rivers makes it easy to hunt, but the main threat arises from destruction or depletion of its specialized habitat. Almost all the country's peat swamp forests have been licensed for timber extraction, and will have been logged by the end of the present decade.

● A proboscis monkey glimpsed through foliage in the Tanjung Puting National Park, central Kalimantan, Borneo. (WWF / Gerald Cubitt)

After logging, the commercially unimportant trees (which are nevertheless sources of food for wildlife) are sometimes poisoned – a practice that renders the forest uninhabitable for wildlife –

before being developed for agriculture. Mangroves have been similarly exploited, mainly for the Japanese pulp and chipboard markets. Virtually all the mangrove forests in Kalimantan and most of those in Sabah, Sarawak and Brunei have been assigned to Japanese timber concessionaires. The clear-felling of mangroves leaves so few mature trees that regeneration is almost nonexistent, and is as detrimental to the local people as it is to wildlife. If the needs of both people and wildlife are to be met, mangroves should be exploited only on the basis of sustainable yield. Mangroves are also being cleared for commercial aquacultural schemes – few of which are viable.

Few of the protected areas in which the proboscis monkey occurs are sufficiently large to sustain viable populations. If the species is to survive it is essential that existing protected areas be enlarged and others created – though as the coastal lowlands, river mouths and deltas forming its primary habitat also carry the country's densest human population, this will not be easy. A survey is needed to provide data on which precise recommendations can be made for setting up an effective system of sanctuaries.

# AUSTRALIA'S TROPICAL RAINFOREST

LOCATION: Extreme northeastern Queensland between 11° S and 22° S, reaching from Townsville in the south to the northern tip of Cape York Peninsula. The World Heritage Site is at 15°39′–19°17′ S, 144°58′–146°27′ E.

AREA: Queensland's tropical rainforest covers a total of 10,516 km²/4,060 sq mls. It consists of two principal blocks of land separated by a tract of open eucalypt forest. The larger, between Townsville and Cooktown, covers 7,900 km²/3,050 sq mls, and the smaller, lying farther north, 2,600km²/1,000 sq mls. The latter consists of a scattering of forest outliers – the largest lying between the McIlwraith and Iron Ranges – linked by fingers of forest reaching along streams, gullies and mountain tops.

ALTITUDE: From the coastal plain to 900 m/3,000 ft on the mountain ranges and tableland lying parallel with the coast, and reaching to the summit of Mount Bartle Frere (1,612 m/5,289 ft), the highest mountain in northern Australia.

CLIMATE: Mean daily temperatures range from a maximum of 31 °C /87.8 °F to a minimum of 23 °C /73.4 °F on the coast, with winter temperatures averaging about 5 °C lower. Above 700 m/2,300 ft temperatures drop low enough to bring sharp frosts. Mean average rainfall is from 4,000 mm/147 in at the coast to 1,200 mm/47 in along the western edge of the forest. In the ranges it can be substantially higher: e.g., 9,140 mm/360 in on the summit of Mount Bellenden Ker. Occasional periods of intense rainfall, sometimes lasting for a week at a stretch, are a feature of the region, as are cyclones, which cause extensive damage. Humidity averages 78% along the coast in summer, but frequently exceeds 90%. Rainforest occurs only on the eastern slopes. The western slopes lie in a rainshadow, causing the rainforest to give way to dry schlerophyll (eucalypt) forest.

In a country as dry as Australia the occurrence of a tract of tropical rain forest, however small, gives it a level of interest and importance out of all proportion to its size. Such an area occurs in the extreme northeast of Queensland, where a mosaic of rainforest interspersed with other types of vegetation – from mangrove swamps to schlerophyll (eucalypt) forest – make for a wealth of unusual plants and animals.

Queensland's rainforest escaped the widespread tree-felling that accompanied the expansion of the dairy industry in other parts of Australia, largely through the State of Queensland's foresight in setting aside most of the forest in the Townsville/Cooktown region as national parks and state forests. Deforestation took place mainly in the lowland areas – principally to clear land for growing such crops as sugarcane and maize – and at the higher elevations, especially on the Atherton Tableland, to establish pastureland for cattle.

In the unprotected areas, the prac-

tice of burning off the sugarcane fields inevitably resulted in fire spreading to the forest. Fires started by Aborigines sometimes resulted in the rainforest being replaced by open grassland bearing occasional eucalypts. This tall, open, savanna-like land is critically important to the conservation of three species of mammals that are restricted to this particular type of habitat: the brush-tailed bettong, yellow-billed glider and swamp rat. In some instances such open patches have in the fullness of time reverted to rainforest, the occasional relict eucalypt growing incongruously amidst rainforest trees providing evidence of the reversion.

Ancient affinities with the prehistoric continent of Gondwanaland are reflected in Australia's flora and fauna, links which are to some extent shared with New Caledonia, Indonesia and above all New Guinea. Queensland's position at the periphery of the great Malesian rainforest has invested it with a strong Malesian flavour. Long isolation of ancient floras

has resulted in an exceptionally high level of endemism. Despite being less rich in plant species – a 1988 estimate gives a total of 1,328 plant species in 534 genera, as compared with 8,000 in Peninsular Malaysia, and 25,000 in Southeast Asia as a whole – Queensland's forest nevertheless contains an unusually interesting flora, including more than 240 species of ferns and their allies, the richest concentration in Australia, one of the most striking being the endemic fan palm *Licuala ramsayi*, which favours poorly drained lowland soils.

Despite occupying only about 0.1% of Australia's overall surface area, Queensland's rainforests nevertheless contain 30% of Australia's total marsupial species (37 marsupials and two monotremes), 30% (47 species) of the frogs and 62% of the butterflies, as well as a high proportion of bats, birds, reptiles and other forms of life. The marsupials include both Australian tree kangaroos, the musky rat kangaroo and the spotted-tailed quoll, a carnivorous marsupial, whose main

● Buttress roots help to support some of the taller species of tree in the shallow soils of the tropical rainforests. (WWF / Fredy Mercay)

range lies over 2,000 km / 1,250 mls away. This species together with the ring-tailed possum and the cassowary is now locally extinct, the blame attributed to increasing fragmentation of the forest and to the spread of the introduced cane toad, native to South America. Carnivores such as the quoll appear to be vulnerable to the poison contained in this toad's skin.

Feral pigs have also contributed to the decline of the native fauna.

Northern Queensland has the most diverse selection of birds in Australia. More than 370 species have been recorded, including 13 endemics, most of which are confined to the uplands. Among them are such species as the Australian fernwren, tooth-billed catbird, bridled honeyeater and golden bowerbird.

The insect fauna, particularly in the upland rainforests, is of great interest; it includes a number of primitive relict species isolated from their nearest relatives by more than 1,600 km / 1,000 mls. The closest relatives of the large stag beetle *Sphaenognathus queenslandicus*, for example, which is found only on Mount Lewis and Mount Windsor Tableland, live in South America.

Endemic moths and butterflies are a notable feature of the Queensland forest fauna, among them the brilliantly coloured *Aenetus monabilis* with a wing span of up to 18 cm / 7 in, the even larger Hercules moth *Coscinocera hercules* – up to 25 cm / almost 10 in, and Cairns birdwing, Australia's largest butterfly.

Since the 1970s, public pressure has been steadily mounting for the remaining rainforest to be fully protected. In 1988, rainforest in the Townsville/ Cooktown region was designated a World Heritage Site, a decision that was initially opposed by the Queensland Government but later accepted. Covering 8,979 km² / 3,467 sq mls, the World Heritage Site incorporates 41 national parks, 43 state forests, 15 timber reserves, as well as the Yarrabah Aboriginal Reserve, which retains the only Aboriginal rainforest culture in Australia, with an oral tradition reaching back 10–15,000 years. But forest in the Cape York region is not included.

Queensland's rainforest adjoins the Great Barrier Reef, also a World Heritage Site. The association between the coastal rainforest and the fringing reef is of unusual interest and possibly unique.

Although still reliant mainly on primary production, tourism is playing an increasingly large part in the economy of the region. The blend of tropical rainforest, perfect sandy beaches and fringing reefs between Daintree River and Cedar Bay is probably unequalled, and combines to make an area of outstanding natural beauty. It has been estimated that 80% of the original rainforest cover remains. The conflicting demands made on the rainforest between timber production and biological conservation suggests that the biosphere reserve concept that allows forests and farmlands to be managed in harmony with protected areas would be the most appropriate system to follow.

● Tropical coastal rainforest at Cow Bay, Cape Tribulation National Park, Queensland.
(Auscape International / Ian D. Goodwin)

# COSTA RICA'S RAINFOREST

LOCATION: Most of Costa Rica's remaining rainforest is contained within the Amistad Biosphere Reserve, formed of a complex of reserves and protected areas with various categories of legal protection, including 2 national parks – Cordillera de Talamanca (1,905 km$^2$/735 sq mls) and Chirripo (501 km$^2$/193 sq mls) – 2 biological reserves, 7 anthropological reserves, a protection forest, and a forest reserve. The Amistad Biosphere Reserve covers the foothills and uplands of the Cordillera de Talamanca, the highest and least spoiled non-volcanic mountain range in Central America, and containing the largest tracts of virgin rainforest remaining in Costa Rica. 8°44'–10°02' N, 82°43'–83°44' W.

AREA: 5,845 km$^2$/2,257 sq mls.

ALTITUDE: 50–3,920 m/164–12,861 ft (Cerro Chirripo, the highest point in southern Central America).

CLIMATE: Temperatures vary from 25 °C/77 °F at sea level to below zero on the higher peaks. Mean annual precipitation ranges from about 2,000 mm/79 in near the Caribbean coast to more than 6,000 mm/236 in high in the mountains.

The convergence of the floras of North and South America, together with various climatic and edaphic (soil and topographic) factors, enable their protected area to sustain a diversity of plants that is possibly unequalled in any other reserve of comparable size in the world. Of Costa Rica's 12 life zones, eight are represented in the reserve. They range from lowland tropical wet rain forest to cloud forest, and include four communities – sub-alpine *paramo* forest, stands of pure oak, lakes of glacial origin, and high altitude bogs – found nowhere else in Central America. The stands of *paramo* on Mounts Chirripo and Kamuk are among the most variegated in the entire Talamanca Range, and the only examples in Costa Rica to show no sign of human degradation.

The fauna contains an admixture of both North and South American forms. A total of 215 mammal, 560 bird, 250 reptile and amphibian, and 115 fish species have been recorded in the reserve. All the Central American felines are represented, including ocelot, jaguarundi, little spotted cat and jaguar. Among other species are the Central American tapir, Central American squirrel monkey and Geoffroy's spider monkey. Of Costa Rica's 16 endangered mammal species, 13 are found in the reserve, as are nine of the 11

endangered bird species. Among the latter are the resplendent quetzal, harpy eagle, crested eagle and orange-breasted falcon. The high-altitude viper *Bothrops negrivisidis*, rarely seen or collected, is known to occur in the reserve.

The Indian tribes originally inhabiting the Talamanca Range were devastated by the Spanish conquest and the diseases introduced by the conquistadors. By 1960, only about 6,000 Indians remained in Costa Rica. Since then, their numbers have gradually increased to about 10,000, most of whom now live within the bounds of the Amistad Reserve, where they have managed to retain their own culture, language and life style. Their presence in or close to the reserve, and the traditional way of life they follow, is compatible with the purposes for which the reserve was created. But oil exploration, copper mining, highway construction, illegal land squatters, forest clearance, watershed degradation, poaching and the looting of archaeological sites are some of the threats with which the reserve authorities and their small staff have to contend.

At the time of its discovery by Europeans in the early 16th century, virtually the whole of Costa Rica was covered with dense forest. Clearance, at first slow – for only the finest timber was exploited – gradually increased in line with population growth and the establishment of permanent farms. Initially, agricultural development concentrated in the more fertile areas, where the climate was benign. Until 1900, less than 10% of the country was being used for agricultural purposes. But the pace of deforestation subsequently quickened, and in the last four decades more than half the forest has gone. If destruction continues at the current rate, virtually no forest will exist outside the national park system by the year 2000.

Deforestation is attributable to the combination of official expansionist policies, unusually liberal land tenure laws and high population growth. As in many other countries in the region, land ownership in Costa Rica is inequitable: about 36% of the land is in the form of large estates owned by 1% of the people. Costa Rica's basic problem is a rapidly expanding human population outstripping the capacity of the essentially agrarian economy to support it.

All tree-cutting is supposed to be authorized by the Forest Service but, in practice, the Forest Service concentrates on issuing permits for timber exploitation and ignores deforestation for agricultural or settlement purposes. In 1980, for instance, permission was given to cut 220 km²/85 sq mls of forest, but this was only about one-third of the area actually cut in Costa Rica that year. Much of the unauthorized cutting is attributable to squatters or colonists who fell and burn forests as a means of establishing claims to unoccupied land. A squatter can gain title to a piece of forest land by 'improving' it – that is, simply cutting down the trees – and occupying it for six months. After six months he cannot be ejected but must be bought out and paid for his 'improvements'.

Many squatters are bought out by cattle ranchers, who have converted large tracts of Costa Rica's tropical forest into pasture. The ranching of cattle on land of such low carrying capacity is economically feasible only because of the high price of beef. Even though only 8% of the terrain is suitable for cattle ranching, 35% is used for this purpose. Indeed, over 60% of Costa Rica's land is suitable only for forestry; if used more intensively the soil quickly deteriorates. Although the rate of deforestation has recently slowed, this is probably due more to the falling price of beef on the international market than to improved forestry practices.

# BRAZIL'S ATLANTIC FOREST

LOCATION: Extending in a narrow strip from the state of Rio Grande do Norte southwards to the Rio Grande do Sul.

AREA: In Columbus's day the Atlantic forest covered about 1 million km²/385,000 sq mls, since when it has been progressively reduced to today's total of 13,000 km²/5,000 sq mls – equivalent to little more than 1% of its original size.

ALTITUDE: From the coastal plain to a maximum of 2,797 m/9,177 ft at Agulhas Negras.

CLIMATE: Although receiving only a moderate rainfall – e.g. 1,120 mm/44 in at Rio de Janeiro, and seldom more than 1,400 mm/55 in anywhere in the region – this is compensated by frequent fogs, a high level of humidity, a brief dry season never lasting more than two months, and by rainfall being spread over the whole year. Rain occurs mainly in autumn and winter – unlike most tropical countries where the wet season is generally in the summer – an anomaly caused by a northerly movement of cold air from Antarctica meeting the Trade Winds.

The Atlantic Forest's geographical location on the east coast of Brazil ensured that it was the first place to be colonized. The prime object of colonization was timber. The exclusive right to exploit the *braza* tree *Caesalpinia echinata* was granted to Fernando de Noronha, who, in 1519 alone, shipped 5,000 trees to Portugal. It was the arrival of this consignment of timber that led to the country being called Brazil. Exploitation of the more valuable trees was followed by clearing the land, and in particular the fertile coastal plain, for agricultural production, much of it for sugarcane. The discovery of gold and diamonds towards the end of the 16th century led to a heavy influx of prospectors, and to further clearance of the forest on which to grow food for the rapidly increasing population.

Within a century the mines were exhausted. Many of the miners turned to farming, and more forest was cleared for coffee, banana and rubber plantations. During the Indian wars, it was common practice to set fire to the forest to drive out the Indians.

Since the end of the Second World War, the rate and extent of deforestation have accelerated, and what was once the Atlantic Forest has become a centre for agricultural and industrial development, as well as containing many big cities, among them Rio de Janeiro and São Paulo. With 43% of Brazil's 148 million people compressed into this region, the Atlantic Forest has been reduced to a few remnants isolated from each other, and almost all privately owned.

The few national parks and reserves that have been set aside cover less than 0.1% of the original forest and their conservation is hampered by shortage of funds, ineffective management and inadequate enforcement of such protective laws as exist. In view of the virtual impossibility of any more forest land becoming available for parks and reserves, the level

of management of what already exists is a matter of crucial importance.

Despite its reduction in size, the fragments of forest that almost miraculously still remain contain a host of interesting plants and animals. Some 2,124 species of butterflies - two-thirds of Brazil's entire butterfly fauna - 913 of them endemic, are to be found there. Of the 21 primate species (belonging to six genera), 17 are endemic, at least 13 of which are regarded as endangered and three as vulnerable.

Heavy demand for Brazilian rosewood *Dalbergia nigra* (for furniture making) has reduced this tree to the point of being officially listed as endangered. No attempt has been made to propagate it in plantations, although the possibility of developing methods of cultivation is being examined.

The woolly spider monkey or muriqui, South America's largest primate, is a species of the forest canopy, seldom descending to the ground - which helps to explain why so little is known about this animal. Intensive hunting for food and sport and for the pet trade and, above all, loss of habitat - primary forest being essential to its survival - have combined to reduce this primate from a total

population that once stood at 400,000 to a mere 400.

The golden lion tamarin, a species of the coastal lowlands, is disappearing equally fast and for broadly similar reasons. Only two groups totalling 250 individual animals are thought to have survived in the wild state.

Of all the Atlantic Forest animals none is more endangered than the maned sloth; very few remain. This highly specialized animal feeds on leaves, and deforestation has greatly reduced its source of food. Attempts at keeping it in captivity have invariably failed. A reafforestation programme designed to benefit the golden lion tamarin will also help the maned sloth.

With Brazil following the tradition adoped by most Latin countries of shooting even the smallest birds, the native birds - which include such species as the rusty-margined guan - have been hunted almost out of existence. Songbirds and birds of bright plumage are also menaced by the cage bird trade. Few Brazilian houses are without caged pets. Their popularity is vouched for by the prominence with which they are displayed in any street market. Many species are becoming rare as a result.

# DEFORESTATION IN BRAZILIAN AMAZONIA

LOCATION: The Amazon Basin in northern Brazil, extending from the Atlantic Ocean to the Andes and reaching along many of the Amazon's tributary rivers.

AREA: The world's largest tropical forest, covering 6 million km$^2$/2.3 million sq mls.

ALTITUDE: From its source at almost 5,000 m/16,000 ft in the Andes, the Amazon drops steeply to little more than 60 m/200 ft where it crosses the international border between Peru and Brazil.

CLIMATE: Annual rainfall averages 2,000–3,000 mm/80–120 in. Humidity high. Average monthly temperatures range from about 23 °C/73 °F to 28 °C/82 °F. Sudden southerly squalls periodically reduce temperatures to below 10 °C/50 °F.

Although the amount of rainforest clearance in Brazil is small in relation to the whole of Amazonia, the explosive surge in deforestation in recent years means that huge areas are likely to disappear very quickly. Most badly affected are the states of Mato Grosso, Rondônia, Acre and southern Pará, although the reasons for their deforestation vary. Newly built roads have encouraged immigration into Rondônia and eastern Acre; elsewhere forest clearance is attributable mainly to the establishment of large cattle ranches.

Construction (financed by the World Bank) of the Belem–Brasilia Highway in 1960 and the now notorious BR–364 in 1967 and, more importantly, their subsequent paving, opened southern Pará, northern Mato Grosso and Rondônia to deforestation. Until 1960, Rondônia was uninhabited except for a few Amerindians and rubber tappers but, by the late 1970s, about 5,000 people were arriving in the area every month. Many of the immigrants were landless people from the south of Brazil, moving under a government sponsored programme. From 1975–80 deforestation increased at an annual rate of 37%.

The planting of pasture by smallholders and landowners alike is a central cause of deforestation in many parts of the Amazon. The settler clearing forest to take an annual crop can expect only one or two harvests before deterioration of the soil structure and nutrient content causes a decline in yields which makes continued planting less attractive than the option of opening a new area. At that stage, he can either abandon the original clearing or plant it with grass. Conversion to pasture has the advantage of producing some income from cattle, and the grass adds to the land's value when it is resold. Real estate speculation is a powerful motive for deforestation, and has been encouraged by a variety of governmental inducements and financial incentives. As in Costa Rica, the position is exacerbated by a small number of landowners owning most of Brazil's productive agricultural land. These wealthy landowners wield considerable political influence; most are not interested in conserving the land, only in making more money from it.

The increase in logging poses a major threat to the Amazon forests. The forests of Southeast Asia currently supply most of the world's tropical hardwoods but, as these are being rapidly destroyed, the large lumber companies are likely to transfer their attention to the Amazon. Charcoal making is also a potential threat. The iron-smelting industry in the Grande Carajás, containing by far the world's largest iron ore deposits, alone has the capacity to absorb vast areas of forest to power its industrial plants.

The huge increase in the number of fires in the Amazon is a further cause for concern. They take place on such a scale that Brazilians talk of three seasons: the dry, the wet and the burning season. Satellite data are being used to monitor the number and position of fires and attempts made to extinguish them before too much damage is done. In September 1991 alone, between 50,000 and 88,000 fires were reported, causing extensive destruction.

A start has been made in Brazil on studying the effects of deforestation on climate. Automatic weather stations that transmit data by satellite are being installed in the country. The stations will be grouped in pairs, one on cleared land and the other in the forest canopy. The data they transmit relating to atmospheric change, temperature, wind direction, air pressure, evaporation and the like should help improve knowledge of the climatic consequences of deforestation.

● Tropical rainforest in the Volcan Poas National Park, Costa Rica. (WWF / Hernan Torres)

## MANAGEMENT OF GHANA'S FORESTS

Ghana's tropical rainforests have declined from an estimated 88,000 km² / 40,000 sq mls at the turn of the century to about 15,000 km² / 5,800 sq mls today. Most of the forest is in the southwestern third of the country and here are found such important timbers as utile *Entandrophragma utile*, wawa and African mahogany. Forestry and logging currently provide employment for about 70,000 people and, since the 1970s, have accounted for 5–6% of Ghana's gross domestic product.

Following the collapse of Ghana's economy in the early 1980s, the World Bank injected large sums of foreign capital into increasing timber production. By 1987, the export of logs and processed wood, mostly to Britain and West Germany, had risen dramatically, bringing in approximately US$ 100 million in foreign exchange earnings. One unfortunate consequence, however, was the establishment under World Bank auspices of fraudulent concessionaires and process-ing companies who plundered about 250 km² / 100 sq mls of Ghana's tropical moist forest, costing the country millions in lost revenue.

The combination of excessive logging, illegal exploitation, fire damage, a growing demand for fuelwood and, above all, shifting cultivation accounts for about 70% of the loss and has left little forest standing except in reserves. Even there, and despite legal protection, the forest is scheduled to be logged.

Over the past decade, a detailed inventory of Ghana's forests has been drawn up. This has been followed by the injection of US$ 64.6 million into a programme aiming at sustainable management of the forest. As well as drawing up management plans for all forest reserves and conservation areas, the project places special emphasis on working with local people in the development of forestry resources outside the reserves. Complementing this project is a range of local initiatives seeking to conserve forest resources. Friends of the Earth-Ghana has assisted ten Ashanti villages to set up agroforestry programmes; the Ghana Association for the Conservation of Nature has helped to establish a number of locally managed reserves in areas serving as 'sacred groves', burial grounds and watersheds; while the citizens of the twin villages of Boabeng-Fiema have protected an area of forest and its incumbent wildlife through traditional beliefs and local by-laws.

These national and local initiatives have cast Ghana in a leading role in terms of conserving Africa's forests at the local level. This is particularly commendable when one considers the pressures the country is under to exploit the forests for short-term economic gain and to convert the land to other forms of use. Most African countries have given in to these pressures, and are now in the position of needing to restore their degraded forests.

# THE FORESTS OF ZAÏRE

LOCATION: Central Africa in the basin of the Zaïre River and its vast network of tributary rivers.

AREA: The Central African rainforest, extending from Cameroon and Gabon on the Atlantic coast to Uganda, Kenya and Tanzania, covers an area of 3.8 million km² / 1.5 million sq mls, of which 1 million km² / 386,000 sq mls are in Zaïre.

ALTITUDE: Averaging about 900 m / 3,000 ft on the central plateau, rising to 3,500 m / 11,500 ft or more in the highlands.

CLIMATE: Mean average temperature remaining fairly constant at 25–27 °C / 77–80 °F. Rainfall 1,500–2,000 mm / 59–79 in, evenly distributed throughout the year. Humidity 80–95%.

In contrast with most other African countries, Zaïre retains almost 70% of its original tree cover. One reason for the continued existence of this huge block of forest is that the Zaïre River, which could otherwise have provided a major highway for human penetration into the centre of the country, is interrupted by two sets of impassable rapids. Railways now bypass both rapids, but the time and cost involved in the transshipment of goods have inhibited exploitation and development of the interior.

Zaïre is well known for its biological wealth. It has more species of birds (1,086) and mammals (409) than any other country in Africa. This is partly due to its large size and the consequent range of climate, geology and topography. In addition, eastern Zaïre was one of the main centres of forest survival during the last Ice Age. These so-called forest 'refugia' are rich in species, with a high ratio of endemics.

The forests in Zaïre are generally semi-evergreen with most areas experiencing a pronounced dry season. The seasonal rainfall makes the forest very sensitive to disturbance, especially to fire. Shifting cultivation, even on a small scale, can allow fire to enter the forest, causing damage out of all proportion to its size. Large tracts of grassland and secondary vegetation have thus replaced the original forest in some areas. Fortunately, central Zaïre has always been sparsely populated; lack of roads and navigable rivers means that most people live around the forest edge. Indeed, with the present poor international market for the few agricultural commodities that the region produces, there continues to be a movement of the population away from the centre to the more prosperous southern, eastern and western areas.

Pygmies live in the forest, but their traditional hunter-gatherer life style has had little, if any, impact on it. They use forest produce mostly for their own consumption, but sometimes to barter for iron implements and other merchandise. But the growing number of people living around the forest margins inevitably increases the level of exploitation. Pygmies have taken to commercializing their

hunting, killing and selling animals to traders from the cities who commission them to hunt on their behalf.

Logging companies, too, have been discouraged from operating in the remote central region as the high cost of extraction allows only the most valuable types of trees to be exploited profitably. The timber industry has thus concentrated on forests near the coast. This has resulted in the removal of virtually all the commercially valuable timber from the Mayombe region. Farmers have then encroached on the logged sites and completed the process of deforestation.

Although Zaïre contains more than

half Africa's forests, it produces only 3.4% of the continent's overall timber exports. The Tropical Forestry Action Plan for Zaïre proposes raising the annual off-take by a factor of eight. This has provoked criticism from many conservation groups, but is proportionately lower than many other African and Asian timber-producing countries. Extraction is nevertheless certain to increase. Despite the government's commitment to increasing protected areas from the present 4% or so to 12–15% of the national territory, there may well be difficulties in retaining Zaïre's forests and their irreplaceable biological riches.

# COASTAL TEMPERATE RAINFOREST

LOCATION: Temperate rainforest, which includes both evergreen and deciduous trees, is found between 32° and 60° latitude, where it is restricted to the coastal margins of western North America, New Zealand's South Island, Tasmania, Chile, parts of Japan, northwest Europe and the Black Sea coast of Turkey and Georgia.

AREA: Originally covering no more than 300,000–400,000 km²/115,000–150,000 sq mls, over half this area has now been eradicated by logging or conversion to agricultural use.

ALTITUDE: Extending from coastal plain into coastal mountain ranges.

CLIMATE: These forests require at least 2,000 mm/79 in annual precipitation and to be free of extended drought in the summer months. British Columbia, for example, has mild winters and cool summers, with temperatures of about 4 °C/40 °F in January and 16 °C/60 °F in July. The Rocky Mountains act as a climatic divide in the region. Warm air moving in from the Pacific has a moderating influence on climate, creating a warm, humid oceanic climate in a narrow coastal strip. The barrier effect of the mountains is very marked, precipitation declining from 5,100 mm/200 in on Vancouver Island to over 2,500 mm/98 in in the coastal ranges, to 1,400 mm/55 in at Vancouver and to 900 mm /35 in in the lower Frazer Valley.

Coastal temperate rainforest is an uncommon category of forest. Due to a unique combination of environmental factors, these forests contain a large number of species and have the highest standing biomass of any terrestrial ecosystem. The propitious climate enables trees to reach tremendous sizes, and to produce very superior types of timber which are much in demand on the world market, a factor that is pushing them rapidly to vanishing point. South America's alerce, *Fitzroya cupressoides,* and monkey puzzle, *Araucaria* spp., and North America's Sitka spruce, *Picea sitchensis,* and yellow cypress *Cha-*

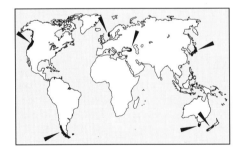

*maecyparis nootkatensis,* all of which are the product of temperate rainforests, are among the world's most highly prized coniferous timber trees.

The temperate rainforests of North

America have been heavily exploited, generally starting in the easily accessible lowlands and working up-slope. Not a single rainforest watershed of any size now remains in the lower states of North America. Only Alaska and British Columbia still retain large undeveloped tracts.

Of around 59,000 km²/22,780 sq mls of coastal temperate rain forest in British Columbia, 40,000 km²/15,500 sq mls are considered 'productive' and thus scheduled to be logged. Productivity is the highest in the world, and forest-related industry generates more than half the province's income. Timber mills are numerous, those sited at the mouth of the Frazer River being the largest in the world.

Only 3,000 km² / 1,158 sq mls are reserved in provincial and national parks; even there the forest is not necessarily safe as provincial park designation does not guarantee complete protection. The logging of these forests is destroying some of the oldest, tallest and most awe-inspiring living things on earth, yet their destruction passes virtually unnoticed.

Present policies allow everything to be felled. The logging companies justify this by claiming to replace the mature forest with new trees, replanting about two-thirds and allowing the rest to regenerate naturally. But these man-made plantations are very different from the natural forests they replace. Logging activities also have the damaging effect of causing landslides, ruining salmon streams and polluting coastal waters.

The disappearance of British Columbia's forests is a matter of great concern to some Canadians, yet causes little international outcry. Most people in the United Kingdom are indifferent to the destruction - if, indeed, they are even aware of it - despite the fact that around a third of Britain's paper pulp comes from Canada, and the UK takes more than half British Columbia's plywood exports.

# AUSTRALIA'S TEMPERATE RAINFOREST

LOCATION: Australia's last great temperate rainforest, and one of the few anywhere in the world, is in southwestern Tasmania. 41°35′–43°40′ S, 145°25′–146°55′ E.

AREA: 4,560 km2 / 1,760 sq mls.

ALTITUDE: Sea level to 1,617 m / 5,305 ft (Mount Ossa, Tasmania's highest peak).

CLIMATE: Coming under the influence of the Roaring Forties, southwestern Tasmania lies in the wettest part of Australia. As well as high rainfall, the climate is characterized by cool temperatures and a high incidence of cloud cover. Annual rainfall in the Gordon-Franklin basin ranges from 1,800 mm / 70 in at the headwaters of the Franklin River to over 3,400 mm / 134 in at the Serpentine Dam.

Australia's temperate rainforest differs from tropical rainforest in having a smaller number of tree species - rarely more than eight - and in the absence of lianas and epiphytes (except for moss and lichens). The vegetation has affinities with New Zealand and South America as much as with Australia. The most characteristic tree is myrtle beech; others include the Huon pine, which lives for 2,000 years or more, and King Billy pine. The forest also carries exceptionally fine stands of pristine tall forest, in which eucalypts such as mountain ash - among the tallest flowering plants in the world - form a canopy 60-90 m/200-300 ft high, over a closed understorey 10-20 m/30-65 ft above the ground. In places, eucalypts - either messmate stringybark or Smithton peppermint - are emergent, giving rise to eucalypt forest with rainforest understorey.

The combination of closed temperate rain forest, open eucalypt forest, buttongrass moorland and alpine communities forms a mosaic of Australian and Antarctic plants that is unique. These diverse habitats sustain a flora and fauna that include a number of archaic forms providing a link with the prehistoric continent of Gondwanaland, which survive only in Tasmania or whose closest surviving relatives are located in South America, Africa and India. King Billy pine is one among several relict Gondwanan conifers. Both plants and animals are represented by an unusually high number of endemics. Fewer exotic species have been introduced into Tasmania than to mainland Australia, with the result that Tasmania's native flora and fauna have been less severely affected than other parts of Australia.

The fauna includes four endemic marsupials: the thylacine, which although officially regarded as extinct is still the subject of occasional sighting reports, and the Tasmanian devil, the next largest surviving carnivorous marsupial.

More than 150 species of birds have been recorded, among them one of Australia's rarest birds, the orange-bellied parrot, found on moorland dominated by buttongrass, a habitat it shares with the ground parrot and the rare broad-toothed rat. Of Tasmania's reptile fauna, four species are endemic, including the Pedra Branca skink, confined to the little rocky island of Pedra Branca. Two of the six species of frog are also endemic, among them the Tasmanian tree frog, while the 15 species of freshwater fish include four endemics, two of them - the Lake Pedder galaxias and swamp galaxias - are regarded as endangered. The introduced trout is believed responsible for the decline of species of native fish.

Of unusual interest is the burrowing freshwater crayfish, another Gondwanan relict, which lives in burrows beneath the tussocks of buttongrass. Its burrows are colonized by a number of endemic invertebrates, among them two primitive crustaceans, *Allanaspides helonomus* and *A. hickmani*, both of which are confined to a very small range in the vicinity of Lake Pedder.

The region's alpine areas sustain a

specialized fauna of great interest, including the rare endemic dragonfly *Archipetalia auriculata*, the most archaic member of the Neopetaliidae, an ancient family having strong affinities with Gondwana-land.

The little development that has taken place in this wilderness area has been largely confined to hydroelectric schemes at Scott's Peak, Mount Arrowsmith, and Lake St Clair, to the inundation of Lake Pedder and its extraordinary quartzite beach, and above all to the construction of the Middle Gordon hydroelectric scheme in the early 1970s, which necessitated damming part of the Gordon River. Subsequent plans to flood the lower reaches of the Gordon and Franklin rivers met with strong opposition from both the national and international communities. The intervention of the Australian Federal Government was necessary before the project was finally abandoned. Opposition, both national and international, was strengthened by inscribing much of the area on the World Heritage List in 1982. Four national parks, two state reserves, two conservation areas and a number of state forests, totalling 10,813 km² / 4,175 sq mls, were incorporated into the Tasmanian Wilderness World Heritage Area. The validity of this action was challenged by the Tasmanian Government, but was dismissed by the High Court of Australia in 1983.

# FIORDLAND, NEW ZEALAND

LOCATION: Southwestern South Island, New Zealand.

AREA: Most of Fiordland lies within the Fiordland National Park, New Zealand's largest national park, covering 12,523 km² / 4,835 sq mls. About two-thirds of the park is forested.

ALTITUDE: Sea level to 2,746 m / 9,010 ft (Mount Tutuko).

CLIMATE: Fiordland is one of the wettest and windiest places on earth; prevailing westerlies sweep in from the Southern Ocean and Tasman Sea bringing blustery wet conditions, and gales are frequent. Mean annual precipitation at sea level (Milford Sound) is 6,200 mm / 244 in. Temperatures show little daily or annual variation. Mean annual temperature 9–10 °C / 48–50 °F, ranging from 14 °C / 57 °F in January to 5 °C / 41 °F in July.

Fiordland is amongst the wildest country in New Zealand. Parts of it are virtually inaccessible except by air. The entire region has been shaped by glacial action. Numerous sheer-sided U-shaped valleys, in places more than 500 m / 1,600 ft deep and separated by narrow dividing ridges, are a characteristic feature of the terrain: along the coast these valleys submerge into the Tasman Sea to form a series of 14 magnificent fiords (or sounds) from which the region derives its name. One of the most impressive is Milford Sound, from whose depths Mitre Peak rises sheer out of the water to a height of 1,700 m / 5,580 ft. Hundreds of glacial lakes are scattered throughout the area – the largest being Te Anau (347 km² / 134 sq mls) and Manapouri (142 km² / 55 sq mls) – set amidst magnificent mountain scenery, topped by Mount Tutuko (2,746 m / 9,010 ft) and Mount Madeline (2,554 m / 8,380 ft). Among countless waterfalls the highest are the Sutherland Falls, which plunge 580 m / 1,900 ft in three cascades. Lying on the boundary between the Pacific and Indo-Australian tectonic plates, and forming part of the 'Pacific Ring of Fire', seismic activity is greater in Fiordland than in almost any other part of New Zealand, though few of the tremors cause significant damage.

Up to the snowline the forest cover is predominantly silver beech and southern rata, the latter a blaze of scarlet during the summer. Mountain beech and red beech are well represented also. Most of the forest trees are stunted and misshapen, and all are festooned with epiphytes; a dense mat of ferns, mosses, liverworts, and numerous flowering plants covers the forest floor. Along some of the river flats sedge-bogs have been formed. On these bottom lands rimu, miro and kahikatea are widespread, many of them up to 800 years old and all components of the great podocarp forests believed to bear a close resemblance to the immense swamp forests of the Mesozoic era. Tree ferns and the curious grass tree are found throughout much of the area. In the alpine zone trees give way to snow-tussock grasses and alpine shrubs such as red mountain heath, mountain totara and alpine toatoa. Many wild flowers, among them the mountain buttercup, the largest of all the buttercups with flowers 5 cm / 2 in across, the Fiordland mountain daisy, the Maori bluebell, and the South Island edelweiss. Twenty-four plant species are endemic to the national park, and an equal number have very restricted distribution centred on Fiordland.

Apart from two species of bats (one an inhabitant of Fiordland), only marine mammals – notably the fur seal, of which there are about 50,000 – have established themselves in New Zealand. The native terrestrial fauna is dominated by birds. Fiordland's bird fauna is remarkably varied and distinctive; it includes representatives of three families that are unique to New Zealand: the Apterygidae or kiwis; the Acanthisittidae or New Zealand wrens; and the Callaeidae or 'wattled crows'. The Fiordland representatives of these families are respectively the South Island brown kiwi, the little spotted kiwi and possibly the great spotted kiwi; the South Island race of the New Zealand bush wren, and the South Island kokako. The two last named are rare and restricted in distribution to a few parts of Fiordland. Two of New Zealand's rarest birds – the takahé and the kakapo, the world's largest parrot, a type of bird generally associated with tropical forest but here living in an alpine zone – have found sanctuary in Fiordland. Among the rare endemic birds now restricted to Fiordland are the southern race of the New Zealand laughing owl or whekau, a relict ground-nesting bird; and the southern race of the piopio or New Zealand 'thrush'. There have been no confirmed records of either bird for a number of years; both are believed to have succumbed to the stoat and other introduced predators; but, if they survive, Fiordland is likely to be their last refuge. Other uncommon or distinctive birds include the Fiordland crested penguin, the New Zealand brown teal, the paradise duck, the blue duck, the South Island weka rail, the parrot-like green kaka, its near relative the kea, and the bellbird, one of the 'honeyeaters'.

The main danger to the endemic Fiordland fauna derives from naturalized animals introduced either intentionally or inadvertently into the region, or that have immigrated from other parts of New Zealand. Alien animals include 43 species of birds, among them the blackbird, thrush, sparrow, starling, goldfinch, Canada goose, mallard, and black swan; and 32 mammal species, including the red deer, wapiti, opposum, rat, and, above all, the stoat. These animals either prey on the native fauna, compete with it for food, or bring in diseases and parasites against which it can offer little or no resistance. Attempts are being made to reduce the numbers of exotic animals by trapping and shooting. The need for these measures is underlined by the fact that 11 of New Zealand's endemic bird species have become extinct since the beginning of the 19th century. But in such rugged country effective vermin control is almost impossible.

Some of the most interesting vegetation in this area, notably certain ferns and undergrowth orchids, has been eradicated on the mainland and survives only on islands in Lake Manapouri. If the level of the lake were raised – as at one time proposed – these islands would, of course, be submerged and their vegetation lost. Flooding would also result in the loss of a considerable area of podocarps, whose berries constitute the basic food for several species of birds. The feeding and nesting areas of about 20 species, among them the crested grebe, would be disturbed or destroyed.

The aim of management is to preserve native plants and animals while eradicating alien species. Helicopters were first used for hunting red deer and wapiti in 1975, since when the vegetation has recovered well. The public has freedom of entry and access, except into four specially protected zones which are either takahé breeding areas, islands which have no exotic animals, or are important breeding grounds for fur seals and seabirds.

In 1990, Fiordland and three other national parks – Mount Aspiring, Mount Cook and Westland – were incorporated into the South-West New Zealand World Heritage Area (Te Wa'hipounamu), in effect a Gondwanaland Heritage Site. Extending along the coast for 450 km / 280 mls, and inland for 40–90 km, / 25–56 mls, it covers 26,000 km² / 10,000 sq mls between latitudes 43° 00′–46° 30′ S and longitudes 166° 26′–170° 40′ E.

● Fiordland, on the southwest coast of New Zealand's South Island, is one of the wettest places in the world, with annual rainfall averaging some 6,200 mm / 244 in. In such conditions temperate rainforest thrives.
(Auscape International / Jean-Paul Ferrero)

# GRASSLANDS

*If to do were as easy as to know what were good to do,*
*chapels had been churches and poor men's cottages princes' palaces*
Shakespeare, *The Merchant of Venice*

In the present state of the earth's climate, grassland is the natural vegetation of nearly one-quarter of the world's land surface, if wooded savanna and shrub steppe are included. While large areas of grassland remain in a relatively natural state, others have been modified to varying degrees by human intervention – a very recent occurrence in relation to the evolution of grassland ecosystems. Many grasslands are secondary formations – that is, were originally forests or wetlands that were subsequently converted to arable land or modified by frequent burning, livestock grazing, mowing, or cultivation. Much grassland has been degraded to wasteland and virtual desert. The aim of this chapter is to examine briefly the present state of the world's major grasslands and man's impact upon them.

The distribution of most natural grasslands is determined by climate – primarily by the quantity and seasonality of precipitation. Soil type is also an important factor but is itself the product of climate acting upon parental rock and influenced by topography, drainage and plant life. Such natural grasslands, termed *edaphic* by ecologists, occur where the availability of soil water falls below the minimum required by forests, at least for part of the year, but where in most years precipitation or flooding is sufficient to support a high proportion of grasses. Some grasslands occur where shallow soils lie on rock or calcitic hardpans, leaving insufficient depth for tree roots. Large browsing herbivores, particularly elephants in tropical Asia and Africa, and deer in temperate regions, contribute to the dynamic balance of wooded grasslands by inhibiting tree growth.

Grasses can be divided into two main categories – temperate and tropical. In the tropics, most natural grasslands occur in the climatic belts between forests and deserts, as in Africa south of the Sahara. They include both the savannas and the seasonally flooded plains such as the central *llanos* of Colombia in South America. In temperate regions they occur mainly in the rain shadows of the great mountain ranges, as in the central Asian steppes to the north of the Tibetan Plateau, and the prairies to the east of the Rocky Mountains in North America. Grassland ecosystems have their close counterparts in every continent. Important differences between them appear to be due to their varied histories of occupancy by very different animal communities and especially to their colonization by man and his domestic animals. Human influences upon the grasslands are oldest in Africa south of the Sahara, dating back to the origins of the human species, and more particularly to the time, probably more than 50,000 years ago, when man first lit grassfires as an aid to hunting.

Since the earliest domestication of wild cattle, sheep and goats, and the first cultivation of cereal crops in southwestern Asia, the Old World grasslands of Eurasia and North Africa have been occupied and exploited by pastoral people and their livestock. In South America, apart from the effects of grass fires, human influence upon the grasslands probably began with the domestication of the llama and the alpaca between 4,000 and 2,500 BC. Sheep, introduced to the New World by the Spanish in the 16th century, were later followed by cattle, both of which have largely, but not entirely, replaced the llama and the alpaca. On the North American prairies, on the other hand, pronghorn antelope, elk and bison, congregating in huge numbers, were not domesticated but hunted by the indigenous Amerindians. The bison especially represented an apparently perpetual and self-sustaining supply of meat and hides. With spear, bow and arrow, and long-acquired traditional skills, native hunters made little impression upon the millions of bison and pronghorn on the plains and elk in the woodlands. It has been argued reasonably that something akin to a symbiotic relationship existed between the native Indians and the bison. The Plains Indians depended upon the bison which, in turn, benefited from the maintenance and even the extension of their habitat through the grass fires lit by the tribesmen.

## The Smaller Grasslands

In addition to the major tropical edaphic grasslands of Africa, Australia and South America, there are many others of importance but of relatively small extent. Examples are the semi-arid wooded grasslands of western and southern Madagascar; the *toich* seasonal floodplains bordering the Sudd swamps of the River Nile; and the floodplains of Lake Bangweulu and the basin of the Kafue River in Zambia. The seasonal floodplains of the tropics, which frequently form an inter-zone between savanna and wetland, are the habitat of grazing species adapted to exploit the specialized grasses of this type of habitat – in Africa, for example, *Vossia*, *Paspalum*, *Oryza* (wild rice) and *Echinochloa* (antelope grass) – by following the advance and retreat of the floodwaters. In tropical Africa this niche is occupied by a group of antelopes, notably the lechwes, kobs, pukus and waterbucks. Africa's extensive floodplains are also used by several other wild ungulates, especially buffalo, as well as by pastoral people. Each year, as the receding waters of the Nile uncover the *toich* grasslands, Shilluk, Nuer and Dinka tribesmen herd their cattle back on to the renewed and apparently indestructible pastures.

In southwestern Tanzania, the steadily rising waters of Lake Rukwa have, in recent years, inundated the former floodplain and continue to rise seasonally into the fringe of the surrounding woodlands. This has probably been caused by several years of above-average rainfall and reduced evaporation due to heavy cloud cover. A small area of floodplain remains in the upper estuary of the Rungwa River, which flows into the northern end of the lake. In years to come, when the lake waters again recede, the seasonal grassland will probably become re-established and the common grazing mammals return – except for the formerly isolated population of puku at the south end of the lake which no longer exists. In the meantime, the cattle of the Sukuma people, recent immigrants from the densely populated region to the south of Lake Victoria, are forced to subsist on poor grazing in the woodlands until the flood-plain is again uncovered.

Despite the generally deleterious changes – including widespread decrease in biological diversity and productivity – which are affecting most of the world's tropical grasslands as a result of human intervention, they have not suffered the complete destruction that has befallen enormous tracts of native temperate grassland in North and South America, Eurasia and Australia. With exceptions, such as the extensively degraded grasslands of the Sahel, the Horn of Africa, the Kalahari, Madagascar, the South American Chaco and much of India's secondary savannalands, the present state of most tropical grasslands gives some hope for their survival as relatively natural ecosystems.

## Secondary Grasslands

Secondary or successional grasslands have arisen from the removal of the original forest vegetation, and are maintained by burning, grazing or mowing. Much of the rougher moorland and downland grass in Europe (including Britain) and Japan, as well as the recently created pastures resulting from the clearance of the Amazon forest for cattle ranching, fall into this category. The highland grasslands in tropical and equatorial regions have replaced montane forest after felling and burning, and are maintained by annual grassfires and by grazing.

The present savanna landscapes and 'cultivation steppe' which typify the greater part of the Indian subcontinent are the result of intensive land-use which, over a period of several millennia, has replaced the original sub-humid and dry forests. In the last four centuries the secondary savanna grasslands have become increasingly degraded by overgrazing and cultivation as the human population has expanded (to its present average density of approx. 200 persons per km²/520 persons per sq ml). Ecologists estimate that 43% of the Indian subcontinent is eroded and in process of change from savanna to desert.

## Grassfires and Successional Change

Most grasslands are adapted to occasional fires. Under natural conditions wild fires are frequently started by lightning, although they are more likely to be deliberately started by people, to the extent that fire has become a dominant factor in grassland ecology. Many grassland communities are maintained by periodic burning, which tends to suppress shrubs and trees that would otherwise grow into dense woodlands or forests. These 'fire-climax' or derived grasslands occur in places where the rainfall is normally sufficient to support the growth of forests or sub-humid woodlands. Frequent fires in these regions, many in upland and highland areas, produce a characteristic pattern of grassed ridges and raised plains dissected by narrow gallery forests bordering streams and rivers. This pattern is exemplified in such places as the *llanos* of Venezuela, the sub-montane grasslands of the Imatong Mountains in southern Sudan and the Nyika Plateau in Malawi. Year after year, fires envelop the ridges and plains, suppressing tree growth and progressively pushing back the forest edge, although this process may be retarded by the curtain of relatively fire-resistant shrub and creeper growth which commonly forms the forest boundary.

The high rainfall regions produce tall grass, 2–4 m/6–12 ft high, which burns fiercely each dry season. The humid savanna grasslands, bordering the Ituri rainforest in northeastern Zaïre and extending into southern Sudan and Uganda, are burned every year by agriculturalists and hunters. Burning takes place in the short dry season, in January and February, when

the widespread fires create a pall of brown smoke which darkens the sky and is frequently carried by westerly winds as far as Kenya and Ethiopia. The aftermath of a grass fire is the characteristic blackened landscape with plumes of smoke rising from the smouldering remains of fallen trees and slowly burning piles of elephant dung. Seedlings and young saplings are commonly either killed or burned down to ground level, but the mature savanna trees, though scorched, withstand the fires. Very hot dry season fires burn off the dry aerial parts of the perennial grasses but otherwise cause them little or no harm. This is because, by that time of the year, most of the nutrient material accumulated during the growing season has been transferred underground to be stored in the elaborate root system, ready to initiate green growth when the rain returns.

Enormous areas of semi-arid savanna in Africa, South America and Australia are subject to localized successional changes from shrubland and woodland to open grassland and back again, a cycle that takes place over several decades or, in the case of the taller woodlands, perhaps a century or more. At any one site, woodland and shrubland regeneration may occur only infrequently, depending upon the conjunction of several favourable conditions, including a sequence of relatively good rains and absence of grass fires. One special circumstance which may determine the incidence and intensity of grass fires, and thus their influence upon the growth of young trees, is the impact of grazing by wild or domestic ungulates. Heavy grazing pressure may keep grass so short that there is little or no fuel to burn. This commonly occurs in grasslands grazed by cattle, but can also be seen in some national parks in Africa, where wild ungulates, especially hippopotamuses, congregate near water. In these circumstances, young trees in tropical savanna grasslands get the chance to grow beyond the fire-vulnerable stage. The resultant 'bush encroachment' may seriously reduce the grazing value of the pasture as, for example, on the shores of Lake Baringo in Kenya, where grasslands heavily grazed by cattle have been replaced by dense acacia thicket.

In the managed pastures of modern cattle ranches, fire becomes a tool for use in controlling bush encroachment. A rotation can be established in which part of the ranch is left ungrazed for at least one out of every three or four growing seasons. Sufficient material accumulates during this period to generate a fire hot enough to kill or suppress young trees and shrubs.

Where management strives to maintain a rough balance of mixed grassland and woodland – as in the Serengeti National Park – considerable effort has to be made to exclude and control the frequent grass fires originating outside the park boundaries and swept in by the wind. In the last three decades, the Serengeti and Tarangire National Parks in Tanzania have experienced over 75% reductions in tree and shrub cover, largely caused by frequent uncontrolled grass fires but also by the browsing of herds of elephant seeking refuge from harassment and habitat destruction. This has resulted in former woodlands and thickets being changed to wooded and open grassland.

● Dinka pastoralists of the southern Sudan have long exploited the seasonal floodplain bordering the Sudd swamps of the River Nile. As the floodwaters recede, they herd their livestock on the newly exposed grasslands
(WWF / Kes Fraser-Hillmann)

Fire control can reverse these changes over a similar period, and any loss of aesthetic quality may be only temporary. The short-term loss of habitat diversity may have more serious consequences for the rare woodland wildlife for which the parks are refuges.

When dry savanna grasslands become overgrazed by livestock – as they have in the past over large parts of southern Australia and the southern USA, and as they continue to be in the Middle East, the steppes of northern China, much of the Indian subcontinent and the semi-arid regions of Africa south of the Sahara – they undergo successional changes which may be reversible in the early stages of degradation but lead to virtually permanent damage if high stocking rates are maintained. Similarly, grain cultivation in arid lands, such as those on the north coast of Africa and along the southern borders of the Sahara, causing rapid loss of topsoil by wind erosion, is a major cause of desert encroachment. Research on soil erosion and its causes has shown that the main damage to soil surfaces in grassland is caused by occasional heavy rainstorms washing away soil loosened by the hooves of grazing animals.

In the course of human expansion, accompanied by the development of

sophisticated systems of management, the world's major grasslands have been radically altered. Palaeobotanical and archaeological evidence indicates that the grasslands of much of southwestern Asia and northern Africa, including the Sahara and the Arabian Peninsula, were reduced to their present desiccated condition between 6,000 and 2,000 years ago, at a time when pastoral and agricultural expansion coincided with a dry phase in the long-term climate pattern. Natural steppe grasslands, established in the previous wet climatic phase, provided conditions suitable for the first domestication of sheep, goats and cattle – which probably took place in the Middle East between 11,000 and 9,000 years ago. It seems probable that the steppe vegetation could not tolerate the stresses caused by the combined impact of a drying climate and heavy grazing by domestic livestock. With the spread of traditional pastoralism and, more recently, of commercial farming and ranching, grazing by domestic livestock has become the most important factor influencing the ecology of grasslands around the world.

Severe overstocking with cattle, commencing in the second half of the 19th century, was responsible for the south-

ern prairies in the USA becoming virtual desert – and creating the notorious 'Dust Bowl'. Sheep farming similarly degraded the grasslands of southern and central Australia in the last century. The arid and semi-arid wooded grasslands of the Sahel region south of the Sahara undergo accelerated desert encroachment with each periodic drought as marginal agriculture, excessive grazing by livestock and the removal of wood for fuel, outstrip the regenerative capacity of the vegetation. Under these circumstances, although the indigenous herb and woody vegetation may have been depleted or removed, intermittent rainfall may permit the growing of grain crops in some years.

Natural grasslands have co-evolved with their native herbivores as self-perpetuating ecological systems. The substitution of livestock (herded or fenced) with access to permanent as distinct from seasonal water, in place of free-ranging and migratory wild ungulates (or kangaroos), imposes stresses upon natural grasslands which they cannot sustain. Where heavy or excessive grazing by domestic stock has taken place, it has commonly caused serious and even permanent degradation towards desert conditions. After severe reduction in vegetational cover, depending upon soil properties and slope, the removal of topsoil through erosion caused by animal trampling, wind and water, may make land rehabilitation difficult or, in poor countries, too expensive to attempt – particularly as there is no guarantee that the process will not be repeated. Even the partial recovery that might be expected within a period of several decades will depend upon the prolonged exclusion of grazing animals, a measure which may be difficult or impossible to achieve where subsistence pastoralists are already short of grazing land.

Each continent's main climatic zones support ecosystems with ecological equivalents on other continents, frequently with broad physiognomic char-

acteristics which make them virtually indistinguishable to the unpractised observer. Both South America's and Australia's natural grasslands have evolved in parallel with their eco-climatic counterparts in Africa. Owing to their relative genetic isolation, however, and their very different histories of colonization by plants and animals since their origins and evolutionary radiation at the beginning of the Oligocene (c. 40 million years ago), the composition of their plant and animal communities differs considerably. The most obvious difference lies in the identity, diversity and abundance of their wild grazing animals. On every continent the grasslands have evolved as grazing ecosystems each with its complement of indigenous consumers. Tropical Africa has 91 wild ungulate species of which over 30 are grazers and partial grazers. In tropical America there are today 21 wild ungulates of which only three deer species (which are primarily browsers) can be regarded as components of the savannas. The only living large indigenous grazing animal is the capybara, a large rodent. South America's two wild camelids, llama and vicuña, inhabit upland and mountain grasslands.

## Range Management

Although the causes of most grassland degradation processes are understood and the grazing regimes necessary to prevent or mitigate them are known, demographic and socioeconomic constraints and traditional practice may inhibit the exercise of the necessary management, even in developed countries. The lessons learned in the US fol-

lowing the Dust Bowl disaster led to the establishment of the new discipline of *range management*. This is concerned with applying basic ecological principles and rational methods to the maintenance and rehabilitation of grasslands and bushlands (collectively called 'range'). Among the best known management principles associated with range are those relating to *carrying capacity* and *stocking rates*.

Subsistence pastoralists are no less subject to the principles of range management than commercial ranchers, but they frequently experience the added difficulty of occupying communal grazing lands over which they have neither individual control nor responsibility. In attempting to maximize their own herds, owners compete for the available forage, creating conditions in which over-exploitation of the grazing land is almost inevitable. Nevertheless, there are accounts of traditional management systems under which natural grasslands have been protected from overgrazing. The *hema* system, once widespread in the Arabian Peninsula, was established by Islamic authority to protect important fodder plants and provide reserves of forage for nomads' livestock. In 1953 the Government of Saudi Arabia opened the reserves to free grazing, with the result that they have become indistinguishable from the surrounding land, denuded of vegetation by destructive grazing and the uncontrolled cutting of trees and shrubs. Loss of fertility and transformation to man-made desert have thus overtaken some of the last reserves on the arid steppes of Arabia.

● The two African species of rhino, the white rhino and the black rhino, are grassland dwellers, although the former is primarily a grazer and the latter a browser. Both species have been heavily poached for their horns. (WWF / Y.J. Rey-Millet)

# THE EURASIAN STEPPE

LOCATION: A northern belt stretching eastwards from the shores of the Black Sea, passing south of the Urals, north of Lake Balkash, through Mongolia into northern China; approx. 45°–50° N and 40°–120° E.

A southern belt running eastwards from central Turkey, south of the Caspian, through Iran, Afghanistan and Pakistan to north-western India; approx. 30°–40° N and 30°–75° E.

A narrow central belt from the Hindu Kush running eastwards between the Tibetan plateau and the Tarim Pendi and Gobi Deserts; approx. 35°–37° N and 75°–110° E.

AREA: Approx. 4.5 million km² / 1.6 million sq mls.

ALTITUDE: Between 500 m / 1,640 ft and 2,000 m / 6,500 ft.

The world's most extensive natural grasslands lie in the temperate zone, represented by the prairies of North America and their ecological counterpart, the Eurasian steppes. Both are bounded on the north, roughly along the 53rd parallel, by coniferous *taiga* forest.

In North America, the prairies reach almost as far south as the semi-arid and sub-desert grasslands of Mexico, while the Eurasian steppes meet or merge on their southern borders with a variety of sub-desert and mountainous habitats, including the Pamir and Tien Shan ranges. At their western limit the steppes occupy the plains of Kazakhstan to the north and east of the Caspian Sea. They extend eastwards for some 4,500 km / 2,800 mls along a belt approximately 1,000 km / 600 mls wide through Mongolia to within 600 km / 375 mls of the Sea of Japan.

Another more southerly and discontinuous belt of steppe grasslands stretches from the plains of Turkey, Syria and Iran, through Afghanistan, the Hindu Kush, and around the northern borders of Tibet into central China. It merges with sub-desert grasslands and desert on its northern and southern borders along the greater part of its 5,500 km / 3,400 mls length. The steppes also originally extended into western Europe, but all that now remains is a scattering of relict grasslands, modified by cultivation and intensive grazing, across southern Ukraine, western Turkey, Hungary and Romania, Macedonia, Greece, Italy, Spain and the islands of the Mediterranean.

A zone running roughly along the northern edge of the Eurasian steppes eastwards as far as the mountainous borders of Kazakhstan, receives enough rainfall to support a mixed agriculture of animal husbandry and cereal cultivation, in which respect it resembles the prairies of North America. In the drier, more southerly, parts of the steppes, huge irrigation projects permit further extensive agriculture, especially cotton growing, where deep ploughing and heavy applications of pesticides have destroyed the indigenous plant and animal life over much of Kazakhstan and Uzbekistan.

The vast grasslands of Mongolia and China are still inhabited by semi-nomadic pastoralists who graze their livestock as they have for several millennia. But, in the last century, the character of the steppes has changed greatly with the near-extermination of the great herds of wild animals which once roamed these rolling grasslands. Several species have been hunted almost to extinction and others reduced to relict populations through competition with livestock for the limited pasturage. Recent assessments by international agencies of the condition of steppe and other grazing lands in China, Bhutan, Nepal, Pakistan, India and Afghanistan are almost wholly pessimistic. Nearly all grasslands in Asia are judged to be overgrazed and badly degraded by livestock. Inner Mongolia, where the rangeland on government farms is reported to be in good condition, is one of the few exceptions.

The general decline is evidently re-

cent. Large parts of the steppe grasslands of Asia appear to have withstood the pressure of intensive use by domestic grazers for several thousand years, although it is uncertain how they differ from the natural grasslands of the past. Considerable efforts have been made in the former Soviet Union and eastern Europe, as in North America, to preserve and restore the flora and fauna of natural grassland ecosystems by establishing national parks and other refuges. However, since very little land remains in Europe, Asia or North America which has not been modified by human influences, it is difficult to ascertain the original natural condition of the grasslands.

The character and origins of the relict indigenous grasslands of eastern Europe have been the subjects of study by plant ecologists, who have broadly divided them into two categories: the continental steppe grasslands (dominant, *Festucetalia valesiacae*) and the sub-Mediterranean grasslands (dominant, *Brometalia erecti*). After the last Ice Age, these communities spread across Europe from the dry steppe regions of southern Russia and the mountains of the northern Mediterranean respectively, assisted by man's clearance of forest and by livestock grazing. The two communities are intermixed in Germany between the Rhine and the Weser. Within them vegetation ecologists recognize several types determined by their origins, climate, altitude, soils and topography: most of them include the grass genus *Stipa* and fescues. The Mediterranean communities are characterized by brome grass, *Bromus erectus*, and, like most steppe grasslands, by a variety of other grass species, beautiful orchids and many wild flowers.

The largest tracts of semi-natural steppe remaining in Europe lie north of the Caucasus Mountains, from the northeastern shore of the Black Sea to the grasslands of the lower Volga and the plains north of the Caspian Sea, with a mean annual precipitation of less than 450 mm/18 in. In the same climatic zone, they extend west of the Black Sea into the Pannonic region of southeastern Europe, including the Great Plain, covering over 10,000 km²/3,850 sq mls – equivalent to 12.7% of the surface area of Hungary. Across this region the former steppe woodlands were replaced, mainly in the 18th and 19th centuries, by open grassland, the *puszta*, which is maintained by fire, livestock grazing and, in the original floodplains of the Danube, by drainage. Two-thirds of the Hungar-

ian Plain is intensively managed for stock raising, while a quarter, lying to the east of the Danube, is grazed more extensively and is considered an important conservation zone: some 1,400 km²/540 sq mls are included in protected areas.

As in Kazakhstan to the east, the indigenous large mammals of the European steppe have mostly been hunted to extinction, though rodents and other small mammals remain abundant. They include hamsters, voles, wood and field mice, and the steppe mouse. The medium-sized rodents of the Eurasian steppes are ground squirrels – marmots and sousliks, which live colonially in burrows, their equivalents on the North American prairies being the prairie dogs. Formerly very abundant from eastern Europe to western China, the bobak or steppe marmot came near to extinction in Europe in the 1940s and 1950s and is still declining in Asia due to the encroachment of agriculture and intensive livestock grazing of the steppe habitat. It

survives only in broken terrain unsuitable for ploughing. But in the Ukraine, the steppe marmot has made a spectacular recovery, partly due to reintroduction and protection, but also to its recently adopted role as colonizer of abandoned farms and villages in depopulated rural areas. The souslik is extinct over the greater part of its former range on the European steppes but remains locally abundant in the upland grasslands of what were formerly Yugoslavia and Czechoslovakia. The brown hare occurs throughout the temperate grasslands of Eurasia and North America. The rabbit is an important agent in preserving grasslands. A native of Iberia and North Africa, it spread across Europe during the Middle Ages, colonizing the new grasslands created by woodland clearance. It was abundant and widespread until the 1950s when the virus disease myxomatosis greatly reduced its numbers. Growing resistance to the disease has enabled it almost to recover.

# THE AFRICAN SAVANNAS

Africa's tropical grasslands occur as a series of climatic zones extending across the greater part of the continent, and range from the sub-desert regions of the Sahel in the north, and the Kalahari sands in the south, to the humid tall-grass savannas of the Guinea Zone of central Nigeria and their southern hemisphere counterparts bordering the forest in southern Zaïre and northern Angola. Both northern and southern tall grass zones are characterized by scattered, short-boled, broad-leaved, fire-resistant trees 12–20m/40–65 ft high, and tall, tussocky, perennial grasses 2–4 m/6–12 ft high adapted to having the dead growth burnt off annually. Among the commonest grasses are the tall blue stem *Andropogon tectorum*, elephant grass *Pennisetum purpureum* and guinea grass *Panicum maximum*.

The drier Northern Guinea Zone of West and Central Africa is a belt of sub-humid deciduous savanna woodlands with a well-developed grass layer occurring mainly on heavily leached, relatively infertile sandy and gravelly soils. It is represented in the northern hemisphere by a wide belt of woodlands, dominated by the broad-leafed tree genus *Isoberlinia*, which extends through the border country of Mali and the Ivory Coast in the west to Sudan's Equatorial Province in the east. In the southern hemisphere it takes the form of a huge area of the closely related *Brachystegia* 'miombo' woodlands of Tanzania, Zambia, northern Zimbabwe and Angola.

The sub-humid deciduous savanna woodlands support a relatively low density of indigenous animal life, except locally in the grassy drainage lines, river valleys and flood plains. The woodlands are the habitat of the common tsetse fly,

the carrier of the blood parasites which cause 'sleeping sickness' in man and cattle but to which indigenous mammals are resistant. The combination of tsetse infestation and the soil's unsuitability for any form of agriculture ensures that enor-

mous areas of miombo remain very little used by people. They are nevertheless subject to human influence through the grass fires lit by hunters and honey-gatherers and the logging of *Pterocarpus angolensis*, one of the most fire-resistant

● Przewalski's horse lived on the steppes of Mongolia and China until the mid-20th century, but is now extinct in the wild state. (WWF / Henry Ausloos)

miombo trees and a source of high-quality timber. Forming a layer 1–2 m/3–6 ft high, the grasses include such common species as African foxtail *Cenchrus ciliaris*, love grass *Eragrostis* spp, blue stems *Andropogon* spp, needle grasses *Aristida* and russet *Loudetia* spp.

The *Brachystegia* and *Isoberlinia*-dominated woodlands are characterized not only by leached, infertile soils and unpalatable vegetation, but also by their relatively impoverished fauna. A few bird and mammal species are particularly associated with the miombo community – for example, Arnott's chat, the Tabora grass warbler and Lichtenstein's hartebeest. Large mammals inhabiting the grassy river valleys and the boundary zone between them and the woodlands include elephant, greater kudu, eland, roan and sable antelopes, and zebra. The last three species are grazers and tend to move into the miombo early in the rainy season to feed on the nutritious new grass growth, but may avoid it for the rest of the year when the commonest grasses – blue stems, thatching grass, *Hyparrhenia* spp, russets and lemon grasses, *Cymbopogon* spp – become tall, fibrous and less nutritious. Most of this grass remains unused and provides fuel for the dry season fires which sweep through the woodlands nearly every year, inhibiting dense growth. Although natural fires do occur, most grass fires are lit

by people and, to this extent, many wooded grasslands are modified fire-climax formations which would otherwise become dense woodlands and thickets.

As a wildlife habitat, the miombo is not uniformly impoverished. Within the *Brachystegia* woodlands, where there are extensive river valley, lake shore or floodplain grasslands, a few 'islands' of exceptional plant and animal diversity have been protected in national parks and reserves. The Selous Game Reserve (50,000 km²/19,000 sq mls), located in the *Brachystegia* woodland zone in Tanzania, is among the largest wildlife sanctuaries in Africa. It has one of the most diverse and abundant large mammal communities, including a population of elephants which numbered some 110,000 until the early 1970s, when the disastrous outbreak of ivory poaching in East and Central Africa reduced it to about 30,000. The Selous Game Reserve is a mosaic of forest, woodland and grassland communities with several permanent rivers, small lakes and seasonal floodplains, providing a unique variety of mammal and bird habitats and long habitat boundaries.

Another extensive climatic belt of wooded grasslands, the Sudan Zone, stretches from Senegal to Ethiopia. In the southern hemisphere it is represented by a belt through southern Angola, Zimbabwe and Mozambique, northern Botswana and the Transvaal, with an extension southwards into Swaziland. The greater part of both northern and

southern zones has been inhabited for a long time and the vegetation has consequently been profoundly modified by cultivation, cutting, grass fires and cattle grazing. Although climatically distinct from the Guinea and Sahelian Zones on either side of it, its tree and grass layers show botanical affinities with both. It owes its origin to the climatic changes that have repeatedly affected Africa since the Pliocene epoch and have caused the sub-humid and semi-arid vegetation zones bordering it to advance and retreat on several occasions.

In contrast with the relatively unnutritious and unused grass layer of the miombo woodlands, the grasslands of the East African semi-arid thorn savanna are very productive within the limitations of their variable rainfall. A high proportion of the grass biomass is normally consumed by a variety of animals, including wild and domestic ungulates (hoofed mammals), rodents, grasshoppers and termites; where sufficient material remains, it commonly fuels grass fires in the dry season. This vegetation zone encompasses the Serengeti region, the Masai Steppe of Tanzania and the dry *nyika* bush of Somalia and eastern and northern Kenya. It is a southward extension, in climatic and ecological terms, of the Sahelian Zone, and is epitomized by

● Water holes in the Tsavo National Park, Kenya. (WWF / Henry Ausloos)
● Plains zebra in the Ngorongoro Crater, Tanzania. (WWF / G.W. Frame)

## SERENGETI PLAINS

The Serengeti provides an excellent illustration of the functioning of a self-sustaining natural grazing ecosystem. Natural grasslands have evolved together with their native herbivores. Grazing by the full complement of herbivorous fauna, including rodents, ungulates (or the marsupial equivalents of both in Australia), grasshoppers, earthworms and termites (in the tropics), is a major influence in determining the species composition, physiognomy and succession of the vegetation. On the Serengeti Plains, the main consumer component of the ecosystem is the 16 species of herbivores constituting the ungulate community. (In the Guinea savanna of equatorial Africa earthworms form a correspondingly high biomass.) The short grass plains of the Serengeti, growing on soil deriving mainly from wind-blown ash from an extinct volcano in the Ngorongoro massif, support a plant community which includes ten common grasses, of which the dominant species are *Sporobolus marginatus*, *S. verdcourtii*, *Digitaria scalarum* and one sedge, *Kyllinga alba*. This vegetation is remarkable for its tolerance of frequent defoliation resulting from grazing during the growing season; it can lose up to 75–80% of its total production, a loss so high that, in other types of grassland, it would cause overgrazing and lead to soil erosion. There is evidently a delicate balance between soil, vegetation and animal impact on the short grass plains. *Andropogon greenwayi*, the dominant grass of the intermediate zone of the plains and endemic to the volcanic soils of the Serengeti region, has successfully adapted to intensive grazing by becoming unusually recumbent.

the many thorny trees, and shrubs of the genus *Acacia*. The umbrella thorn, *Acacia tortilis*, a tree 10-15 m/30-50 ft high, the typical flat-topped acacia tree of Africa's arid and semi-arid grasslands, is distributed from the Middle East to Namibia and is the only large tree occurring in desert *wadis* in the Sahara and the Arabian Peninsula. The grass layer of the acacia savannas varies from the sparse, feathery annual needle grasses of the sub-desert, with mean annual rainfall of about 100 mm/4 in, to the community of some 25 perennial grasses in the acacia woodlands at 500-700 mm/20-28 in rainfall. Common species include African foxtail, horse-tail *Chloris roxburghiana*, finger grasses *Digitaria* spp and African timothy grasses *Setaria* spp. But the most characteristic grass of the acacia savanna is red oat-grass *Themeda triandra*, a fire-adapted species which becomes dominant over huge areas as a result of regular annual burning and, late in the rains, colours the landscape pale brick-red. Perennial grasses are favoured by late fires, which burn off the dead hay after the transfer of nutrients to the storage roots. They are vulnerable to heavy grazing and trampling by livestock, which cause their replacement by annuals, as has occurred over the greater part of northeastern Africa, where the formerly dominant perennial,

the golden beard grass *Chrysopogon aucheri*, has become rare and localized.

The acacia savannas of eastern and southern Africa are well known as the habitats of the greatest assemblages of large wild mammals in the world. As wildlife habitat, the region is exemplified by the open plains and woodlands of the Serengeti ecosystem, which includes the Serengeti National Park and Ngorongoro Conservation Area in Tanzania, and the adjoining Masai Mara Game Reserve in Kenya. During the dry season some 2 million wild ungulates make an annual migration in order to take advantage of the highly nutritious wet season grazing on the eastern Serengeti Plains and the open woodlands in the wetter regions to the north and west. The wildebeest calving season coincides with the period when the short grasses of the plains are at their most nutritious.

The major consumer component of the acacia savanna ecosystem is a remarkable complex of specialist grazing animals: the African buffalo, the wildebeest, hartebeest, topi, impala, two gazelles, kob and waterbuck, two reedbucks, zebra and, in the dry savanna of northern Kenya and southern Ethiopia, Grevy's zebra. The African elephant is a mixed feeder in the acacia savanna, consuming large amounts of grass as well as the foliage of trees and shrubs.

The remarkable diversity of large thorn savanna herbivores is almost matched by the diversity of the large carnivores which prey upon them: lion, cheetah, African wild dog and spotted hyena. (The leopard is more of a woodland and forest predator.) Cheetah and wild dog are typically hunters of the open plains, running down their prey, the former relying on short bursts of speed (up to 120 kph/75 mph) and the latter on endurance and pack cooperation. In East Africa spotted hyenas are pack hunters as well as scavengers. Three jackals, black-backed, side-striped and common or Asiatic, overlap in eastern Africa but generally frequent different habitats (black-backed in open woodland, side-striped on hillsides and common on open plains) where they scavenge and hunt small prey. Lions, living and hunting in prides, are the only socially cooperative cats in the world – probably as an adaptation to hunting in savanna grasslands with little cover. Several smaller cats, including caracal, serval and African wild cat, inhabit savanna grasslands, preying upon small mammals and birds, while small predators – mongooses and civets, weasels, polecats and ratel – are also well represented. Rodents, of which numerous species occur in the savannas, are the main prey of these small carnivores.

The African acacia savanna supports a rich bird fauna, with 70-85 species breeding in wooded grasslands. Typical resident grassland birds include the large Kori bustard, several francolins, both helmeted and vulturine guinea fowl and secretary bird, while Caspian plover, Montague's and pallid harriers and steppe eagle are common winter visitors and passage migrants on the African plains.

The ostrich – the largest living bird – is as characteristic of Africa's grasslands as the zebra and gazelle with which it frequently associates for mutual vigilance. Its defence against predation is its running speed, although it can also deliver a powerful kick. The ostrich is one of several large flightless birds inhabiting the world's grasslands.

Wildlife and habitat research, carried out in the national parks and reserves, is one of many facets of conserving the African savannas. It is a remarkable story of the vision, dedication and efforts of the small number of founders, trustees and managers of the parks and wildlife organizations, meriting more than the brief reference that can be made here. While nearly all grasslands outside protected areas are being modified to varying degrees – and some severely degraded – by human intervention, Africa's national parks have managed to conserve many significant habitats in virtually natural condition, despite the problems imposed by developing (and struggling) economies and pressures on the land from rapidly expanding human numbers.

It is evident that the long-term survival of examples of natural grassland, like that of other major pristine ecosystems and landscapes, will come increasingly to depend on protected areas, the national park system in particular. Fortunately, with growing public understanding of environmental issues, the parks movement is now gaining strength in most countries, with some financial support from international sources. The growing tourist industry enables many African national parks to become at least partly self-financing. Although tourism is a necessary and, normally, a relatively benign use of the national parks, it is nevertheless an artificial intrusion which can become damaging. The presence of visitors inevitably causes disturbance to plant and animal life. Leaving aside the question of fire, the greatest potential for damage to vegetation within protected areas lies in the destructive effects of vehicle wheels upon grasses and other small plants. The proliferation of wheel tracks on grassland, many of them leaving virtually permanent scars leading to erosion, is a serious problem in some East African national parks. The logical solution – confining vehicles to prepared tracks – has been introduced in some parks but is long overdue in others.

The presence of wild ungulates is an important feature of many grasslands used by livestock. Over the greater part of the Masai Steppe in Kenya and Tanzania, indigenous grazers such as zebra, wildebeest, Thomson's gazelle and oryx can frequently be seen grazing close to domestic herds. The continued survival of wildlife in these areas has been largely due to the tolerance shown by the pastoralists, with whose cattle the wild grazing animals may compete for forage and water, and to their exclusion of hunting tribes from the region.

The grasslands of eastern Africa are the home of several pastoral tribes, including the Masai of Kenya and Tanzania, the Turkana of northern Kenya, the Boran of southern Ethiopia and the Karamojong in Uganda. They are the descendants of Cushitic and Nilotic peoples who moved into the region from the Nile valley and

## ECOLOGICAL RESEARCH

The ecology and behaviour of mammals, birds and many other animal species have been the subject of studies by biologists working individually or attached to the three research institutions which were established in East African national parks in the 1960s: the Uganda Institute of Ecology (formerly the Nuffield Unit of Tropical Animal Ecology) in Uganda's Queen Elizabeth National Park; the Tsavo Research Project in Kenya; and the Serengeti Research Institute in Tanzania. Each has taken advantage of the exceptional opportunities arising from its location in a national park to study the structure and functioning of natural savanna ecosystems, their research concentrating mainly on issues most relevant to the parks in which they were based.

In Tsavo National Park studies of elephant population dynamics, movements and feeding ecology were carried out in the 1960s and 1970s in response to the severe impact of a very large elephant population on trees and shrubs. The resulting change towards open grassland, compounded by the increasing incidence of grass fires, was reducing the habitat of the woodland mammals for which the Tsavo Park was the most important sanctuary in East Africa: black rhino, lesser kudu and gerenuk. Drought, causing high mortality among young elephants, appeared as a partial solution to overpopulation, but the outbreak of poaching, which reduced Tsavo's elephants from 45,000 in 1965 to 6,000 by 1985, removed the problem — it also removed the opportunity to learn how elephant populations are regulated under natural conditions — a question that may now never be answered.

Uganda's most pressing problem in the early 1960s also related to excessive numbers, this time of hippopotamus in the Queen Elizabeth National Park, and the effects their overgrazing was having on the grasslands bordering the lake shore and on other grazing animals.

The Serengeti Research Institute owed its origins partly to the serious drought that overtook the East African savannas in 1960 and 1961, causing large numbers of cattle to die and resulting in accelerated degradation of the grasslands of the Masai Steppe. The Director and Trustees of Tanzania National Parks were concerned that the Serengeti grasslands might be similarly degraded by the large herds of wild grazing animals. As very little was known about the dynamics of grazing ecosystems at that time, a research project was instituted to study the wildebeest and zebra, and their interactions with the grasslands. Although no indications of damage to the Serengeti grasslands were found, the research drew attention to the lack of information on the ecology of the park, and on tropical grazing ecosystems in general.

Since its establishment in 1966, the Serengeti Research Institute's main aim has been to gather scientific information relevant to park management. During that time, a considerable body of knowledge — the work of some 60 scientists — has been published. Their studies cover a wide field: grassland and woodland ecosystems, including the interactions of climate, geology, geomorphology and land form, soils, vegetation and animal populations, land use near the borders of the park and human influences (poaching, tree felling, tourism) within it. Among the topics to have received attention are: vegetation dynamics under the influence of climate, animals and fire; standing crop biomass and primary production in the grass layer in relation to climate and grazing; qualitative and quantitative consumption by wild grazing animals; woodland dynamics in relation to browsing animals and fire.

north-eastern Africa between 800 and 300 years ago, introducing cattle, sheep, goats, donkeys and camels into the drier areas. In the relatively short time since their arrival, they have exerted an influence on the ecology of the semi-arid East African grasslands out of all proportion to their numbers.

The system of traditional nomadism adopted by the pastoral tribes has enabled them to continue to use the sparse grazing of arid and semi-arid grasslands over a period of several centuries. The extent to which degradation of the former savanna grasslands of the Sahara and Arabian regions, and their reduction to desert in the last 6,000 years, has been due to pastoralism or to climate change is uncertain. Whatever the cause, the change has been relatively slow. Throughout the Sahel and north-eastern Africa the increasing trend towards settlement of nomadic pastoralists in permanent villages – generally encouraged by governments for administrative and political reasons, and for the convenience of establishing famine relief centres – combined with growing numbers, has led to foci of accelerated degradation spreading out from the villages.

# THE AUSTRALIAN GRASSLANDS

LOCATION: Western grasslands, approx. 21°–33° S and 115°–123° E. East-central grasslands, approx. 20°–33° S and 132°–145° E. Northern grasslands, approx. 13°–23° S and 123°–145° E. Northwest and central desert grasslands, approx. 18°–32° S and 123°–133° E.

AREA: Western grasslands, approx. 980,000 km² / 377,000 sq mls; East-central grasslands, approx. 1,265,000 km² / 485,000 sq mls; Northern grasslands, approx. 320,000 km² / 123,000 sq mls; Northwest and central desert grasslands, approx. 2,010,000 km² / 730,000 sq. mls. Open and wooded grasslands cover over 90% of Australia.

ALTITUDE: Between 100 m / 328 ft and 500 m / 1,640 ft.

● The emu is the Australian representative of a group of large, fast-running, flightless birds ideally adapted to grassland habitats. Other examples include the ostrich of Africa and the rhea of South America. The related cassowary of northern Australia and New Guinea is a rainforest dweller. (WWF / Fredy Mercay)

Natural grasslands, both wooded and open, occupy a broad zone spanning tropical and temperate regions between Australia's central desert to the west and Great Dividing Range to the east. Another wide belt of grassland is located between the desert and the southwest coast. Narrower belts of grassland border the desert on the north and south. They include a variety of communities which, in the tropical north, correspond ecologically to South American and, more particularly, to African savannas (with which they share several grass and tree genera) – although the term 'savanna' is not normally used in Australia. Like South America, Australia has neither indigenous bovids nor any other wild ungulates. The natural grazing niches are occupied by the large red and grey kangaroos, the combined numbers of which are estimated at 19 million. They are distributed, in varying densities, over almost all the continent's natural grasslands. Australia's commonest grazing mammal is the introduced rabbit; in some regions it consumes more grass than cattle.

Australia's grassland birds include the flightless emu, which is common in nearly all habitats except the driest deserts and densest forests. The large Australian bustard is widespread and common away from human settlement. The drier grasslands are the home of the budgerigar, immense flocks of which frequently congregate at waterholes.

Australian Aborigines were hunter-gatherers and did not cultivate the land. Their only domestic animal was the dog. The sheep and cattle introduced by British colonists had far-reaching, and locally disastrous, consequences for the more fragile grasslands and the native fauna associated with them, but have less effect in other regions as management improves.

Reaching in a discontinuous arc from the Kimberley Plateau in the north-west, across the Northern Territories into Queensland are the Mitchell Grasslands. Dominated by perennial tussock grasses *Astrebla* spp. (Mitchell grass), this semi-arid grassland community is unusually resilient and, unlike other Australian grasslands, has proved well suited as

pasture for commercial livestock production, particularly following recent advances in range management and animal husbandry practices.

The semi-arid woodlands lying to the west of the Great Dividing Range in New South Wales and extending into South Australia were formerly wooded grasslands. Heavy grazing and the mistaken suppression of fire, to which the indigenous grasses are adapted, resulted in the loss of the grass layer and its replacement by woody growth, leaving the surface of the soil almost bare. Range ecologists doubt whether the damage can be economically repaired.

The Mitchell Grasslands and semi-arid woodlands each occupy about 8% of the overall land surface of Australia, while 22% – mainly in the west and centre of Australia and including all the major deserts – is covered by 'hummock' grass, a name deriving from the rounded tussocks of the dominant perennial, evergreen, grasses of the genera *Triodia* and *Plectrachne*. This 'spinifex' vegetation is adapted to a harsh and variable climate, poor soils and frequent fires. The spinifex

bird (an Old World warbler) and the spinifex pigeon are characteristic birds of the hummock grasslands. Otherwise, it sustains few mammal and bird species; but it possesses the richest lizard fauna in the world. Among the most remarkable is the moloch (harmless despite its name), a 20 cm/8 in spiny, ant-eating agamid (lizard), the skin of which appears to absorb water like blotting paper – probably as an adaptation to the dry environment where dew is the main source of moisture. Water dropped on to it is not actually absorbed but quickly dispersed along minute channels between the lizard's scales. Once the water reaches its lips, the moloch drinks.

Officially designated as unoccupied Crown land, most of the region has been set aside as an Aboriginal 'reserve', or for conservation. Cattle are grazed in some areas at very low densities. In Australia there are no parks or reserves designated specifically for the conservation of natural grassland, although they receive incidental protection in some national parks as, for example, Ayers Rock–Mount Olga National Park (1,325 km²/511 sq mls).

# THE SOUTH AMERICAN SAVANNAS

LOCATION: *Llanos* of Venezuela and Colombia, approx. 3°–10° N and 63°–75° W; *cerrados-campos* tropical grasslands of Brazil approx. 5°–23° S and 40°–60° W; temperate sub-humid *pampas* of southeastern Brazil, southern Paraguay, Uruguay and northeastern Argentina, approx. 25°–35° S and 50°–60° W; semi-arid *pampas* of Argentina, approx. 30°–40° S and 57°–68° W.

AREA: *Llanos* approx. 1.3 million km²/500,000 sq mls; *cerrados-campos* approx. 1.5 million km²/580,000 sq mls; *pampas* originally 430,000 km²/166,000 sq mls, now very much reduced.

ALTITUDE: *Llanos* sea level to 1,000 m/3,281 ft; *cerrados-campos* 200 m/656 ft to 500 m/1,640 ft; *pampas* sea level to 1,000 m/3,281 ft.

● The giant anteater of the South American pampas is frequently shot on sight by the local people, even though it is not a pest, nor are its pelt or meat of any value. (Matthew Hillier SWLA)

Tropical South America (where the term 'savanna' originated) has large areas of natural and formerly natural grassland. They include the equatorial *llanos* plains of the Orinoco basin in Venezuela and Colombia, as well as the *cerrados* (humid savanna woodlands) and *campos* (sub-humid wooded grasslands) which occupy the upland plateau of central Brazil, to the south and east of the Amazon rainforest. To the south lie the dry temperate *pampas* of Argentina, Uruguay and southeastern Brazil, and the sub-desert grasslands of Patagonia; to the west the arid montane grasslands of the Andes from Ecuador through Peru, Bolivia, Chile and Argentina, reaching to Tierra del Fuego.

Although the South American continent is rich in animal life, most of the indigenous fauna is associated with forest. The savanna grassland fauna is comparatively impoverished. The semi-aquatic capybara, although much reduced in numbers, is still scattered throughout the seasonally flooded grasslands of South America. Besides being hunted for meat, this largest of living rodents (weighing as much as 80 kg/175 lbs) is kept in thousands on ranches, especially in Venezuela. As an animal naturally adapted to floodplain pastures, it has several advantages over cattle. At a stocking rate of almost one per hectare and with each female giving birth to two litters of four or five young a year, capybara herds produce over 60 kg of meat per ha/53 lbs per acre – at least four times as much as cattle.

The jaguar, ocelot and cougar are widespread feline predators in all habitats. Three indigenous canids occur in the savannas: bush dog, zorro and maned wolf. Formerly widespread on the savannas and northern pampas, the maned wolf is now largely confined to the Brazilian *cerrados*. With long legs enabling it to see above the vegetation, it is particularly well adapted to life in grassland habitats. The zorros, on the other hand, are short-legged, fox-like animals, all uncommon, of which two species – the Argentine grey fox and hoary zorro of Brazil – inhabit savannas.

Among the typical grassland ground birds are representatives of three South American endemic families, the primitive partridge-like tinamous (Tinamidae), the long-legged seriamas (Cariamidae) and the large, flightless rheas (Rheidae). The common rhea is the most characteristic bird of the temperate pampas.

As in Africa and Australia, the complement of common plants in each South American savanna community may be in the order of 50–200 species, if trees, palms, shrubs, grasses, sedges and other herbs are included. The Brazilian *cerrados* are assumed by ecologists to have derived through fire and clearance of the original *cerrado* dry forest, of which only relicts remain. They bear many physiognomic and ecological similarities to the African savannas, but tend to be richer in tree and shrub species. Unlike the wooded savannas of Africa, in which a few tree genera are dominant – notably *Acacia*, *Combretum* and *Brachystegia* – and those of Australia – *Eucalyptus*, *Acacia* and *Melaleuca* – the savannas of South America are characterized by more woody species, in roughly equal abundance. This is explained by more intensive human influence in Africa over a far longer period of time.

Every savanna is likely to contain 20–30 common grass species. In the *cerrado*, the grass cover is largely composed of tall perennial species that are fibrous or woody and of low forage value, the commonest genera being *Andropogon*, *Axonopus*, *Icanthus* and *Paspalum*. As in the African tropical savannas, the grasses tend to be tallest (3–4 m/9–13 ft) in the most humid regions adjacent to the rainforest and shorter in the drier areas. Where the Orinoco's swift-flowing, heavily silt-laden, tributaries are checked by the abrupt change of gradient as they emerge from the ravines in the eastern Andean foothills, they deposit their sediment on the flatter land. The obstructed rivers divide into branches (distributaries), spreading debris to form alluvial fans upon which the main area of Colombia's extensive *llanos* flood plain grasslands have developed. There the commonest grasses include *Trachypogon* spp. and *Axonopus*.

Towards the south of the Mato Grosso, in the transition zone between the tropical and the temperate region, the *cerrado* is being progressively replaced by *campos limpos* – grassy shrub savanna occurring on poor soils of leached sands or compacted clay. The vegetation, formerly rich in tropical species and valued for livestock forage, has been degraded by frequent burning and heavy grazing. Although the region retains most of its tropical flora, it is being colonized by grasses characteristic of the temperate pampas of Uruguay and Argentina to the south, *Stipa*, *Bromus* and *Aristida pallens*. In recent years natural savannas in Brazil have been extensively replaced by pastures of artificially introduced African guinea-grass, *Panicum maximum*; forest clearings in Amazonia, Parana and Mato Grosso have also been planted with this species to provide forage for cattle. Other African grasses – *Hyparrhenia rufa*, *Digitaria decumbens* and *Brachiaria* spp. – are planted on poorer soils in the *cerrados* and *campos limpos*.

The basin of the Paraguay River and its tributaries forms a vast depression in the centre of South America that includes two major biogeographical areas – Pantanal and Chaco – each containing

## SOUTH AMERICAN DEER

Three native deer species, white-tailed deer, swamp deer and pampas deer, inhabit the South American savannas. The white-tailed deer, widespread in North and Central America, occurs in the region north of Brazil and is the only deer to be found in the *llanos*. The pampas deer has a wide distribution from the Brazilian *cerrados* to the Argentinian pampas, overlapping that of the swamp deer, which is restricted to the southern *cerrados* and the Brazilian and Uruguayan pampas. As browsers, all three deer species are associated more with trees and shrubs than with grass. In San Luis province of central Argentina, in one of the few remnants of true dry pampa, about 150 pampas deer of the rare Argentinian subspecies *Ozotoceros bezoarticus celer* – half the surviving wild stock – live on a 30,000 ha/74,000 acre *estancia* where cattle are grazed at low density. This *estancia* also possesses a representative tract of the almost equally rare temperate grassland biome of South America.

## FLIGHTLESS BIRDS

Among the birds of the pampas, the common rhea, the largest bird on the American continent, is sometimes called the South American ostrich. It is much smaller than the African ostrich, to which it bears only a superficial resemblance. The ostrich, the rheas and the emu of Australia, although unrelated, show evolutionary convergence in their adaptation to similar ecological niches. Together with the forest-dwelling cassowaries (Casuariidae) of Australia and New Guinea and kiwis (Apterygidae) of New Zealand, they are grouped as 'ratites' (flightless birds with no keel on the breastbone). The grassland ratites of the three continents are long-legged, nomadic birds living in small flocks (occasionally congregating in large numbers). They escape from predators by running, and feed on mixed diets of insects, reptiles, and small mammals as well as leaves, fruits and seeds. They exhibit very similar reproductive behaviour, with the males, which are polygamous, incubating the eggs and shepherding the chicks (although the female ostrich shares in these tasks).

The flightless ratites might be expected to be particularly vulnerable to human persecution and it is perhaps surprising that they have survived in all three continents despite considerable disturbance from hunting and habitat destruction (mainly due to agriculture). Ostriches were exterminated in the Arabian Peninsula as late as 1940 and have disappeared from the Egyptian and Nubian deserts.

They are protected in the national parks of eastern and southern Africa. Emu are common and widespread in Australia, despite determined campaigns to eliminate them from farmland. Rheas formerly roamed the Brazilian and Argentine pampas in great flocks, but the advance of agriculture and intensive hunting for food and sport decimated them. Both the ostrich and the rheas tame easily and have been raised commercially for their meat and feathers; ostrich skin makes a fine soft leather.

Modern man has exterminated many large flightless birds, among them two groups of ratites, the last representatives of which died out between 700 and 500 years ago. Twenty-two species of moa (Dinornithidae) in New Zealand and eleven species of elephant bird (Aepyornithidae) in Madagascar and South Africa have been identified from remains in the form of subfossil bones, feathers and eggs. The largest moa was 4 m / 13 ft tall and the smallest about the size of a turkey. The elephant bird, which lived in Madagascar, stood only 3 m / 9 ft high, but was the heaviest bird known, weighing about 450 kg / 1,000 lbs. Its enormous eggs, with a capacity of 9 litres / 2 gallons may have given rise to the legend of the 'roc' of the Arabian Nights. It seems probable that some of the larger elephant birds and moas frequented grasslands but, as the people who hunted them left no written or pictorial records, we cannot be sure.

---

distinctive grassland systems. The Pantanal occupies approx. 130,000 km² / 50,000 sq mls of low-lying alluvial plains on the borders of Brazil, Bolivia and Paraguay. Here Amazonian rainforest outliers in the north meet the dry forests of the Chaco to the west and the *cerrados* in the centre. Much of this region is subject to annual or irregular flooding and supports grasslands (including the dominant grasses *Paspalum intermedium* and *Sorghastrum agrostoides*) with scattered tree formations in which *Copernicia alba* and *Tabebuia nodosa* are common. Periodic drought, as occurred between 1962 and 1971, resulting in greatly decreased flooding, is responsible for increased tree

densities and the spread of woodlands into formerly open grassland. Swamp areas are occupied by giant papyrus-like sedge *Cyperus giganteus* and aquatic monocotyledons, including the indigenous water hyacinth *Eichhornia*, which has become a major weed pest in tropical lakes and rivers around the world. In the Pantanal, flooded grasslands and the margins of swamps are grazed by cattle and locally occupied by swamp deer and capybara.

The Chaco, lying astride the Tropic of Capricorn, covers approx. 800,000 km² / 310,000 sq mls of ancient land surface (peneplain) in Bolivia, Paraguay and Argentina. The climate, which features a

variable annual rainfall (500–1100 mm / 20–43 in) and a pronounced dry season, supports *chaqueana* vegetation – originally dry forest covering large parts of northern Argentina and Paraguay, but subsequently greatly modified by clearance and burning. This has given rise to woodlands of *Schinopsis lorentzii* and *Aspidosperma quebracho*, with an impoverished grassy herb layer (grazed by livestock), and a variety of shrub formations without herbaceous understorey, but whose foliage and fruits are seasonally important as forage. Grasslands occur in seasonally flooded areas or result from burning; but while widespread overgrazing limits the spread of fires, it is responsible for the expansion of dense shrublands at the expense of grass besides causing serious soil erosion. The Chaco region presents a picture of progressive degradation.

### The Pampas

Very little remains of the South American temperate grasslands, which originally extended over 430,000 km² / 166,000 sq mls. As in the tall-grass eastern prairies of North America, the higher rainfall grasslands of eastern Argentina have been ploughed for cereal cultivation. The present-day pampas, which have replaced the natural dry grasslands of western Argentina, are secondary grasslands. Three hundred years of cattle grazing has suppressed the dominant native species and permitted colonization by many grassland plants from adjacent highland and sub-tropical habitats. With over 400 species of grasses, the large, well-managed *estancias* have a greater diversity than the original grasslands. But recent economic pressures and changes of ownership, with subdivision of properties and increased cattle stocking rates, have impoverished large tracts of pampas.

No true pampas grasslands are included in protected areas, although the eastern part of Nahuel Huapi National Park (4,758 km² / 1,837 sq mls) on the slopes of the Andes in southern Argentina, includes approx. 1,000 km² / 385 sq mls of dry, semi-natural pampa plains in which temperate grasses – fescues *Festuca* and meadow grasses *Poa* – occur with sub-desert needle-grass *Stipa speciosa*. In Nahuel Huapi Park small herds of guanaco graze the lower and montane grasslands and vicuña the alpine meadows.

● Seasonal flooding in the Chaco of Paraguay. The vast Chacó region of Paraguay, Bolivia and Argentina has been progressively degraded by tree felling, burning and overgrazing. (WWF / Hartmut Jungius)

● The maned wolf is now largely confined to the Brazilian *cerrados*, the humid savanna woodlands to the south and east of the Amazon rainforest, although it was once widespread across all the South American savannas and even the northern pampas. (Planet Earth Pictures / Richard Matthews)

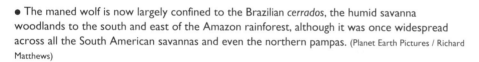

● The most characteristic species of the North American prairelands was the bison. Once numbered in millions, the plains bison was almost wiped out by the end of the 19th century. Careful management in reserves such as Canada's Wood Buffalo National Park, enabled numbers to increase. This photograph was taken in the Grand Teton National Park in Wyoming. (Planet Earth Pictures / Franz J. Kamenzind)

# THE NORTH AMERICAN PRAIRIES

LOCATION: Approx. 30°–50° N and 85°–110° W.

AREA: Approx. 1.47 million km²/575,000 sq mls.

ALTITUDE: Between 600 m/2,000 ft and 1,675 m/5,500 ft.

The first Europeans to enter central North America early in the 19th century found much of the region a vast natural grassland. Lying between the Rocky Mountains in the west and the Great Lakes in the east, it stretched from the *taiga* forests in the north to the semi-arid Mexican plateau in the south. The colonization of the North American prairies by Europeans, the near-extermination of the bison population, variously estimated at between ten and one hundred million animals, the displacement of the native Indians from their tradi-tional homelands, and the virtual de-struction of a remarkable and highly suc-cessful human culture must be viewed as one of history's great human and envi-ronmental tragedies. But it was only one episode in a saga of pillage and plunder that practically destroyed the grasslands and forests, together with their native fauna.

The bison was the emblem of the American frontier. Like all the other apparently limitless resources of the west, including the grasslands themselves, it was also a resource to be exploited. The great buffalo slaughter began at the time of the American Civil War with almost every element of colonial society participating. The army wanted to starve the Plains Indians into submission; cattlemen wanted forage for their livestock; hide hunters wanted hides and tongues to trade; sportsmen killed for trophies and pleasure. Buffalo hunting was for a time the main occupation on the plains and a few men achieved fame as prodigious buffalo hunters, contributing to the saga of the 'conquest' of the prairies. As the massacre continued, there was concern that the bison would be completely annihilated. But General Sheridan proclaimed to the Texas legislature that the buffalo hunters were more effective in subjugating the Plains Indians than the army had been in 30 years and that the extermination of the buffalo was the only way to bring lasting peace and civilization to the region.

So thorough was the slaughter that when, in 1887, the American Museum of Natural History tried to procure specimens, they had great difficulty in finding any. Fortunately, some remnant bison herds in Montana, Wyoming and southern Canada were overlooked by the hunters. Their descendants survive in a few wildlife refuges in the west – living symbols of the catastrophe which the colonists brought upon the grasslands and forests of North America.

Like the eastern pampas of Argentina and the northern steppes of Central Asia, the wetter tall-grass areas of the eastern North American prairies have been transformed into arable croplands, much of it growing cereals – cultivated grasses grown as monocrops and dependent upon intensive management. The damage to the original grassland ecosystems goes a great deal further than the obvious visible changes. Tillage completely destroys the natural plant cover and, in combination with the use of chemical fertilizers and pesticides, also changes the microbiological components of the native soils, depleting the organic content which has accumulated over the centuries.

The drier prairie, steppe and savanna grasslands which cannot sustain agriculture have survived as grazing lands, though modified and managed to varying degrees and supporting differing regimes and intensities of animal husbandry. They include the semi-arid grazing lands of the traditional nomadic and transhumant pastoralists of eastern Africa and the Mongolian steppes, the sub-desert grasslands grazed by the sheep, goats and camels of Bedu and Berber nomads in the Middle East and North Africa, and cattle ranches in the US, Argentina, Australia and southern Africa. In all these exploited grassland ecosystems, perhaps the most conspicuous feature of their transformation from the natural state is the replacement of their original diverse native faunas by a few domestic animals, above all by cattle.

Parts of the North American prairie grasslands, mainly in the north and west, although modified to varying extent, have retained many of their native grasses and herbs. Vegetation ecologists have recognized three main types of grasslands which are related to climatic conditions.

## The Mixed-Grass Prairie

The mixed-grass prairie is a very large region which extends roughly down the centre of North America from southern Alberta and Saskatchewan in Canada to northern Texas, lying between the Rocky Mountains in the west and approximately along the 100th meridian in the east. In the southwestern part of this region, mainly in the states of Colorado and New Mexico, the mixed prairie merges in the west with the drier short-grass prairie and, in the east, from Manitoba through Missouri and Oklahoma to central Texas, with the wetter tall-grass (true) prairie.

In the tall-grass, which has largely been replaced by arable land, three open grass communities are recognized in which the almost ubiquitous genus *Andropogon*, blue stem grasses, is dominant, with *A. gerardi*, *Stipa spartea* and the fescue *Festuca scabrella* among the commonest species. The tall-grass also extends as the understorey into the western margins of the broad-leafed forests of the eastern US which, today, are mostly secondary formations.

The mixed-grass prairie features several associations, some of which are closely related to the tall-grass communities. In the south, *Andropogon scoparius* is a common dominant, a role filled towards the north by the wheatgrasses and needle-and-thread grasses. Blue grama and buffalo grass are common throughout the mixed-grass prairie.

In the US and Canada cattle ranching was established in the mixed-grass prairie by the 1870s, stimulated by the building of the railways which brought the early settlers. The farming of wheat and other cereals began in the eastern grasslands in the 1860s and expanded westwards until by the end of the century the original grasslands had virtually disappeared, together with the vast herds of bison and pronghorn associated with them. The remaining rangelands were grazed by domestic cattle. One outcome of arable farming and cattle ranching was a serious increase in the numbers of several species of rodent. Overgrazing by cattle produced conditions favouring the pocket gopher and ground-squirrel populations. The black-tailed prairie dog was particularly abundant: in 1905 there were estimated to be 800 million in Texas alone. These rodent irruptions – which were accompanied by a corresponding

## RE-CREATING TALL-GRASS PRAIRIE

By the beginning of the 20th century little, if any, of North America's natural tall-grass prairie remained. The importance of conserving examples of the prairie was first recognized by two ecologists, John T. Curtis and Aldo Leopold when, in the mid-1930s, they established a small prairie restoration experiment at the University of Wisconsin Arboretum in Madison by planting native species in simulations of natural communities. Today, at that site, now called the Curtis Prairie, 24 ha / 60 acres of grassland have been established with over 350 indigenous plant species. In the mid-1940s, the botanist Henry Greene began restoring a second site at the Arboretum, now 16 ha / 40 acres of prairie and considered an outstanding example of restored natural grasslands.

increase in the numbers of coyote – were countered by systematic poisoning and trapping campaigns, encouraged by government bounties. So successful were they that several once-abundant species were reduced to scarcity and became very localized.

## The Short-Grass Prairie

The short-grass prairie, lying along the eastern foot of the Rocky Mountains in Colorado and New Mexico and just reaching northwestern Texas, is dominated by blue grama and buffalo grass. These short grasses are favoured by heavy grazing and tend to predominate on ranches in the mixed-grass prairie as well. Precipitation in the short-grass region is normally too low for agriculture but suffices for stock raising. Its greatly depleted wildlife communities are similar to those of the mixed-grass prairies.

## Other Types of Natural Grasslands

Vegetation ecologists recognize four further natural grassland communities in North America:

*the shrub steppe*, situated mainly in the dry inland valleys of Washington and Oregon states which includes wheat grass, fescue and sagewort shrubs;

*annual grassland* in California, containing many introduced annual species suited to the Mediterranean type climate, and deriving, through grazing pressure, from the former indigenous perennial grassland;

*desert grassland* in Texas and New Mexico where the dominant grasses *Bouteloua* and *Hilaria* grow among xerophytic shrubs such as *Larrea divaricata*, *Prosopis* spp. or *Yucca* spp;

*montane grassland* below the tree-line on the Rocky Mountains, in which wheat grasses and fescues are the dominant grasses in meadows and mountain wild flowers grow in abundance.

## What Future have Grasslands?

Despite the degradation of most of the earth's natural grassland regions, there is room for some optimism about their future. The grasses, as a single family, constitute a very high proportion of the total living material on the land surface of the earth. They show great resilience under conditions of heavy use and stress.

## PRINCIPAL GRASSLAND TYPES

### Natural (edaphic) Grasslands

| Tropical | Examples | Temperate | Examples |
|---|---|---|---|
| floodplain | Kafue Nat. Pk. (Zambia) | floodplain | Neusiedler See (Austria) |
| sub-humid medium grassland | Selous Game Reserve (Tanzania) | mixed-grass prairie and steppe | *Prairie Nat. Pk. (Canada) |
| semi-arid perennial grassland | Serengeti Nat. Pk. (Tanzania) | short-grass prairie and steppe | Rocky Mountain Nat. Pk. (USA) |
| semi-arid annual grassland | Sibiloi Nat. Pk. (Kenya) | semi-arid grassland (pampa) | Nahuel Huapi Nat. Pk. (Argentina) |
| arid annual grassland | *Kalahari Gemsbok Nat. Pk. (Botswana) | tundra | Kandalakshskiy Reserve (Russian SFSR) |
| alpine (afro-alpine) | Ruwenzori Mts. Nat. Pk. (Uganda) | alpine | Vanoise Nat. Pk. (France) |

### Derived and Successional Grasslands

| Tropical | Examples | Temperate | Examples |
|---|---|---|---|
| montane | Mount Kenya Nat. Pk. (Kenya) | montane | Abruzzo Nat. Pk. (Italy) |
| humid tall-grass | *Garamba Nat. Pk. (Zaïre) | tall-grass prairie and steppe | *Curtis Prairie (Restored) (USA) |
| wooded successional grassland | Tarangire Nat. Pk. (Tanzania) | forest secondary grassland | Wood Buffalo Nat. Pk. (Canada) |
| managed grassland | Laikipia Plains (Kenya) | managed grassland | Temperate rangelands (E. Australia) |

* Protected areas set aside primarily to safeguard grassland ecosystems.

They tolerate, and are adapted to, more physical damage, including repeated defoliation and burning, than any other group of plants. Under intensive use they are capable of tolerating great changes in their community structure and functioning, and many species are amenable to cultivation. Humans are more dependent upon the grasses than on any other plant family. Most people's diet includes a large proportion of the seed of three cultivated grasses – rice, maize and wheat – while 11 other grasses produce important, but less universally consumed grains and one – sugarcane – supplies more than half the world with sugar. About 70 million people are directly dependent on grasslands for their livelihood or way of life.

As the necessity for conserving the earth's biological diversity becomes more generally understood, it must also become evident that preserving the integrity of the grasslands is at least as important as preserving the forests and wetlands.

# WETLANDS

*All offices of nature should again*
*Do their due functions.*
Shakespeare, *Cymbeline*

Wetlands are found in every part of the world and in every type of climate. Covering about 6% of the earth's land surface, they range from huge estuaries and lagoons to tiny mountain tarns and hot springs. Their waters can be either static or flowing, fresh or brackish. They include coastal wetland and freshwater marsh, lake and pond, peatland, bog and fen, floodplain and swamp forest, to mention only some of the better known categories.

The earliest cities were established along the floodplains of the Tigris and Euphrates and other great rivers, and a high proportion of the world's population continues to live in river valleys, along coasts or is in other ways dependent on wetlands. Yet, despite their importance, wetlands have generally been belittled. Most people look upon them as pestilential sumps, breeding grounds for mosquitoes and other noxious insects that are carriers of diseases such as malaria, yellow fever and bilharzia. Indeed, their association with disease and waste has long encouraged the view that wetlands are fit only to be drained, dredged, or in-filled to make way for some more acceptable use, and means that they are among the first targets of the developer. Reclamation of wetlands is seen as a beneficial form of development and thus often attracts government subsidies. Price support schemes, by means of which artificially high prices are paid for growing certain crops on wetlands, provide further inducement. But perceptions are gradually changing,

and traditional attitudes are increasingly being questioned. There is a growing awareness that wetland communities, if properly managed, can be more productive in terms of plant matter than conventional agricultural land.

Rice, a wetland plant, is the principal food for more than half the world's population. Oil palm, originally from the wetlands of West Africa, is an exceptionally important source of edible vegetable oil, giving a higher yield than any other plant. Sago palm, the 'breadfruit tree', native to the swamps of Southeast Asia, is the high-yielding source of sago flour, the staple food of the Papuans of New Guinea, the Amerindians of the Amazon/ Orinoco floodplain, and many other peoples. Given that it can yield about 7– 9 tonnes of starch per hectare, it would make more sense to grow sago on wetlands instead of draining them to grow conventional crops with a much lower yield.

Some important inedible plants are also wetland products. Reedbrakes are among the most biologically productive habitats on earth. Reeds and palm fronds

are widely used for thatching, matting and in manufacturing paper, and willows are used for basket weaving.

## Mangrove Swamps

One of the most important wetland plants is mangrove, of which there are about sixty species, ranging from the red and black mangroves of equatorial South America, that reach a height of 40–50 m/ 130–160 ft, to shrubs less than 1 m/3 ft high. Mangroves grow in the intertidal zone and are salt tolerant. The greatest concentration is in the Indian Ocean– West Pacific region, and especially in the area extending from Vietnam, Thailand and Malaysia to Sumatra, Java and Borneo. Indonesia has more managed mangrove swamps than any other country in the world, estimated at 38,000 km² /14,670 sq mls, over 75% of which occur in Irian Jaya (western New Guinea). The largest unbroken stand of mangroves – covering almost 10,000 km²/3,860 sq mls – lies in the Ganges delta, along the Bay of Bengal. In areas such as this where cyclones, storms and flooding are frequent, mangroves have a particularly important protective role in stabilizing coastal lands.

Mangroves have long been a valuable article of commerce in tropical countries. The wood is highly durable and used for such purposes as building poles, charcoal and firewood; the bark has a high tannin content, which makes it suitable for use in the manufacture of leather. Many people in Asia, Africa, Latin America and elsewhere follow a way of life that is

● Floodlands at the north end of Lake Baikal, Siberia. Lake Baikal is, by volume, the largest body of fresh water in the world, and contains many unique species, including a freshwater seal. (WWF / Marek Libersky)

● Male and female fiddler crabs on mudflats in Western Australia. One of the male's claws is greatly enlarged, and is waved in a characteristic way to attract females. (Planet Earth Pictures / Purdy and Matthews)

entirely dependent on harvesting mangrove products.

Hundreds of species of plants and animals live in association with mangroves and the mudflats that are an integral part of them. As well as fish, crustaceans (shrimps, prawns and crabs) and shellfish (oysters, clams and mussels), they provide ideal conditions for birds and mammals of many kinds. Mangrove forests are thus a rich source of food, ranging from meat and birds' eggs to honey and other forest produce. Indonesia's swamp forests, for example, produce valuable tropical timbers. In the early 1980s, one species of dipterocarp, *Ramin*, alone accounted for almost half (45%) of the country's sawnwood exports, valued at US$119 million a year.

Overexploitation has destroyed many mangrove forests. Agricultural development (mainly in the form of rice paddy and wood chip production) accounts for much of the loss. As mangroves are mainly located in some of the world's poorest countries, the temptation to exploit them to destruction for short-term gain often proves irresistible. Haiti is one of a number of Caribbean islands that has already lost virtually all its mangroves. The destruction of mangrove forests deprives people not only of the mangroves themselves but also of important nurseries for fish and crustaceans, and the ecological base that sustains them. The effect of their loss is reflected in the breeding cycles of species inhabiting seas and oceans far from the forest itself. Extensive clearance of mangroves in Ecuador has affected shrimp reproduction to the extent that there are no longer sufficient larvae to stock shrimp ponds; almost half the ponds are now empty.

## Wooded Wetlands

Wooded wetlands – woodlands growing in association with wetlands – occur in many parts of the world. The largest, the Pantanal, shared between Brazil, Paraguay, and Bolivia, is a vast complex of seasonally flooded wetlands and interconnecting lakes, covering up to 130,000 km²/50,000 sq mls. Deforestation and the expansion of agriculture taking place along the Paraguay River some distance upstream have altered the flood regime, thereby affecting the Pantanal's ecology and reducing its productivity. An international management plan, agreed by the three countries concerned, is vital for the future of this exceptionally important wetland.

## Peatlands

Peatlands are tracts of dead plant matter in which conditions of low temperature, high acidity, low nutrient intake, waterlogging and oxygen deficiency combine to slow down the normal processes of decomposition, thereby allowing the organic matter gradually to accumulate in the form of peat. Deposits of peat often form in association with swamps and marshes, floodplains and coastal wetlands; but peat also occurs in a variety of distinctive types of wetland such as fen, moor, bog and muskeg. Peat is found on every continent, occurring in layers frequently 10 m/33 ft or more deep. Though rare in Africa, they are widespread in Southeast Asia, particularly in Borneo, Sumatra and Irian Jaya. They also occur in parts of the Amazon basin and on some Caribbean islands. All told peatlands are estimated to cover 5 million km²/2 million sq miles.

As well as sustaining many commercially valuable trees, peat swamp forests are important as watershed areas, serving as natural reservoirs which, by absorbing and storing excess water, help reduce flooding. In many northern countries peat is the traditional domestic fuel, and is being increasingly exploited for horticultural and industrial purposes. Third World countries are being encouraged to use it in order to reduce their reliance on fuelwood and on imported oil and coal. But although stocks are large, peat is nonetheless a finite resource. Mechanized peat cutting, together with reclamation and drainage for agricultural and other purposes, is greatly depleting these important wetlands. Britain's peat bogs are almost exhausted – only 4% are still intact – and it is claimed that if commercial extraction continues at its present rate they will disappear within 20 years.

## Floodplains

Floodplains are temporarily flooded lands commonly found on the lower reaches of rivers throughout the world. After the flood waters have receded, they leave permanent or semi-permanent areas of standing water. Seasonal flooding sometimes takes place on a titanic scale as, for example, in the Lower Mekong basin.

Despite their impermanence and shallowness, floodplains are important dry season refuges, not only for fish, birds and wild herbivores but also for pastoralists. By serving as dry season grazing areas, they are a vital component of the nomads' grazing strategy. As the floodwaters recede, the newly exposed grasslands provide rich grazing for large numbers of animals, both wild and domestic, at a time when grazing elsewhere is sparse. Many wild animals depend on the floodplain to see them through the dry season. Even agriculturalists have to adapt to the rhythm of the floodplain cycle, planting rice as the river rises, harvesting it when the waters recede.

In the heavily industrialized regions of Europe and North America, intensive development of riverine wetlands has resulted in the disappearance of all but a few floodplains. While human needs are responsible for the destruction of some wetlands, they have caused others to be created, such as dams, reservoirs, canals, even sewage farms and abandoned gravel pits. Though these artificial wetlands cannot compare with natural wetlands, there are places, particularly in the vicinity of densely populated urban areas, where they are almost the only viable wildlife habitat that remains. They are also important for recreational purposes.

## Coastal Wetlands

River deltas and estuaries, coastal wetlands and salt marshes are the breeding grounds and nurseries for all manner of marine life, including many economically important species of fish. Two-thirds of the world's fish catch begins life in tidal wetlands. Their still waters protect both eggs and fry from strong currents and turbulence, while the dense aquatic vegetation provides cover from predators. With some species – shrimps, for example – spawning takes place at sea, the larvae migrating to wetland nurseries along the coast. With others, the adult fish move into the wetlands to spawn. While some marine species are content to go no farther than an estuary, others migrate over long distances, making their way up large rivers such as the Amazon, the Niger or Mekong.

Coastal wetlands can thus be seen as an integral part of the marine system, with a vital role in the life cycles of aquatic organisms forming the base of marine food chains; they sustain not only fisheries but an abundance of other forms

## THE RAMSAR CONVENTION

The Ramsar Convention – the Convention on Wetlands of International Importance Especially as Wetland Habitat – named after the Iranian town close to the Caspian at which the convention was adopted in 1971, provides a framework for cooperation at the international level in the conservation of wetland habitats, and is the principal global forum for promoting conservation of wetlands throughout the world. The main purpose of the convention is to halt the decline of wetland habitats and to ensure that they remain ecologically viable. The Convention places contracting states under an obligation to safeguard designated wetlands under their control and, in particular, those on the List of Wetlands of International Importance. The Ramsar Bureau, established in 1988, is the medium through which the aims of the convention are pursued and international collaboration enhanced.

By 1992, a total of 546 sites in 65 countries covering a total of 329,140 km²/127,090 sq mls had been listed under the Ramsar Convention as wetlands of international importance. Impressive though this figure may be, it represents only a small proportion of the more important wetlands. Moreover, it includes some areas that, despite their protective status, are inadequately managed.

An important advance is the establishment of the Ramsar Wetland Conservation Fund to assist Third World countries in conserving suitable wetland sites.

of wildlife. Fiddler crabs, for instance, occur in astonishing numbers – densities of 19.75 million per ha/8 million per acre have been recorded on an island off the Atlantic coast of North America. As fiddler crabs are among the principal prey of birds, raccoons and other marshland inhabitants, they are an important constituent of the food chain and of marshland ecology.

### Freshwater Wetlands

Freshwater wetlands are also, of course, highly productive. Some of the largest are to be found in Indonesia, where more than a quarter of the land surface area consists of wetland of one kind or another. As well as natural wetlands, Indonesia has some of the most extensive artificial wetlands in the world, in the form of rice paddies and fish ponds, mostly on densely populated Java.

Pisciculture – fish farming – is one of the most fruitful sources of food, yielding far more than land of comparable size, and has been practised for centuries. The monastic fishpond was as indispensable a feature of European medieval culture as the dovecote and the deer park. Fishponds continue to be an important source of food, particularly in the Third World. More sophisticated methods of fish husbandry are being developed for example in Scotland, where the dearth of wild salmon has given impetus to the intensive 'farming' of salmon, although this has led to problems and criticism by some ecologists.

Aquaculture – covering a variety of species, from shrimps and oysters to turtles and crocodiles – is also becoming increasingly popular. Alligator farming is a growing industry in the southern USA. Multi-cropping systems are being developed which combine crayfish farming with rice production, for example, or timber and protein production. Having seen the advantages of aquaculture, many Third World countries are intensifying production as a means not only of feeding their own people but as a valuable source of foreign exchange.

### Faunal Diversity

Wetlands are dynamic ecosystems, subject to change, and greatly influenced by factors beyond their boundaries. Ecological diversity derives from the differing vegetational zones resulting from variations in water depth and nutrient levels. As vitally important life support systems, they sustain spectacular wildlife communities, ranging from algae and insects to birds and mammals, and are important refuges for both aquatic and terrestrial animals, among them valuable furbearers such as beaver, coypu, muskrat and otter. An indication of their value can be gained from the US$43 million turnover of Canada's fur trapping industry in 1976, while in Uzbekistan and Turkmenistan muskrat trapping is also an important industry.

Other characteristic aquatic or semi-aquatic mammals include such species as the great Indian and Javan rhinoceroses in Asia, hippopotamus and lechwe in Africa, and capybara in South America. Mangroves and wooded wetlands are the chosen habitat of the proboscis monkey in Borneo, the spotted cuscus in Papua New Guinea, and of course crocodiles, caimans and alligators. Mangrove forest in the Ganges delta is the most extensive habitat remaining to the Bengal tiger, and the floodplain of the Pantanal carries the largest number of jaguars in Latin America. Estuaries and shallow coastal waters in some tropical regions are inhabited by the aquatic plant-eating mammal, the manatee, while P'o-yang Lake, in southern China, is the wintering ground for more than 90% of the world population of the Siberian crane.

Destruction of certain key species can have widespread, and often unexpected, ecological consequences. Elimination of the Nile crocodile, for example, seriously affected the biological equilibrium in waters where they were previously common. Crocodiles generally prefer to eat the coarser types of predatory fish: the economically important species, such as *Tilapia*, are seldom included in their diet. Immature crocodiles eat freshwater crabs, predatory giant water beetles and many other creatures that prey on fish, fry and eggs, while adult crocodiles take a heavy toll of fish-eating animals such as otters, marsh mongooses, snakes and monitor lizards, as well as cormorants, darters and other fish-eating birds. They also keep the shores and waters clean by acting as scavengers, performing functions similar to those of hyenas and vultures on land, and equally essential. Loss of crocodiles results in a rise in the numbers of cannibal fish, such as barbel and lungfish. The numbers of omnivorous crabs and other invertebrates also increases, taking a heavy toll of eggs and fry. Similar discoveries have been made in respect of the two Amazonian crocodiles, the black caiman and yacare caiman. Both feed mainly on fish, but their disappearance from some areas has been followed by a sharp decline in fish populations. Crocodile excreta has been shown to be a key element in the food chain.

## Wetland Birdlife

Wetlands are rich in birdlife, particularly waterfowl, which are the most dependent on wetlands for survival. Birds are seasoned intercontinental, even inter-hemispheric, travellers. If it were not for migration, the size of bird populations would be limited by the amount of winter food on the breeding grounds. As most migratory birds are insect-eaters, the decline of their food supplies, induced by the onset of winter, forces them to move to warmer regions where insects are abundant. Waterfowl which breed on the arctic tundra overfly many countries along centuries-old flyways to their wintering grounds.

● Mangroves in the Daintree National Park, Queensland, Australia. Mangroves grow between the low and high tide marks in many tropical and subtropical areas. Their aerial roots help to trap silt, and play an important role in stabilizing coastlines. (WWF / Paul Trummer)

The wetlands of the Eurasian steppe are the principal breeding grounds for huge numbers of migrant waterfowl, birds from western Eurasia wintering in Africa, southwestern Asia, the Mediterranean and southwestern Europe, while those from eastern Eurasia winter in southern Asia. Although most of the numerous shallow lakes, both fresh and brackish, scattered across the steppe are individually small, their collective area is substantial. The most important lake complex on the steppe is in the Turgai region, where 5,000 lakes totalling 4,856 km²/1,875 sq mls are used by waterfowl both for breeding and moulting, and as a concentration zone prior to migration. Among several similar lake complexes on the steppe, the southern Ishim region has 2,700 lakes with a surface area of more than 4,047 km²/1,562 sq mls. One of the richest is the Kurgaldjin-Tengiz complex in central Kazakhstan containing 3,900 lakes, including Lake Kurgaldjin (360 km²/139 sq mls) and Lake Tengiz (1,497 km²/578 sq mls). Besides being frequented by large numbers of wild-

fowl, the world's most northerly flamingo colony, numbering about 12,500 birds, is located on Lake Tengiz.

A little farther north, the forest-steppe zone has been described as the richest wildfowl area in Russia. Four places are of particular interest: the Baraba region, between the rivers Ob and Irtish, with about 2,500 lakes, including Lake Chany (3,339 km²/1,289 sq mls); the Ishim region, between the Irtish and Ishim rivers, with about 1,600 lakes covering 2,023 km²/781 sq mls; the Tobolsk or Kurgan forest-steppe, between the Ishim and Tobol rivers, with 1,500 lakes; and the Trans-Ural region with numerous lakes both on the plain and in the foothills. These wetland systems comprise the main global breeding and moulting grounds for many species of waterfowl. With bird populations fluctuating widely according to lake level, no comprehensive census has been taken, but they can certainly be numbered in hundreds of thousands. In recent years, several species have tended to decline, partly because of increased hunting pressure

but mainly because of agricultural development, including the reclamation of bogs and marshes, and irrigation projects. Deep ploughing of virgin lands is affecting surface water run-off and causing lake levels to fall.

While carefully protected on their Siberian breeding grounds, waterfowl are subjected to persistent harassment as soon as they cross the Russian frontier. Russian ornithologists are understandably concerned that the success of their work in building up wildfowl populations is being undermined by indiscriminate hunting beyond their frontiers. As an international asset, birds require international protection. It is of critical importance that key wetlands along traditional flyways should be set aside as true sanctuaries for birds of passage, and the hunting of birds on international flyways strictly controlled. Size is by no means the only criterion: location is all-important. The Azraq oasis, for instance, covers no more than $12 \text{ km}^2$ / 4.6 sq mls, but is the only place in the Jordan desert with permanent water, and so is a key sanctuary for huge numbers of migrant birds on passage from southeastern Europe across Asia Minor to the Red Sea.

Canada, the United States and Mexico have shown the way by collaborating over a North American Waterfowl Management Plan which provides the framework for protecting breeding and nursery areas and wintering grounds for millions of ducks and geese. The scheme is funded from the sale of Federal Duck Stamps which all duck hunters are obliged to purchase. The yield from this source – about US$50 million a year – is used for extending the system of refuges.

## Exploitation and Degradation

There is hardly a country in the world that is not exploiting its wetlands, sometimes in ways that are fraught with environmental and ecological imponderables. Since 1850, for instance, more than 90% of Switzerland's fens and bogs have been drained. Most of the remaining river courses are now artificially constructed. The former Soviet Union planned to pump water from the River Ob over a distance of 2,500 km / 1,500 mls for irrigation purposes. If this scheme is ever implemented it will substantially reduce that river's flow, with incalculable ecological consequences for the Ob's Arctic floodplains and Siberia's coastal wetlands.

## NILE PERCH IN LAKE VICTORIA

Attempts at improving fisheries have not always been successful. The introduction of the Nile perch into Lake Victoria, Africa's largest lake, in 1960, was an unmitigated ecological disaster. The proposal, made by the Uganda Game and Fisheries Department, was strongly opposed by scientists of the East African Fisheries Research Organization, but their opinions and warnings were ignored.

If the introduction of this exotic species was intended to increase the lake's productivity, the idea badly misfired, for the Nile perch, Africa's largest freshwater fish, which reaches a length of 2 m / 6½ ft and weighs up to 80 kg / 175 lbs, is a voracious predator. The first signs that all was not well appeared 10 years after the initial introduction, when perch began to be netted in increasing numbers.

Within a few years the Nile perch had virtually wiped out the endemic cichlid fish and destroyed the fishery. Of the 300 or more varieties of *Haplochromis* endemic to the lake, almost two-thirds were destroyed and the remainder reduced to a remnant. The Horniman Museum, in London, is holding what is believed to be the only surviving colony of *H. pyrrhocephalus*.

The local people do not rate the Nile perch highly as food: they prefer smaller fish such as *Tilapia*, which they have habitually caught. They are, moreover, accustomed to sun-dry fish, but the oily nature of the Nile perch makes sun-drying impossible: perch have to be smoked. So more and more trees are being cut down for fuel with which to do the smoking.

The introduction of the Nile perch has dramatically altered Lake Victoria's complex ecosystem, which not only produced an important and abundant source of high-quality protein, but also served as an outstanding natural laboratory for the study of evolution and speciation.

---

The USA and Canada are examining an equally ambitious scheme for bringing water from Alaska and Canada over thousands of kilometres to irrigate land in the southwestern US and Mexico. China plans to abstract water from the Chang Jiang (Yangtze) to irrigate some $40,000 \text{ km}^2$ / 15,500 sq mls in the arid north, and has earmarked more than $66,000 \text{ km}^2$ / 25,500 sq mls of wetlands for reclamation by the end of the present decade. Flood prevention measures include substantial modification to the banks of the Chang Jiang, thereby reducing the area of floodplain. The riverine lakes – including the country's largest, P'o-yang Lake, and its satellite wetlands – absorb a large proportion of the floodwaters brought down by that great river. Most of them will now be isolated by sluice gates that enable water to be retained during the dry season for irrigation and fish culture. As these lakes are important nursery areas, fish stocks are likely to be reduced. The programme includes reclaiming $6,000 \text{ km}^2$ / 2,300 sq mls in the delta of the Huang He (Yellow River) in Shandong Province, for agricultural use and for fish and shrimp ponds. The delta, one of China's largest estuarine systems, is of great importance as a fish nursery, for wintering waterfowl and for birds of passage using the Asia / Pacific flyway.

Almost half of South Korea's birds are dependent on wetlands, especially wintering ducks, geese and swans (40 species), wintering cranes (three species) and passage shorebirds (52 species). Yet more than half of South Korea's coastal wetlands, including seven sites of international importance, are threatened with reclamation. This could pose a threat to the country's fishing industry, currently valued at US$1.2 billion a year and employing nearly 400,000 people.

## Reclamation and Dams

Reclamation for agricultural development is the biggest threat to wetlands. Development is frequently misguided in the sense that greater productivity would result from leaving wetlands alone and managing them properly than from reclaiming them. Ignorance and prejudice both play a part. Mistaken western attitudes have been introduced into the Third World, often through foreign aid schemes, supported by international agencies whose planners neglect to take ecological and environmental factors into account. The construction of hydroelectric dams at the Kafue Gorge and farther upstream, for example, to generate power for Zambia's copper industry has radically altered the flood regime of the Kafue floodplain.

Dams and barrages for hydroelectric and irrigation purposes are greatly favoured by governments bent on achieving cheap power and on bringing marginal lands into production. Dams have the added attraction of being regarded as prestige projects. But dams frequently reduce the flow of rivers and lower the water-table for some distance downstream, often at the expense of fisheries and agricultural production, and the lives and livelihoods of large numbers of people.

Soil erosion, arising from poor agricultural practices upstream, causes many dams to silt up. By lowering the floodplain's capacity to store flood water, siltation increases the likelihood of flooding farther downstream as well as decreasing the dry season flow. This problem is particularly severe along the lower reaches of the Huang He, described as 'the greatest earth mover'. Siltation is having the effect of raising the river bed at the rate of 75-150 mm / 3-6 in a year, at the same time as reclamation is reducing the size of the lakes alongside the river, thereby lowering their ability to store floodwater, and thus increasing the likelihood of downstream flooding.

The Aswan High Dam in Egypt is one of the most notorious examples of the unexpected side effects of dam building. Construction of this huge man-made lake, against all informed scientific opinion, resulted in a reduction in the amount of silt being brought down the Nile, and a consequent drop in the quantity of nutrients and sediments reaching the Nile delta, causing a serious decline in the fish catch (particularly sardine) in the eastern Mediterranean, which in 1962-68 fell from 38,000 tons to 14,000 tons.

Attempts are being made to increase the amount of Nile water available to the Sudan and Egypt by building the Jonglei Canal through the Sudd. If this canal is ever completed, it is likely to compound the damage already done by the Aswan Dam by dramatically altering the ecology of a large region, in the process seriously affecting the lives of about 400,000 pastoralists of the Dinka, Shilluk and Nuer tribes, whose culture and way of life are based on conforming with the Sudd's seasonal flood regime. Given the evidence, Sudan's interests would best be served by leaving the Sudd untouched.

Large numbers of wetlands have already gone; and, once gone, they are extremely difficult to restore. Many more are threatened. Under its transmigration programme, the Indonesian Government is settling millions of people from the overpopulated island of Java – and some of the other 13,000 islands that constitute the archipelago – in the forests and wetlands of Kalimantan (southern Borneo) and Irian Jaya. But the environmental conditions are unsuited to the type of agriculture to which the colonists are accustomed. Inappropriate land-use methods have already led many settlements to fail; they have also given rise to devastating fires which, in 1983-84, destroyed 35,000 km²/13,500 sq mls of vegetation and peat in what was described by IUCN as one of this century's worst environmental disasters. Much of Kutai National Park (2,000 km²/772 sq mls) in East Kalimantan was devastated.

## Chemical Damage

Wetlands in the industrialized countries are being increasingly damaged by acid rain. More than 20% of Sweden's 85,000 large or medium-sized lakes are now acidified, while in Asia about 5,000 km² / 2,000 sq mls of wetlands are being lost each year.

The runoff of chemical fertilizers from agricultural land, and the introduction of industrial effluents and raw sewage are responsible for the loss or degradation of many wetlands. Forty-two of peninsular Malaysia's major rivers have been declared biologically dead, primarily through the introduction of effluents from rubber and oil palm production, sewage and industrial wastes. Fish, shellfish and crustaceans are no longer found in these rivers, and their waters are considered unfit either for drinking or washing. Of Asia's 191 wetlands of international importance, 69 (36%) are under moderate to severe threat, despite being accorded protective status.

About 95% of fish caught along the US Atlantic coast spend part of their life-cycle in coastal wetlands. But this has not prevented wetlands on the eastern seaboard of the USA from being replaced by dump-and-fill, factories and houses; large areas are now buried under asphalt and concrete, while chemicals and sewage pollute the wetland that remains.

Conservation of many important wetlands is frequently complicated by control being in the hands of a number of countries. Measures taken in one country can seriously affect another. The nation located upstream can exert a powerful influence, leaving downstream states dependent on its goodwill. The catastrophic flooding of the Ganges delta in Bangladesh, for example, is due to deforestation in the Himalayas. Eroded soil from northern India and Nepal is washed downriver and deposited in the delta. The problem can be resolved only by halting the deforestation. But pressure of population is forcing the Nepalese to move higher and higher into the mountains to find land, and to till hillsides too steep for cultivation.

There is general failure to understand the natural values of wetlands, the contribution they make to human welfare, and the importance of retaining them in their own right. The first essential is to dispel the prevalent ignorance and prejudice through an educational programme designed to impart a better understanding of the values of wetlands and of the benefits they bring mankind. It should be aimed not only at school children and the public at large but also at landowners, governments and agencies with ultimate responsibility for planning and policy. International bodies providing both the funds and the expertise should understand the ecological implications and be made aware of the consequences of their actions.

Politicians and planners are readily persuaded that wetland reclamation will increase food production and help raise living standards, and must therefore be in their country's best interests. Huge sums of money are spent on conventional development schemes. Without careful assessment of all the factors involved, the end result can have very different long-term effects from those envisaged. The challenge facing wetland scientists and decision-makers alike lies in devising systems of utilizing the diversity and natural productivity of wetlands on a sustainable basis. This calls for greater knowledge of the ecological complexities of wetlands and greater understanding of their functioning; it also requires a coordinated interdisciplinary approach to ensure their rational use.

# THE EVERGLADES

LOCATION: Southern tip of the Florida peninsula, bounded on the west by the Gulf of Mexico, and on the south and southeast by the Florida Keys. The biosphere reserve includes Fort Jefferson National Monument, a group of seven coral reefs – the Dry Tortugas – and the shoals and waters around them. 24°50′–25°55′ N, 80°20′–81°30′ W.

AREA: 5,858 km²/2,261 sq mls, of which the Everglades National Park covers 5,668 km²/2,188 sq mls, and Fort Jefferson National Monument 190 km²/73 sq mls.

ALTITUDE: Sea level to 2 m/6.5 ft.

CLIMATE: Annual precipitation often exceeds 1,270 mm/50 in; temperatures moderate: up to 23–35 °C /73–95 °F in summer; winters mild. The region is prone to hurricane-strength storms.

Everglades National Park is in the form of a shallow basin tilted slightly to the southwest and underlain with limestone. It is a vital recharge area for the Biscayne Aquifer, the sole source of water for Miami and the whole of southeastern Florida. The principal source of water is Lake Okeechobee – the largest lake in North America after the Great Lakes – which periodically overflows, inundating two-thirds of the park.

Lying at the interface between temperate and subtropical America, between fresh and brackish water, and between shallow bays and deeper coastal waters, the park's habitats are exceptionally diverse. This diversity is reflected in the 1,000 species of seed-bearing plants, 25 varieties of orchids, 120 tree species, both tropical (e.g. palms and mangroves) and temperate (e.g. ash and oak) and even desert plants (e.g. cactus and yucca), as well as aquatic vegetation, both freshwater and marine.

The transitional zone from fresh to saltwater (glade to mangrove) is of exceptional biological diversity, particularly of economically valuable crustacea. As well as about 150 species of fish, 400 species of land and water vertebrates occur in the park, 36 of which are threatened. They include the Florida panther, round-tailed muskrat, mangrove fox squirrel, and Florida manatee. Over 323 bird species have been recorded, among them the Everglades kite – equipped with a long, sickle-shaped bill specially adapted for feeding on the giant freshwater snail *Pomacea paludosa* – bald eagle, osprey, roseate spoonbill, wood ibis, ivory-billed woodpecker, and many others. The sixty known species of amphibians and reptiles include both the American alligator and the American crocodile, while among the butterflies are such species as the endangered Schaus swallowtail.

The key to the perpetuation of the Everglades' wildlife is the annual inundation. At that time the park is littered with shallow ponds – known as 'gator holes – which are deepened and enlarged by alligators, and attract large numbers of fish and fish-eating birds. These 'gator holes are supremely important to the ecology of the Everglades, and the alligator thus a key species in maintaining ecological equilibrium. Starting with the summer rains, the runoff flows slowly across the Everglades – moving at the rate of about 500 metres a day – covering the entire area in a thin sheet of water and triggering an explosion of life of all kinds. Plankton, insects, small fish and many other aquatic creatures forming the broad base of the food pyramid proliferate in the nutrient-rich environment.

In their natural state, the Everglades remained inundated for about nine months of the year. Since the 1880s, drainage and flood control programmes have dramatically transformed the face of this great wetland. Reduction of the level of Lake Okeechobee through extraction of water for urban and agricultural use has significantly affected the park. Canals and levees in the north and east of the park draw off water for flood control purposes, emptying much of it into the Atlantic and the Gulf of Mexico.

The reduced flow cuts the flood period to five months and inundates a

## FLORIDA KEY DEER

Smallest of the thirty or so races of the white-tailed deer, the Florida key deer is less than 76 cm/30 in high and weighs no more than 22 kg/50 lbs. This little deer has never been known to exist outside the confines of the chain of keys, or small islands, protruding from the southern tip of the Florida Peninsula. It lives on most of these islands, readily swimming from one to another. The key deer was at one time remorselessly hunted, often by night using packs of dogs and spotlights. Development of the keys for human use, in which fire played a conspicuous part, has dramatically reduced and altered the habitat available to the deer. By 1949 the total population had been reduced to about thirty individuals. The deer now has full legal protection, and regular patrolling by wardens has brought hunting firmly under control, allowing the population to build up to about 300. The establishment of the Key Deer National Wildlife Refuge (27 km²/10 sq mls) in 1953 safeguards the greater part of the animal's habitat. The most serious difficulties currently affecting the deer are shortage of suitable habitat and road kills on the road running the length of the keys.

smaller area than previously. The 'gator holes dry out more quickly, shortening the period of food production and thus reducing the numbers of birds and other animals. In the mid-1930s, the number of wading birds breeding in the Everglades exceeded 1.5 million; by the mid-1940s the figure had dropped to 300,000; and in the period 1950–62 to 50,000. It is even lower today. Alligators are estimated to have declined by 95% since the early 1920s. The reduced flow has also increased the salinity of estuaries, bays, sloughs and tidal flats along the coast, with consequent reductions in fish, shrimps and other organisms. This has impaired the ability of the coastal wetlands to feed as many birds and other

● Kitching Creek, a typical scene in the Everglades National Park. (Planet Earth Pictures / J. Brian Alker)

● Pelicans, gulls and terns form just part of the Everglades' rich array of birdlife. (WWF / Y. J. Rey-Millet)

animals as previously. Disruption of the natural flood cycle and water regime has adversely affected the flora also. Stands of pure sawgrass, for example, have been invaded by poisonwood, buttonwood, willow and other woody plants. Further problems have been caused by the introduction of exotic plants; water hyacinth and Brazilian pepper have proved particularly difficult to eradicate. The change in vegetation has a major bearing on the ecology of the entire region. Exotic fish, notably *Tilapia* and walking catfish, compete with the indigenous species.

Almost all that now remains of the Everglades ecosystem lies within the Everglades National Park. Its survival is dependent on maintaining an adequate flow of water; but control of water is in the hands not of the park authorities but of the US Army Corps of Engineers. It is essential that the park is guaranteed sufficient water to meet the ecological requirements of the Everglades' natural ecosystem. Unless that is done, this unparalleled wetland wilderness will go by default. The Everglades was accepted as a World Heritage Site in 1979.

# THE PANTANAL

LOCATION: Mainly in Brazil, but reaching into Paraguay and Bolivia. Approximately 14°50′–22°50′ S, 55°00′–59°00′ W.

AREA: The most extensive marshland and seasonally inundated savanna in the world, the Pantanal covers 130,000 km² / 50,000 sq mls.

ALTITUDE: About 100 m / 330 ft above sea level.

DEPTH: The Rio Paraguay varies in depth from 4–5 m / 13–16 ft. The waters begin to rise in December, reaching their highest level in May or June, about a month after the rains have finished. By then the Pantanal has been transformed into a vast swamp. When the waters subside, huge expanses of land dry out, and remain dry until the following flood season.

CLIMATE: Annual rainfall 1,000–1,300 mm / 39–51 in, mostly between December and March.

The Pantatal is the largest wetland in the world, and an ecosystem of international importance, characterized by alternating periods of inundation and desiccation, a regime to which the plants and animals have had to adapt. The region is predominantly seasonally inundated marshland and savanna (*cerrado*), and is flooded from December to June. The nature of the habitat inhibits tree growth, but there are occasional infusions of gallery forest, semi-deciduous forest, and *chaco*, interspersed with a number of large permanent lakes, located mainly in the northwest. The Pantanal ranch-owners' practice of regularly burning the grasslands has greatly affected the vegetation, little of which remains undisturbed.

This region is the most important habitat for wetland birds in South America. Among them are such species as the finfoot, a rare relative of the rails, and the roseate spoonbill, which nests colonially in trees, often in close proximity to colonies of wood ibis, one of the most abundant birds of the Pantanal. Besides huge flocks of herons and other marsh birds, the Pantanal is also home to the curious and little known boat-bill, nocturnal in habit and possessing a large spoon-shaped bill – an adaptation to searching in mud for its prey. The open marshland and wet savanna are frequented by immense numbers of water-fowl, both resident and migratory, including the red-winged whistling duck and kamichi, the latter's presence made evident day and night by the music of its trumpeting. The area also serves as an essential stopping place for Arctic shorebirds migrating to and from their wintering grounds in the south. Other notable birds include the rhea and the hyacinth macaw.

Among the mammals living in the Pantanal can be found the giant otter, giant anteater, maned wolf, marsh deer, and South American tapir. Monkeys are uncommon, the most numerous being the black howler. Of the reptiles, the yacare caiman, once abundant, is being rapidly exterminated.

The region has become a major centre for cocaine smuggling and illicit traffic in wildlife. Hundreds of thousands of caiman are poached each year, and their skins flown out or taken over the border into Paraguay and Bolivia. One convoy of seven trucks was recently found to be carrying caiman skins worth US$9 million. The standard bribe for driving through customs is US$15,000 per vehicle. The high market value of hyacinth macaws – US$8,000 a pair in the USA and US$15,000 in Europe – makes smuggling for the pet trade a lucrative business. No more than 3,000 individual hyacinth macaws are estimated to remain in the wild.

Dam construction for power production; deforestation on the periphery of the Pantanal to make way for cattle ranching; charcoal burning; drainage of wetland for agricultural expansion; pollution of rivers by erosion; siltation; pesticide run-off; and toxic effluents from sugarcane processing and gold mining are among the factors impeding natural drainage and altering the water regime, with serious ecological consequences. The reduced productivity of the fishery affects all the animals higher up the food chain.

Since much of the Pantanal is privately owned, ranch owners must clearly play a major part in conserving it. Some appreciate the need to protect their land and have already taken steps to do so; others are showing interest in the possibility of commercial caiman farming, which, if strictly controlled and monitored, might reduce pressure on wild stocks. Since less than 2% of the land is at present protected, it would be useful if the Pantanal National Park (1,350 km² / 521 sq mls) could be extended and other parks created as opportunity occurs. This should be accompanied by an educational programme designed to encourage a greater understanding of the importance of the Pantanal, both locally and nationally, and the need to utilize its natural resources in ways that can be sustained.

## GIANT OTTER

Distribution of the giant otter centres on the Amazon basin, where it was once widespread, but only Guyana, Surinam and French Guiana still possess viable populations; elsewhere only a scattering of small isolated groups of giant otters now remains.

The giant otter is monogamous, pairing for life. More gregarious than other otters, it lives in family units of up to about 30 individuals, close to the holts that it excavates in the river bank. An inhabitant of large rivers and creeks, it spends about four months of every year

in seasonally flooded forest. This otter is strongly territorial, marking its territory with spraints, characteristic droppings deposited where they are most likely to be seen by other otters. Diurnal in habit, it hunts by sight, at which time its speed and agility in water (aided by a tail flattened like a beaver's) are unrivalled, but its thickly webbed feet make travel on land difficult. As well as fish (including piranha) and other aquatic food, the giant otter preys on the paca, the large rodent that is the quarry of almost every carnivore in South America. The reduction of

fish stocks, caused by overfishing, is often blamed on the giant otter and used to justify killing it.

Habitat destruction and river pollution are major contributory factors in the decline of the giant otter; but hunting for the fur trade is the chief cause. Its pelt is extremely valuable, and the animal's diurnal habits, social structure and curiosity make it easy to shoot. In parts of its range it has been hunted to extinction. Cubs are in demand for the pet trade but, being difficult to raise in captivity, usually die.

# KUSHIRO MARSH

LOCATION: Eastern Hokkaido, north of the city of Kushiro. 43°03′ N, 144°24′ E.

AREA: 290 km²/112 sq mls, including 226 km²/87 sq mls of moorland and a 50 km²/19 sq mls Ramsar Site.

ALTITUDE: 1.5–5 m/5–18 ft.

CLIMATE: Cold temperate. The influence of sea fog ensures that spring and summer are generally cold and humid; autumn and winter, on the other hand, are fine and dry. Snow is frequent in winter, with temperatures often falling below −20 °C/−4 °F. Annual rainfall averages about 1,124 mm/44 in.

Kushiro Marsh is a large floodplain that includes several small freshwater lakes, separated from the sea by a strip of coastal dunes about 1.6 km/1 ml wide. The Kushiro River meanders through the marshes for a distance of 67 km/41 mls, creating many oxbow lakes, especially along its lower reaches. The main lakes are Toroko (620 ha/1,532 acres), Takkobunuma (137 ha/338 acres) and Shirarutoronuma (337 ha/833 acres). It is an extensive area of open marsh and peat bogs. The floodplain vegetation consists predominantly of reeds, with large tracts of sedges.

This marsh is one of the largest and most important natural wetlands remaining in Japan. The shallow marshes are particularly important as a breeding and wintering ground for the Japanese or red-crowned crane, the entire Japanese

population of which breeds in eastern Hokkaido, mostly in Kushiro. Large numbers of migratory ducks, geese, swans and shorebirds frequent the marshes seasonally, some remaining through the winter. Numerous birds of prey, among them Steller's sea eagle and white-tailed eagle, either breed or winter on the marshes. Altogether, more than 150 species of birds have been recorded from the area, together with 26 species of mammals, including raccoon dog and red fox. Thirty-two species of fish, 46 dragonfly and 84 species of butterfly have been identified.

Although much of the marsh remains relatively undisturbed, the clearance of forest for agriculture and the straightening and deepening of rivers in the catchment area are having a detrimental effect on the wetlands downstream. Acceler-

ated run-off in the hills causes soil erosion, thus increasing siltation and bringing excessive spring floods and summer droughts. Kushiro's urban sprawl is encroaching into the marsh. Along the Ninishibetsu River alone, about 1,568 ha/3,874 acres of marsh have been destroyed since 1972.

About 50 km²/19 sq mls in the centre of the Kushiro wetland, representing 17% of the total marsh habitat, have been listed under the Ramsar Convention. Kushiro is now designated a National Wildlife Protection Area; it is also protected as a Special Natural Monument under the Japanese law for the Protection of Cultural Properties. Land reclamation, the felling of trees and bamboo, and erection of structures are now strictly regulated, while hunting and fishing are prohibited.

● The inundated floodplain of the Pantanal, the most important habitat for wetland birds in South America. (WWF / Hartmut Jungius)

● The giant otter can grow to over 2 m / 6 ft in length (including tail). (WWF / Hartmut Jungius)

● The Japanese, or red-crowned, crane performing its elegant courtship ritual. The entire Japanese population of the bird breeds in eastern Hokkaido, mainly in the Kushiro Marsh. (WWF / Sture Karlsson)

## LAST RUN OF THE SATSUKIMASU SALMON

The Nagara River, near Nagoya, on the west coast of the main Japanese island of Honshu, is one of only three rivers in the entire Japanese archipelago that has not yet been dammed. It is also the last unobstructed river remaining within the range of the Satsukimasu salmon, which at one time had unimpeded access to numerous streams along the south coast of Honshu.

In 1988, in the face of strong opposition from the Freshwater Fish Protection Association of Japan, the Ecological Society of Japan, and a number of other reputable organizations, Japan's Ministry of Construction authorized the building of the Nagara Estuary Dam, ostensibly because the dam represents 'the only way to prevent flooding' and will have 'no adverse effect on natural ecology'. Such assertions are dismissed by opponents of the scheme, who believe that as well as blocking the salmon migration, the dam will destroy important habitats for clams and other wildlife. Despite mounting opposition by local residents and government assurances to the contrary, no environmental impact assessment has been undertaken.

The Environmental Agency of Japan, the government body responsible for wildlife protection and for management of wetlands under the Ramsar Convention, has conspicuously failed to oppose construction of a dam that will inevitably result in the extinction of an endemic species of fish which the Agency itself officially designated 'endangered' in 1990.

# YANCHENG MARSHES

LOCATION: On the northern and central coast of China's Jiangsu Province, about 65 km / 40 mls west-north-west of Shanghai. 32°32'–34°30' N, 119°27'–120°56' E.

AREA: 2,430 km² / 938 sq mls.

ALTITUDE: up to 4 m / 13 ft.

CLIMATE: Humid temperate, with cool dry winters. Annual rainfall averages about 1,020 mm / 40 in, falling mostly in summer. Temperatures range from a maximum of 39 °C / 102 °F to a minimum of −17 °C / 1 °F. Typhoons occasionally occur in late summer.

The Yancheng marshes are an extensive complex of marshlands and ponds (both fresh and brackish), including grasslands, reedbeds, intertidal mudflats, salt pans, channels and tidal creeks, all extending for 300 km / 186 mls along the coast of central and northern Jiangsu Province. Depositions of silt from the Chang Jiang (Yangtze) are progressively extending the area of mudflats. A wealth of aquatic vegetation exists in the marshes, both submerged and floating, and it is dominated by extensive reedbeds, interspersed with grass and sedge marshes.

These wetlands are important wintering grounds for large numbers of migratory wildfowl. A census made in January 1988 counted 32 species of waterfowl totalling more than 535,000 birds, together with numerous gulls and other shorebirds. The marshes are particularly well known as a wintering ground for the endangered red-crowned crane, whose numbers increased from 200 in 1981 to over 600 in 1987. The marshes are also important for nesting birds, among them the rare Saunder's gull, whose breeding grounds were discovered in the northern part of the area as recently as 1984. A second colony was discovered in the south in 1987. Altogether, 226 species of birds have been recorded. Two interesting mammals are associated with these marshlands: the Chinese water deer (Yancheng is one of the few places where it remains relatively common) and Père David's deer.

The marshes are unfortunately located in one of the most densely populated regions of China. As a result, parts have been drained for agricultural purposes or for use as fish ponds; a substantial reed-harvesting industry has developed; grass is cut for use as fuel for brick making; salt is extracted on a commercial basis; and proposals have been made for constructing a seawall to enclose 400 km² / 154 sq mls of saltmarsh close to one of the Saunder's gull colonies. These activities inevitably cause disturbance. Hunting of Chinese water deer is prohibited but still occurs.

However, the entire marshland is included in the Yancheng Protection Zone. This is divided into three distinct zones: an Absolute Protection Area (100 km² / 38 sq mls); an Experimental Area (470 km² / 181 sq mls), half of it outside the sea-wall; and a No-Hunting Area (1,860 km² / 718 sq mls). Two reserves have been established in the protected area: Yancheng Nature Reserve (400 km² / 154 sq mls) and the Da Feng Milu Reserve (about 10 km² / 4 sq mls) as a site for the reintroduction of Père David's deer.

## PÈRE DAVID'S DEER

Père David's deer, the sole living representative of its genus, has never been known to Western science in the wild state. During the Shang dynasty (1766–1122 BC), the Chihli Plains, in the swamps of which it lived, were brought under cultivation. Thereafter, the species survived for almost 3,000 years in a semi-wild state in specially protected areas. At the time of its discovery by the famous French missionary-naturalist Père David in 1865, the only surviving herd was in the Imperial Hunting Park, the Non Hai-tzu (or Southern Lake), about 60 km / 37 mls south of Beijing. In 1894, flood waters breached the wall (72 km / 45 mls in cir-cumference) surrounding the park. Many of the deer escaped into the surrounding countryside, where they were eaten by the famine-stricken peasants. The few animals to survive were shot by the international troops stationed in the park during the Boxer Rising.

One man's initiative was instrumental in saving the animal. Events in China persuaded the Duke of Bedford to establish a herd at Woburn by bringing together all available specimens from zoos in Europe. From a nucleus of 16 animals, several herds were gradually established in other countries. In 1964, half a century after the deer had been exterminated in its homeland, four specimens from the London Zoo were sent back to China, where they were installed in the Beijing Zoo. A further 20 from the Woburn herd were returned in 1985. The following year, 39 more were flown to China for release into the Da Feng Milu Reserve. In 1991, others were released into a 10 km² / 4 sq mls reserve on the Huang He River in Hubei Province. The climax came with the re-establishment of a flourishing herd in the former Imperial Hunting Park (now known as the Nan Haizi Milu Park).

# MEKONG DELTA

LOCATION: The Mekong delta extends from Phnom Penh, capital of Kampuchea, where the Mekong River divides into two main branches, to the coast of Vietnam on the South China Sea. 8°33′–11°35′ N, 104°30′–106°50′ E. In its passage through the delta the river divides into the nine channels from which the Mekong ('Nine Dragons') derives its name.

AREA: The delta forms an immense triangular plain, extending for about 270 km / 168 mls from its apex at Phnom Penh to the coast of southwestern Vietnam, and covering 49,520 km² / 19,120 sq mls, of which roughly 70% is in Vietnam and 30% in Kampuchea.

ALTITUDE: Almost the whole of the delta lies below 5 m / 16 ft.

For between five and eight months of the year the Mekong floodplain is inundated to a depth of 1.5– 7 m / 5–23 ft.

CLIMATE: Tropical monsoonal. Pronounced rainy season during the southwest monsoon (May– October). Rainfall varies from 1,500 mm / 59 in in the centre and northwest to more than 2,350 mm / 92 in in the south. About three-quarters of the rainfall is concentrated into a four-month period at the height of the rainy season. Mean annual temperature 26 °C / 78.8 °F. Humidity high.

The Mekong delta is one of the most productive regions in Southeast Asia, with an annual yield of about 6.5 million tonnes of rice, besides much other agricultural produce. The delta also supports one of the world's largest and most prolific inland fisheries, yielding 150-200,000 tonnes of fish a year and providing half the region's protein requirement.

About half the delta – 24,740 km² / 9,552 sq mls – is cultivated as rice paddy. Of the balance, about 2,800 km² / 1,080 sq mls are covered with mangrove and *Melaleuca* forest. Mangroves were originally more extensive, but the deliberate destruction by aerial spraying of herbicides during the Vietnam War was responsible for the loss of 400 km² / 155 sq mls of mangrove forest.

The delta sustains 260 species of fish (most of them commercially valuable), 35 of reptiles, six of amphibians, 386 of birds, and 23 of mammals. Reptiles include the estuarine crocodile and the endangered river terrapin. Mangrove and *Melaleuca* forests support huge nesting colonies of cormorants, herons, egrets, storks and ibises. Some 92 species of waterfowl, both resident and migratory, frequent the delta, as well as eight species of kingfisher and many other birds. Endangered birds include the eastern Sarus crane, which vanished during the Vietnam War but has subsequently re-turned, the giant ibis, the white shouldered ibis and the white-winged wood duck. None of the latter three species has been seen since 1980, but the two ibises are believed still to exist in wetlands close to Vietnam's border with Kampuchea. Among the mammals the smooth-coated otter and possibly the small-clawed otter occur in the delta, as well as the fishing cat and the crab-eating macaque. Particular interest attaches to the five species of dolphins associated with the delta. Two – the Malay dolphin and the Red Sea bottle-nosed dolphin – are confined to the estuarine area, but the other three – the Chinese white dolphin, black finless porpoise and Irrawaddy dolphin also frequent the lower reaches of the river itself, the last-named being the only one to occur upstream of Khone Falls in Kampuchea.

The greatest threat to the Mekong delta stems from a rapidly rising human population, which in the century 1880-1980 increased from 1.7 million to 13.5 million. This population explosion has caused increased agricultural development in the shape of a number of wetland reclamation schemes, at the expense of both floodplain and tidally inundated lands. The run-off from pesticides and fertilizers used for agricultural purposes could have serious consequences for fish and wildlife. One of the principal aims of the Committee for Coordination of Investigations of the Lower Mekong Basin (the Mekong Committee) is the control of seasonal flooding by storing flood-water upstream and by constructing levees and dikes throughout the delta region. Flood control, development of hydraulic power, and extensive agricultural schemes are already affecting delta ecosystems. The construction of dams along the Mekong in Kampuchea and Laos will further reduce seasonal flow, limit flooding, curtail deposition of sediment, reduce the level of nutrients in the delta, and alter the hydrology of the region. Together with change in water quality and timing of peak flow periods, this will adversely affect migratory fish using the floodplain for spawning and damage the delta's value as a fishery.

It is in the interests of all the delta's inhabitants that the productivity of this great wetland should be placed on a sustainable basis. Since the end of the Vietnam War, the Mekong Committee has commissioned studies on numerous aspects of the delta. Six reserves have been set aside for conservation, and extensive reafforestation and wetland restoration programmes are being undertaken, including the planting of an almost continuous strip of mangrove forest along the seaward edge of the delta, which extends for 600 km / 230 mls.

● Père David's deer has been reintroduced into the wild in several sites in China, including the Yangcheng Marshes. This group is at the Da Feng Forest Farm in Jiangxi Province. (WWF / Hartmut Jungius)

# PROSPERITY ON THE FLOOD

As the annual inundation of the Nile is the life blood of Egypt, so the seasonal flooding of the Mekong River is crucial to Kampuchea. Indeed, the Khmer empire, which flourished from the 9th to the 13th centuries, owed its prosperity and strength to the flood and to the Great Lake. This lake, the largest in Southeast Asia, lies in a huge shallow basin to the west of the Mekong, to which it is connected by the Tonle Sap River. In late May or June, the Mekong, swollen by melting snow from its headwaters on the Tibetan Plateau and by monsoonal rain collected as it passes through the mountains of Yunnan, discharges into the Tonle Sap. The quantity of water is so great that it blocks the course of the Tonle Sap, forcing it to reverse its flow into the Great Lake. As it fills, overflows and spreads over the land, the lake quadruples in size from 2,500–3,000 km²/ 965–1,160 sq mls to 13,000 km²/5,000 sq mls. At the height of the dry season the

inundated area covers 70,000 km²/27,000 sq mls. By absorbing and storing flood-water and gradually releasing it during dry periods, the Great Lake regulates the flow downstream into Vietnam.

The lake's phenomenal productivity derives from natural fertilization of the land by the deposition of millions of tonnes of sediment brought down on the annual flood. Flooding releases nutrients from the soil, vegetation and inundated organic matter, giving rise to spawning and nursery grounds from which young fish repopulate the Mekong basin. By enriching the waters, inundation also provides conditions conducive to a massive irruption of fish populations – to the great benefit of the 40 million people currently living in the lower Mekong basin.

The clearance of freshwater swamp forest from around the periphery of the Great Lake for agriculture, fish farming and firewood has increased siltation, lead-

ing to fears that the lake may eventually disappear. These fears are based on a steady decline in the fishery of the Great Lake–Tonle Sap inundation zone from a mean annual yield of 139,000 tonnes in the period 1939–51, to 50–80,000 tonnes in the 1970s, and 63,000 tonnes in 1984. To counter the effects of siltation, the Mekong Committee plans to build a barrage to regulate the inflow and outflow of water from the Great Lake via the Tonle Sap River (and to provide water for irrigation).

The construction of large reservoirs in Laos and Kampuchea and the Tonle Sap barrage, as well as levees and embankments around the perimeter of the Great Lake and throughout the delta region would, if fully implemented, almost halve the annual Mekong flood discharge from 40,000 cusecs (cubic ft per second) to 23,000 cusecs, which would reduce the entire high water surface area of the inundation zone by 92%.

# THE SUNDARBANS

LOCATION: Southern part of the Ganges / Brahmaputra delta, in Bangladesh and India (West Bengal). 21°32′–22°29′ N, 88°03′–89°53′ E.

AREA: 10,271 km²/3,965 sq mls. Just over half (56% – about 5,771 km²/2,228 sq mls) of this huge wetland is in Bangladesh and 44% (4,500 km²/1,737 sq mls) in the Indian state of West Bengal.

ALTITUDE: Sea level to 3 m/10 ft.

CLIMATE: Humid tropical maritime (humidity averaging 80%). Monsoon mid-June to mid-September. Violent southwesterlies mid-March to September. Storms, often developing into cyclones, are common in May and October–November. Mean annual rainfall varies from 1,700 mm/67 in in the centre and north, to 2,790 mm/110 in along parts of the coast. In the pre-monsoon period (March–April) temperatures may rise to 43 °C/109 °F. Maximum winter temperature about 24 °C/75 °F.

The Sundarbans lie in the delta of the Ganges, Brahmaputra and Meghna rivers, forming a labyrinth of channels and waterways draining into the Bay of Bengal. Deposits of sand, gravel and clay brought down by these great rivers have created a mosaic of countless small mangrove-covered islands, which are regularly submerged by the spring tides. During the high tides associated with the coming of the monsoon, the greater part of the area is subject to flooding. The combination of seasonally swollen rivers overflowing their banks and frequent cyclones originating in the Bay of Bengal and accompanied by tidal waves can cause widespread devastation. Severe storms are a regular occurrence, inundating thousands of square kilometres and causing heavy loss of life. The Sundarbans is the principal nursery for shrimps along the entire coast of eastern India, and an important spawning grounds for other crustaceans and fish.

● The Sundarbans are one of the most important tiger refuges on the Indian subcontinent, being home to about 350 Bengal tigers. Tigers are strong swimmers, and have adapted well to the semi-aquatic conditions. (WWF / R.H. Waller)

● As a fish-eater, the gharial was regarded as a competitor by the fishermen of the Sundarbans, and was hunted almost out of existence. It now survives in very few places. (WWF / Ron Whitaker)

The Sundarbans of Bangladesh form part of the largest continuous mangrove forest in the world. In the Indian sector the mangrove forests have largely disappeared, and are now confined to islands east of the Matlah River. About one-third of the area is taken up with waterways (varying in width from a few metres to several kilometres), canals, tidal creeks, mudflats and dunes.

The brackish waters of this wetland habitat are well suited to a number of aquatic mammals, among them the Indian smooth-coated otter, which has been heavily persecuted by hide hunters, and the fishing cat, which remains abundant. The tidal waters are frequented by the Ganges river dolphin, the Irrawaddy dolphin, the Indo-Pacific hump-backed dolphin, and the finless porpoise.

The Sundarbans also have a thriving population of about 350 Bengal tigers, which have adapted well to the unusual conditions. Tigers are strong swimmers, readily moving from island to island, and seeming as much at home in water as on land. There is no more important tiger refuge on the Indian subcontinent. The tigers of the Sundarbans have acquired a reputation for man-eating. This is said to be partly attributable to the decline of their natural prey – notably the type of wild cattle known as the gaur, which has been hunted almost out of existence, and the barasingha or swamp deer, which is now extinct in the area. The wild boar nevertheless remains plentiful, while the chital or axis deer, the tiger's principal prey species, is more abundant in the Sundarbans than anywhere else in its range. Its reputation as a maneater makes the tiger a difficult animal to conserve;

but if it were to disappear, deer and wild boar would certainly increase. So, too, would the damage to woodlands, thus affecting the livelihoods of considerable numbers of people.

One hundred years ago both the great Indian rhinoceros and the Javan rhinoceros, as well as the wild Asiatic water buffalo were present in substantial numbers; but by the end of the 19th century all three species had vanished from the area. More recent local extinctions include not only the swamp deer but also hog deer and leopard. Illegal hunting and trapping remain widespread; fishermen and woodcutters are the chief offenders, but government officials and members of the armed forces are implicated.

Three species of crocodiles inhabit the Sundarbans. The marsh crocodile, or mugger, was at one time common along the Sundarbans' inland creeks and waterways, but in recent years has been virtually exterminated by hide hunters. The gharial, a long-nosed species which is harmless to humans but, being a fisheater, is regarded as a competitor, has similarly been all but wiped out: it no longer exists in the Sundarbans, and occurs in very few other places. Even the more formidable estuarine crocodile is now rare in estuaries along the sea coast where it was once plentiful. In 1979, an attempt was made to re-establish the estuarine crocodile by releasing 40 captive-bred animals in India's Sundarbans National Park.

More than 270 species of birds have been recorded in the Sundarbans. At one time it was an important wintering ground for ducks and geese, but only small numbers have been recorded in

recent years. Its importance today is as a staging and wintering area for migratory shorebirds, gulls and terns. As might be supposed of such a watery environment, kingfishers are particularly plentiful: no less than eight species inhabit this immense swamp jungle. Other birds include the swamp francolin and 35 species of birds of prey, among them the white-bellied sea eagle and the crested serpent eagle, both of which prey on the snakes that abound in the Sundarbans, and Pallas's fish eagle.

The seabirds and aquatic fauna of the Sundarbans are under constant threat from oil spillage; oil carried by high tides could also damage the forest itself. The proposed construction of a fertilizer plant at Mathurapur, only 5 km / 3 mls from the Satpukur sluice gate on the edge of the Sundarbans, would result in harmful effluents entering the waterways of the Sundarbans. Siltation is of growing concern: some rivers are showing signs of becoming unnavigable.

Forestry is a major industry in the Sundarbans. The most pressing threat to the mangrove forests comes from over-exploitation. The diversion of more than one-third of the Ganges' dry-season flow through construction of the Farraka Barrage in India has led to increasing salinity in the Sundarbans, affecting the natural regeneration of mangroves and in some places causing die-back, as well as disrupting the migration patterns of many species of fish. These factors, in conjunction with the gradual eastward migration of the Ganges, presage long-term ecological change. It is essential that action be taken to maintain an adequate flow of Ganges water through the dry season.

Besides giving the entire wetland the status of Forest Reserve, Bangladesh has set aside three wildlife Sundarbans sanctuaries – East (54 km² / 21 sq mls), West (90 km² / 35 sq mls) and South (178 km² / 69 sq mls). India has proclaimed the Sundarbans Tiger Reserve (2,585 km² / 1,000 sq mls) in West Bengal, part of which has been given the status of Sundarbans National Park (1,330 km² / 513 sq mls), and was inscribed on the World Heritage List in 1985.

## THE FLOODING OF BANGLADESH

Flood damage in Bangladesh has become a regular occurrence. The people of Bangladesh live on the mudflats of a delta founded on soil washed down from the mountains of Nepal. Rising human numbers have forced Nepal's subsistence farmers to spread on to hillsides too steep for cultivation, resulting in the soil eroding and being washed downstream into the delta. As the drainage channels become choked with silt and the mudflats build up, the water becomes shallower and the land floods more easily. The floods cause heavy loss of life, leaving millions homeless and bringing ruination to the country. In Bangladesh, population pressure forces people to live in and farm areas that are dangerously prone to flooding, yet any attempts to raise living standards must deal with the problem of bringing human fecundity under control. Evasion of this issue will leave flood and famine as the ultimate biological controls.

# ARAL SEA

LOCATION: Central Asia, east of the Caspian Sea, on the border between Kazakhstan and Uzbekistan. 43°–46° N, 57°45′–61°00′ E.

AREA: Since being deprived of its inflow in 1961, the Aral Sea has lost 69% of its water; its surface area has shrunk from 66,458 km²/25,661 sq mls to 36,500 km²/14,000 sq mls.

ALTITUDE: 53 m/174 ft.

DEPTH: Originally averaged 21 m/70 ft, maximum 69 m/226 ft. By the end of 1987, the water level had fallen to 13 m/42 ft. Until 1961, the Aral Sea contained 1,064 km³/255 cu mls of water; replenished by an annual inflow of 56 km³/13 cu mls of river water, average salinity was maintained at 10–11%.

CLIMATE: Subtropical desert; excessively hot, with strong dry winds, sandstorms and a high level of evaporation. Rainfall averages 100 mm/4 in per annum, equivalent to one-twelfth the evaporation rate. Average winter temperature (Jan–Feb) ranges from −12 °C/10.4 °F in the north to −6 °C/21.2 °F in the south; summer (July) averages from 23.3 °C/73.9 °F in the north to 26.1 °C/79 °F in the south.

The Aral Sea is surrounded by four deserts: the Aralian to the north, the Karakum to the south, the Kyzylkum to the east, and the Ust Urt Plateau to the west. The combination of desert, riverine forest and lake shore gives rise to a distinctive flora and fauna, some of which is represented on the island of Barsa-Kel'mes (183 km²/71 sq mls), now a nature reserve.

The Aral Sea is the fourth largest body of inland water in the world and the second largest in Asia. Its scientific interest stems from the fluctuations in area and volume that have taken place in geologically recent times. Within the past two decades it has achieved notoriety as an object lesson in ecological mismanagement.

About 90% of the Aral Sea's inflow comes from two great rivers – the Syr Darya discharging into it from the north and the Amu Darya (the Oxus of the ancients) from the south. These are the only rivers in the region to maintain a constant year-round flow. This is because glaciers in the Tien Shan Mountains continue to melt throughout the summer, thus feeding these two rivers after the sources of other rivers have dried up. Each year evaporation takes out about the same amount of water as the rivers bring in. Since the early 1960s, the water level of the Aral Sea has been systematically reduced by the diversion of water

from both rivers to the collective farms for growing cotton under irrigation. This has involved constructing a complex system of irrigation canals, dams, reservoirs and barrages, while two enormous hydroelectric stations, the Nurek (300 m /984 ft high) and the Rogun (335 m /1099 ft) – the two highest dams in the world – have been erected on the Vaksh River (one of the Amu Darya's main tributaries).

Large-scale irrigation has had the effect of leaching nutrients from the soil. The bureaucratic solution to this problem was to make good the loss with massive applications of chemical fertilizer, to the extent that agricultural production became wholly dependent on chemicals. The problem was compounded by growing cotton year after year on the same land. The weeds and pests resulting from this monoculture were controlled by liberal applications of herbicides and pesticides.

The mechanical harvesting of cotton requires that the leaves should first be removed. This was done by aerial spraying with defoliants, including even Agent Orange (the defoliant used with such appalling consequences in the Vietnam War). After the crop had been harvested, the waste material was fed to cattle, thereby introducing Agent Orange into the food chain, with dire effects on the

## THE BACTRIAN DEER

Among the many animals living in association with the dense *tugai* vegetation found along the course of the Amu Darya is the Bactrian deer. At one time, it inhabited the full length of the Amu Darya as far as the Aral Sea and the delta of the Syr Darya, but systematic eradication of the *tugai* has dramatically reduced its range. Fire remains a perpetual hazard to the few patches of *tugai* that remain. Today, this deer is confined to the vicinity of the Vaksh and Pjandzh rivers, in southwestern Tajikistan, along the border with Afghanistan. Two reserves have been established primarily for its protection: Tigrovaya Balka Nature Reserve (497 km²/192 sq mls) and the Aral-Paigambar Reserve, an island 40 km²/15 sq mls, in the Amu Darya, near Termez, in Uzbekistan. Both reserves are being used for propagating the Bactrian deer with the aim of reintroducing it into suitable parts of its former range as opportunity occurs.

● Dead fish in stagnant water in the former fishing town of Muynak, which is now 50 km / 30 mls from the Aral Sea. The remains of abandoned port buildings stand in the background. (WWF / F. Ardito / Panda Photo)

people living and working in the area: the land they walk on, the water they drink, even the air they breathe have all been saturated with toxic chemicals.

Between 1974 and 1986, none of the Syr Darya's water reached the Aral at all, and from 1982 to 1989 none from the Amu Darya either. This resulted in a sharp lowering of the water level and a threefold increase in salinity. By the end of 1987 the water level had fallen to the extent that the sea was divided into two parts – the Greater and Lesser Aral. Of the 1,064 km³/255 cu mls of water that the Aral Sea once held, more than half has already been lost, leaving a series of saline lakes and ponds. The original seabed is now a desert containing 6 billion tonnes of salt, while a new salt desert, covering 26,000 km²/10,000 sq mls, has formed where the Karakum and Kyzylkum deserts converge. These two sources generate huge dust clouds (up to 100 km/60 mls wide) containing toxic salts that are carried on the wind over long distances.

The ruination of the Aral Sea is epitomized by the town of Muynak. This self-named 'city of fishermen', originally on an island at the southern end of the sea, now stands in the desert, 50 km/30 mls from the sea. The only fish now processed in Muynak's canning factory come from the Atlantic.

The conversion of this once-fertile region into a manmade desert has had a devastating impact on the local people, their lives wrecked by a combination of bureaucratic incompetence and agricultural ineptitude. With drinking water dangerously contaminated by toxic substances and the soil so saline that nothing will grow, the effect on human health and wellbeing has been calamitous: 30 million people are at risk. Infant mortality has soared, with physical and mental deformity commonplace, particularly among children who spend a great deal of time working in the cotton fields, where they are exposed to herbicides, pesticides and defoliants.

The contraction of the Aral Sea and the consequent rise in salinity has had an equally catastrophic effect on wildlife, fish in particular. Sturgeon, carp, bream and roach have been wiped out. From an annual yield of 40–50,000 tonnes, the catch has fallen to zero, and the fishing fleets lie abandoned in the sand. Of the 178 species of vertebrates that once lived around the Aral Sea and its delta, only 38 have survived.

The most disturbing aspect of the destruction of the Aral Sea is that it was deliberately planned. The former Soviet Ministry of Water Resources evidently believed that the sacrifice of the Aral Sea was justified in order to bring 100,000 km²/38,600 sq mls under irrigation. The plan envisaged eventually increasing this figure to 235,000 km²/90,700 sq mls, an aim that could only be achieved by reversing the direction of rivers, such as the Ob, flowing into the Arctic. The scheme has now been discontinued.

The problems remain, however. Even if the enormous sums of money needed to meet the cost of large-scale reclamation were to be made available (at a time when the economies of Kazakhstan and Uzbekistan are in a parlous state), the process of degradation has gone so far that it may well be irreversible.

# THE SUDD

LOCATION: Central Sudan.

AREA: Varies from about 16,500 km²/6,370 sq mls in the dry season to perhaps 32,000 km²/12,350 sq mls in the rains.

ALTITUDE: 200–500 m/650–1,600 ft.

CLIMATE: Equatorial. Rainfall averages over 970 mm/38 in, mostly April–October. Temperature: averages from 26 °C/78 °F during the rains (July–August) to about 29 °C/84 °F in the dry season. Daytime temperatures reach 30–37 °C/86–98 °F throughout the year.

The basin of the upper Nile, occupying much of the southern Sudan and Uganda, forms a complex of lakes, rivers and swamps set among lush grassland and woodland savannas. Of all these swamps, the greatest is the huge, impenetrable wetland of the Sudd, which lies in a region of heavy rainfall and high humidity. Its waters are shallow and slow-moving; papyrus dominates the vegetation, forming a tangle of stalks 4 m/13 ft tall, giving way to the Nile cabbage in more open water. The surface of the water is littered with great mats of floating vegetation, often strong enough to bear the weight of quite large animals, and creating so impenetrable a barrier that for centuries it effectively prevented access by river to the upper Nile. Floating vegetation chokes the river channels, forcing the water to spread over a wide area. This has the effect of regulating the flow: no matter how much water enters the Sudd from the south and west, the flow at the Nile's northern outlet remains more or less constant.

Fringing the swamp are extensive

floodplains which become waterlogged in the rainy season, but provide good grazing during the dry season, when large numbers of animals, both wild and domestic, are drawn in from the surrounding country.

The Sudd's ecosystem is founded upon seasonal inundation. As the Nile's floodwaters spread over the land, the wetlands double in size. Man and animal alike have adapted to this regime, conforming with the floodwater cycle by retreating to higher ground and peripheral areas during the inundation and returning to the floodplain in the wake of the receding waters. During the dry-season phase (from November to May), Nilotic pastoralists – of the Shilluk, Dinka and Nuer tribes – spend five or six months in temporary camps within reach of the river, before the June/July flooding signals the start of the return migration to their permanent villages. This age-old natural rhythm will be profoundly affected – as, indeed, will the hydrological regime and the ecology of the entire region – if plans for the construction of the 360 km / 224 mls Jonglei Canal (temporarily suspended because of civil war) are ever revived. Designed to drain part of the Sudd for agricultural development – in the process reducing the area of the swamps by 21–25%, and the floodplains by 15–17% – the Jonglei Canal project has all the makings of an ecological fiasco comparable with the construction of the Aswan High Dam.

The Sudd supports some of the greatest concentrations of wildlife in Africa.

Two antelopes, the tiang and the white-eared kob, are particularly abundant, assembling in spectacular numbers to undertake their mass migration through the Omo River delta, when they form what are probably the biggest aggregations of large mammals remaining in Africa.

Most animals shun the dense swamp vegetation, but some have adapted to it. The Nile lechwe, for example – an antelope almost entirely confined to the Sudd region, and separated by several thousand kilometres from its nearest relatives in Zambia and Botswana – has elongated hooves that spread the weight of its body and thereby keep it from sinking into the swamp. Large herds graze the waterlogged floodplain, plunging into deeper water if danger threatens. Numbers are believed to be in the order of 30–40,000.

The Sudd's bird fauna is the richest of any wetland in Africa. None is more extraordinary than the whale-headed, or shoe-billed, stork, so called on account of its enormous bill. Despite standing more than 1 metre/3 ft high, it is seldom seen. It lives deep in the swamp, where it spends much of its time standing motionless on floating vegetation waiting, heron-like, for fish or other prey to come within reach.

The swamps of the Sudd mark the division between the Sudan's predominantly Muslim north, where the people (mainly of Arab extraction) are agriculturalists, and the south, peopled by Nilotic pastoralists who since independence in 1956 have been fighting for self-determination. The Khartoum government has declared its intention of converting the 'unproductive' wetlands of the Sudd into an agriculturally productive region. This will result in the Nilotic pastoralists, whose culture and life style are geared to the Sudd's ecological regime, being forced to abandon their traditional way of life and exchange their customary nomadic affluence for the hopelessness and grinding poverty that are the common lot of Third World peasant agriculturalists.

Several game reserves – among them Seraf (9,700 km²/3,745 sq mls), Shambe (620 km²/240 sq mls), Bandingiru and Mongala, together covering more than 10,000 km²/3,860 sq mls – have been established in the Sudd, while the entire Dinder/Rahad floodplain is included in the Dinder National Park (8,900 km²/3,436 sq mls) and the adjoining Rahad Reserve. But protection is only nominal. The status of the reserves needs to be strengthened, their size increased and their management improved, thus recognizing the Sudd's standing as one of the most important natural sanctuaries in Africa and as a wildlife reservoir for replenishing adjacent areas. But there is even more to it than that: the Sudd is of such importance to the stability of the Nile ecosystem and to the riparian countries dependent on its waters, that this greatest of all African wetlands should be permanently conserved in its entirety. For the Sudan there could be no higher conservation priority.

## THE ILEMI TRIANGLE

Adjoining the Sudd in extreme southeastern Sudan is an immense tract of country known as the Ilemi Triangle. During the dry season (from early November to mid-March) the entire area is practically waterless, as bare as a desert and almost devoid of life. But with the coming of the rains the scene changes dramatically. Dry river beds are transformed into raging torrents and the black cotton soil becomes a quagmire. The onset of the rains triggers a southerly movement of wild animals from the Sudd in search of fresh young grass. Many thousands of animals take part in this migration, trekking across the plain in long columns reaching as far as the eye can see.

White-eared kob predominate, but other species are also present. An indication of numbers is given in an account written in 1940, describing the main formation as 'roughly in column of ten and packed so close that noses almost touched tails and flanks rubbed flanks. How long this line was could not be estimated but it continued moving past the camp from sunrise to sunset ...' Another first-hand report, written in 1947, described how 'the whole countryside appears to be on the move. I have seen natives throw spears with their eyes shut and hit an animal every time.'

The reverse migration back to the Sudd starts in October when the land starts to dry out and the grass begins to die back. More than a million kob are estimated to make the 650 km/400 mls round journey from their dry season grazing area on the Upper Nile floodplain to their wet season range in the Ilemi Triangle and back again, rivalling the great wildebeest migrations on the Serengeti Plains and the caribou migrations across the Arctic tundra.

# KAFUE NATIONAL PARK

LOCATION: South-central Zambia, in Central, Southern and Northwestern provinces. 14°00′–16°40′ S, 25°15′–26°45′ E.

ALTITUDE: 970–1,470 m / 3,182–4,823 ft.

AREA: 22,400 km² / 8,648 sq mls. The park adjoins the Kafue Flats Game Management Area (5,172 km² / 1,997 sq mls).

CLIMATE: The year is divided between a cool, dry season and a warm rainy season (November–March). Mean annual rainfall about 1,000 mm / 39 in. Temperatures range from a maximum of 15–27 °C / 60–80 °F in the cool season to 27–35 °C / 80–95 °F in the hot season, dropping sharply at night.

Kafue National Park is located in the mid-reaches of the Kafue River and its two major tributaries, the Lufupa and Lunga, that flow into the park from the north. A perennial swamp in the northwestern extremity of the park drains into the Lufupa via the Busanga floodplain. The park is of outstanding international importance for the conservation of antelopes. Vegetation in the southern part of the park is mainly miombo or *Brachystegia* woodland, with tracts of mopane. In the northern sector, areas of open grassy floodplain ('dambos') are ringed by miombo-termitaria woodland.

The park protects some of Africa's largest populations of typical miombo species, among them roan and sable antelope and Lichtenstein's hartebeest. It also supports locally high densities of such species as wildebeest, waterbuck, puku and red lechwe. Large mammals are well represented, among them elephant, black rhino, buffalo, and lion. Both yellow and chacma baboons live in close proximity. Waterfowl are abundant, one of the most notable being the wattled crane, which has its main breeding ground on the Kafue Flats. The slaty egret and the shoe-billed stork have also been recorded from the area.

The Kafue park is bisected by the main road from Lusaka in the east to

## A FLOODPLAIN ANTELOPE

The lechwe occurs in southern Central Africa. Of the three subspecies, the most widely distributed is the red lechwe, found principally on the Upper Zambezi and Kafue rivers, extending into Angola, the Caprivi Strip and Botswana. The Kafue lechwe is restricted to Zambia's central Kafue Flats, while the black lechwe is confined to the floodplains of Lake Bengweulu and the upper Chambeshi Basin in northeastern Zambia. Overall numbers are estimated at about 100,000: red lechwe 30,000; Kafue lechwe 40,000; and black lechwe 40,000.

The lechwe's movements are closely related to the availability of the floodplain grasses and aquatic plants on which it feeds. The Kafue Flats are subject to regular flooding, starting in December, reaching a maximum in May, and receding to the lowest level by November.

Over the last three or four decades the red lechwe has undergone a dramatic decline. In 1932, numbers on the Kafue Flats totalled 250,000 but, by 1958, they had been reduced to 30,000. The black lechwe fell from 150,000 to 15,000 during the same period. The decline was largely brought about by the *chila*, the traditional tribal hunt involving a large body of men, accompanied by dogs, encircling a herd of lechwe and gradually closing in. Ready markets for 'bushmeat' are found on Zambia's copperbelt and among the expanding population of the capital, Lusaka. Major water control schemes along the Kafue River, starting in 1967 with the construction of a hydro-electric dam at Kafue Gorge to provide electricity for the copper industry, followed in 1977 by a second dam at Itezhitezhi, 250 km / 155 mls upstream, have radically altered the natural flood regime and reduced the area of seasonally inundated floodplain.

Prime responsibility for conserving the lechwe rests with Zambia. The Kafue National Park protects one of the most important lechwe populations which, since 1950, has extended its range into the Kansonso–Busanga Game Management Area. Two further, smaller, national parks, Lochinvar and Blue Lagoon, have been established on part of the Kafue Flats adjoining the Kafue Flats Game Management Area. Most of the black lechwe's range now similarly lies within designated game management areas. Projects are being planned to demonstrate that protection of wetlands on the Kafue Flats and Lake Bangweulu is compatible with development. The National Parks and Wildlife Department is instituting a system of management based on controlled culling to utilize the lechwe's high potential for protein production, with the aim of benefiting the local people in a region where conventional agriculture is greatly restricted.

● Work on the Jonglei Canal – which would have a disastrous impact on the ecology of the Sudd wetlands – has been temporarily suspended because of the civil war in southern Sudan. (WWF / Oswald ITEN)

Mongu in the west. Construction of a dam on the Kafue River at Itezhi-tezhi, just outside the park's eastern boundary, has affected the water regime. A squatter settlement erected by a remnant of the labour force employed on building the dam has been established within the park, and the island in Lake Itezhi-tezhi is permanently occupied by fishermen, despite being within the park. Nitrates, phosphorus residues and other wastes are being discharged into the Kafue River. A national conservation strategy has been drawn up, identifying the main problems facing Zambia and recommending how they may best be resolved.

● Lechwe bounding across the Kafue floodplain in Zambia. The traditional hunt, the *chila,* has been replaced by controlled culling designed to benefit the local people while ensuring the survival of the lechwe. (Jacana / Alain Antony)

# OKAVANGO

LOCATION: Northwestern Botswana. 19°05′–19°45′ S, 22°28′–23°45′ E.

AREA: Varies from 16,000 km²/6,175 sq mls in the dry season to 22,000 km²/8,500 sq mls in the wet.

ALTITUDE: About 900 m/2,950 ft.

DEPTH: Up to 5 m/16 ft in the channels, shallow elsewhere.

CLIMATE: Rainfall about 475 mm/18 in, but subject to wide variation, a good wet season sometimes being followed by years of drought. Temperature ranges from a maximum of 41 °C/105 °F to a minimum of −6 °C/21 °F.

Northern Botswana is one of the last great wilderness areas remaining in Africa, and the Okavango Delta, an oasis set amidst the sands of the Kalahari Desert, is its most spectacular feature. Rising in the highlands of Angola, the Okavango River flows for several hundred kilometres through the Kalahari.

For the last part of its journey it moves swiftly through the 'Panhandle', before fanning out into the Okavango Delta. The Panhandle, about 80 km/50 mls long and 15 km/9 mls wide, forms 12,000 sq km/4,600 sq mls of reedbeds, dominated by papyrus. The delta itself fluctuates in size according to season but, even under severe drought conditions, contains permanent water and provides a haven for water-dependent species.

● The vast swamps and floodplains of Botswana's Okavango Delta. (WWF / Jim Thorsell / IUCN)

The inflow into the Panhandle reaches its maximum in February or March, when the Okavango River overflows its banks and revives the swamps on either side. On reaching the delta the water spreads out and slowly advances, so slowly in fact that it usually takes about five months to cover the entire delta: it does not reach the southern floodplains until the middle of the dry season. After spreading over many thousands of square kilometres of shallow floodplain, the last of the Okavango's waters are carried across the Kalahari by the Boteti River, which flows into Lake Xau, some distance short of Lake Makgadikgadi (37,000 km²/14,285 sq mls), which is the largest salt pan in the world.

Most of the water is carried through the delta in a few reed-lined channels, 20–30 m/60–100 ft wide and up to 5 m/16 ft deep. Except for these channels, the delta is covered by shallow water, forming a complex of floodplain grasslands, seasonal and perennial swamps, lakes and lagoons linked by an intricate web of interconnecting waterways in a constantly changing pattern.

Throughout the delta there are innumerable islands of every shape and size, often dotted with termite mounds and rising sufficiently high above the water to allow trees to become established. Many of the islands are fringed with tall sycamore figs, bearing clusters of fruit throughout the year, and are an important source of food for many animals – including several species of fish that specialize in feeding on figs as they fall into the water. Islands in the southern part of the delta are of a different type. The largest, Chief's Island, covers more than 1,000 km²/386 sq mls, and has vegetation and animals more characteristic of the acacia thorn scrub and mopani of the Kalahari woodlands.

Groves of tall trees lining the river banks and growing on islands serve as vantage points for fish eagles by day and fishing owls by night, as well as providing nesting sites for many kinds of birds. Crocodiles and hippos inhabit the main channels of the Panhandle, leaving the papyrus swamp to its specialized fauna, principally birds such as the pied kingfisher, the little bittern and various herons. Only in the delta itself does the combination of flooded grasslands interspersed with numerous islands provide ideal conditions for supporting substan-

tial concentrations of large mammals. In its lower reaches the delta's lush reedbeds give way to an extensive area of floodplain. It is on these seasonally inundated grasslands, where the delta merges with the Kalahari, that the wild fauna of the Okavango can best be seen.

As the dry weather advances and the water recedes, the floodplain dries out and the larger grazing animals move in. The drying of pans in the northern Kalahari causes increasing numbers of animals to fall back and concentrate on the delta, and up to twenty species of large grazing animals occupy the floodplain at any one time. Nowhere else is there a comparable intermingling of nomadic desert antelopes such as gemsbok and springbok with wetland species such as lechwe and waterbuck, or of plains game such as wildebeest, hartebeest and impala, with woodland species like kudu and sable and roan antelopes.

Among the larger grazing animals of the floodplain, the most common is the red lechwe. This semi-aquatic antelope grazes on grasses and shoots growing in the shallow waters. A common sight in the Okavango – and one that never palls – is of lechwe bounding through the water with characteristic loping gait, kicking up showers of spray as they go. During the breeding season, from January to April, males often position themselves on the tops of termite mounds, using them as vantage points.

The only antelope to inhabit the reedbeds – and to include a high proportion of papyrus in its diet – is the sitatunga, here at the southern extremity of its

range. The Okavango provides ideal conditions for an animal whose entire life is spent deep in the papyrus swamps.

Hippos occupy the lagoons by day, trekking to the islands at night to graze, in the process cutting paths through the almost impenetrable reed thickets. Other animals make use of these hippo paths, as do Bushmen on papyrus rafts and local fishermen in their dug-out canoes, known as *mokoros*.

Water birds – ducks and geese (the spur-winged goose is particularly abundant), herons, ibises and fish eagles – are common on the floodplain, while the swamps are the wintering grounds for migrant ducks from southern Africa moving north to avoid the southern winter. With the coming of summer these birds return south, their place being taken by an influx of Palaearctic (northern Eurasian) migrants.

Water lilies are widespread on the more open waters of the delta. Living in close association with them is the little pygmy goose, which feeds almost exclusively on lily fruit. Pygmy geese usually nest in tree hollows lined with down. The goslings leave the nest shortly after hatching, leaping into space and free-falling to the water.

The Okavango holds Botswana's only permanent surface water, for which there is incessant demand from agricultural and industrial users, notably for augmenting the supply of water to the Orapa Diamond Mine, some 280 km/175 mls from the delta. Diamond extraction – one of the country's principal revenue earners – requires huge quantities of

## THE CATFISH RUN

The extraordinary phenomenon known as the 'catfish run' is unique to the Okavango Delta. Each year, between September and November, when the level of water in the Panhandle is still falling, catfish gather in huge shoals and migrate upstream, slowly making their way to their spawning grounds on the floodplains of the Panhandle. As they go, they indulge in a frenzy of feeding, flushing out small fish lurking among the papyrus roots, and causing the surface of the water to erupt. The noise of their feeding can be heard at a distance, and attracts flocks of fish-eating birds – cormorants, darters, fish eagles, herons and kingfishers. Crocodiles also lie in wait to seize the catfish as they float listlessly past, gorged with food. Pack-hunting and intensive feeding are part of the process of the catfish getting into breeding condition before the arrival of the floodwaters from Angola, usually in February and March.

water. Botswana's livestock industry is given special help by the EC, which guarantees to take a substantial annual quota of beef. This has encouraged the livestock industry to expand, and has led to proposals for eradicating tsetse fly from the delta (using insecticides such as Dieldrin and Endosulphan), to facilitate the introduction of cattle. The Okavango Delta is a stable ecosystem; it is also very fragile, and any interference by people could have far-reaching consequences. Misuse, wether through ignorance, thoughtlessness or greed could easily destroy the finest wetlands remaining in Africa.

Only a small part of the Okavango Delta is protected. In 1962, the Batawana people took a positive step by establishing the Moremi Wildlife Reserve (1,800 km² / 695 sq mls) on their own tribal land – the first tribal wildlife conservation area in Botswana, and one of the few in Africa.

# WADDEN SEA

LOCATION: Extends from the northern Netherlands along the entire North Sea coast of Germany into southwestern Denmark, from Den Helder in the west to Esbjerg in the northeast, and separated from the North Sea by a chain of islands. 53°15′–55°16′ N, 5°15′–8°32′ E.

AREA: 4,788 km² / 1,848 sq mls: Netherlands 2,500 km² / 965 sq mls; Germany 880 km² / 340 sq mls; Denmark 1,408 km² / 543 sq mls.

ALTITUDE: Sea level to 34 m / 110 ft (the height of the tallest dunes).

DEPTH: Averages 3–4 m / 10–13 ft; maximum over 20 m / 65 ft; tidal variation 1.7–3.5 m / 5.5–11.4 ft

CLIMATE: Temperate. The shallow waters are relatively warm in summer with ice forming in winter.

The Wadden Sea is Europe's largest and most important wetland complex, and a vitally important nursery for many species of fish and crustaceans. The shallow waters sustain almost all the North Sea's herring, 80% of plaice, and more than half the sole during some part of their life cycle. The islands in the area support numerous seabird colonies, and are feeding grounds for immense numbers of waterfowl from Siberia, Scandinavia and northeastern Europe.

The region is dominated by a huge complex of tidal mudflats, marshland and low islands, much of it shallow sea which dries out at low tide, leaving large tracts of mud or sand intersected by branching creeks. Habitats range from open seas and tidal zone to river estuary and freshwater ponds, and from sandy beaches, sandbanks and dunes to salt marshes and grass-covered islands.

The Wadden Sea accommodates several million waterfowl, many of which are migratory. Up to 30,000 moulting and 74,000 roosting eider ducks, as well as 200,000 roosting scoters have been recorded on the island of Mana. Trischen Island and the Knechtsand are the moulting ground for almost the entire European population of shelduck. The Sea also supports the largest population of avocet in the world. Other species found there include brent goose, pink-footed goose, spoonbill, oystercatcher, curlew, and bar-tailed godwit. Some 2–3,000 common seal also frequent the area.

Since 1963, 350 km² / 135 sq mls of habitat have been lost through reclamation. Other potential threats arise from mass tourism and the exploitation of gas and oil deposits, including the installation of pipelines to mainland Europe from natural gas fields and oil platforms in the North Sea. Pollution in the form of organic and mineral wastes, chlorinated hydrocarbons and other toxic substances originating from the nearby industrial regions of France, Germany, Switzerland and the Netherlands is carried into the Wadden Sea by rivers, chiefly the Rhine, Scheldt, Meuse, Weser and Elbe.

In 1982, the governments of the Netherlands, Germany and Denmark signed a trilateral agreement for the protection of the Wadden Sea and the region was also designated a Ramsar site in 1985. Substantial parts of the Sea, including a number of islands, have been declared nature reserves. The most important are the Nordfriesische Wattenmeer (1,400 km² / 540 sq mls) and Wattenmeer Knechtsand / Eversand (325 km² / 125 sq mls). An action group, the Society for Conservation of the Wadden Sea, has been set up to increase public awareness of the issues, to lobby politicians and to discourage harmful development.

# ISLANDS, ATOLLS AND CORAL REEFS

*If that the heavens do not their visible spirits
Send quickly down to tame these vile offences,
Humanity must perforce prey on itself,
Like monsters of the deep.*
Shakespeare, *King Lear*

There are three principal kinds of island: inshore, offshore and oceanic. The first comprises all islands situated on continental shelves and which at some stage have been attached to the nearby continental land mass, but have become separated during the processes of uplift and subsidence that shaped the surface of the earth, either by the land sinking or the sea level rising (e.g. the British Isles and Japan).

Offshore islands, on the other hand, are groups of islands - archipelagos - which, although located on the periphery of continental land masses, have had no direct connection with them. Their proximity has nevertheless resulted in infusions of plants and animals from the mainland (e.g. the West Indies).

The third category, oceanic islands, have had no connection with any continental land mass, but are of marine origin. Some are the tops of granitic submarine mountains (e.g. the Sey-

chelles), but most are the product of volcanic activity. They include volcanic islands, formed of accumulations of larva rising from the seabed until the tip of the cone broaches the surface (e.g. the Galápagos and Solomon islands); coral atolls, consisting of submerged volcanic cones capped and fringed with coral growth to form a reef enclosing a lagoon (e.g. Laysan in the Pacific and the Chagos islands in the Indian Ocean) and elevated limestone atolls (e.g. Aldabra, also in the Indian Ocean).

## Life of Islands

Recently formed inshore and offshore islands are characterized by plants and animals similar to those of the continental land mass from which they sprang; but older ones (e.g. Madagascar and New Zealand – which are in effect miniature continents) and old island continents (e.g. Australia) possess faunas that are fundamentally different from those of

the nearest land mass, often because they retain types of animals that have become extinct on the mainland (e.g. Australia's marsupial and Madagascar's prosimian faunas). Oceanic islands owe their floras and faunas to the chance of sea and air currents and violent storms; involuntary immigrants arriving in this manner do not always originate from the nearest land mass.

Such accidental dispersal explains why so few terrestrial mammals (other than bats) and so few reptiles colonize oceanic islands, and why there is a preponderance of sea birds and insects. As a rule only small mammals and reptiles can be successfully transported to outlying islands, probably on natural rafts of floating vegetation, logs, branches, or other debris. There are, however, occasional records of large animals surviving relatively long sea passages as, for example, the adult hippopotamus from the East African mainland that arrived on the is-

land of Pemba – 50 km / 30 mls offshore – in the severe floods of 1961. During glacial epochs, when ocean levels were much lower than today, more land would have been exposed, thus forming a series of stepping stones that made islands much more accessible. About 25,000 years ago, when ocean levels were 150 m / 500 ft lower, the Seychelles, for example, was not a group of small islands but a substantial land mass.

But although the colonization of oceanic islands by both plants and animals is haphazard, the successful immigrants gain certain immediate advantages: competition from other organisms is either non-existent or minimal; some animals find vacant ecological niches and are thus able to exploit a variety of favourable opportunities that were not available on the mainland, a circumstance that permits them to evolve in directions that would otherwise be impossible; others are obliged to adapt themselves to

● The Komodo dragon – in fact a large lizard that can grow up to 3m / 10 ft in length – is found only on a few Indonesian islands, where it occupies the ecological niche – which in other parts of the world is filled by large mammalian carnivores. (WWF / Michel Terrettaz)

● The breath-taking scenery of the Cirque de Mafate on the isolated Indian Ocean island of Réunion. The whole island was once covered with evergreen rainforest, most of which has been cleared, and erosion is now widespread. (Explorer / Luc Girard)

unfamiliar environmental conditions. These factors, accentuated by inbreeding, open the way to adaptive radiation – the process whereby animals spread out from their centre of origin and adapt themselves to whatever ecological opportunities happen to be available, changing in both appearance and behaviour to meet the new conditions. This explains why island floras and faunas develop special characteristics, why island animals often fill ecological niches very different from those occupied by their ancestral mainland forms, why endemic species (species of plants and animals that exist nowhere else) are so frequently found on islands, and why islands provide the best opportunity for studying living examples of evolution and adaptive radiation.

The best-known examples of adaptive radiation are Darwin's Galápagos finches which, in the process of exploiting a variety of specialized ecological opportunities, became sufficiently differentiated to be classified into 13 distinctive species in three separate genera.

Isolation shields animals and plants but leaves them ill-equipped to meet competition. Thus, when aggressive and adaptable invaders arrive in the wake of man, the native species can offer no resistance and are quickly supplanted.

## The Outside Threat

Man's impact on islands has had dramatic consequences. During the course of their epic ocean voyages, migrant Melanesians, Micronesians and Polynesians took with them various staple food plants, such as taro, yams and sweet potatoes, as well as dogs, pigs, chickens and, unwittingly, rats, all of which were small enough to be carried by outrigger canoe. Europeans entering the arena in the Age of Discovery regarded oceanic islands as temporary havens where crews could recover from the rigours of the voyage, food and water could be replenished, and ships repaired.

The early explorers and adventurers brought with them livestock such as goats, sheep, pigs, rabbits, chickens, and, of course, rats and mice. Domestic animals were released, either intentionally, with the aim of provisioning subsequent voyages, or inadvertently. Occasionally temporary or permanent victualling stations were established on islands where crops could be grown. These pioneer voyagers were followed by sealers, whalers, buccaneers and traders, who systematically exploited the resources of the islands they visited.

The effect of introductions has almost invariably been disastrous. Island vegetation, being fragile and usually of limited extent, is extremely vulnerable to disturbance, and the plant/animal associations are easily disrupted or destroyed. Exploitation of the flora frequently leads to the disappearance of the fauna, or its relegation to the few relict patches of indigenous vegetation that remain. The fragility of island faunas can be gauged from the disproportionately high number of insular birds considered endangered. Sixty per cent of birds listed in the *Red Data Book* are island forms, as is the great majority of extinct birds.

Many indigenous animals knew no fear of man and made no attempt at escape or resistance. This is well illustrated by the story of Steller's sea cow. The largest of the sirenians (the order that includes the dugong and the manatees), it attained a length of 7.3 m / 24 ft, and inhabited Bering Island and Copper Island in the Bering Sea. Expeditions wintered on the islands with the express purpose of laying in stores of sea cow meat. Within nine years of man's first visit, no sea cows remained on Copper Island and, by 1768, 27 years after its discovery, the species was extinct.

Apart from the damage caused by people, the alien animals introduced by them preyed on the native fauna. The eggs of ground-nesting birds, tortoises and turtles are particularly vulnerable to predation by rats and rooting pigs (a danger that is greatly reduced among birds habitually nesting on cliff faces); flightless birds (loss of the power of flight is a relatively common phenomenon among island birds) are defenceless against feral cats and dogs (i.e. domestic animals gone wild); while goats and sheep destroy the vegetation.

Earlier generations did not hold a monopoly of manmade changes: modern technology has made exploitation possible on a larger scale and at a quicker rate than ever before. On Ocean Island in the Pacific, for example, strip-mining for phosphates and other minerals has been so extensive that virtually the entire land surface has been removed; such damage is irreversible. This ecological *Rake's Progress* accelerated during the Second World War when the militarization of islands became a feature of the fighting in the Pacific theatre, and the concept of island-hopping to penetrate Japan's protective screen of island bases was adopted as basic US strategy.

The destruction of the war years was followed by the phase of post-war 'reconstruction' that was almost as calamitous. But this time, instead of enjoying the dispensation of acts performed under stress of war, devastation arrived in the guise of 'development'. The pace was accelerated by radical improvements in the means of transport, communications and other technological triumphs. Today, few islands are more than several hours' flying time from major centres of population. Remoteness and isolation have disappeared.

As well as exploitation of the native forests and the spread of plantation agriculture, islands have been taken over for such purposes as airfields and military bases, communications and satellite tracking stations, even for nuclear testing grounds. A more insidious factor has been the spread of mass tourism, which debases not only the islands but sometimes the islanders as well. Mass tourism is in many respects the greatest threat to islands. Airfields, roads, hotels, restaurants, housing, souvenir shops and a wide range of service industries are an integral part of the business. These, together with problems associated with sewage and waste disposal, have completely transformed many islands. The Canaries, Hawaii and the Bahamas are examples of the type of development that can occur. Many other islands in the Mediterranean, Indian and Pacific oceans are moving in the same direction, their natural features debased by vulgarization.

## Saving Islands

Before ruination is complete, measures need to be taken as a matter of urgency to protect islands and islets from further disfigurement by setting aside such unspoiled areas as remain as national parks or protected areas in which exploitation is either prohibited or effectively controlled. And what happens ashore is bound to affect neighbouring marine environments. When rivers laden with silt (arising from poor farming practices) or carrying sewage or other wastes discharge into the sea, the effect on coral reefs and other coastal ecosystems can be profound.

No two islands are alike, so protective measures must be tailored to meet the specific requirements of each island. The fact that most have already been debased makes it all the more important that the few unspoiled islands remaining should be placed under the control of the international scientific community, as has happened with Aldabra.

## Coral Reefs

Largely restricted to shallow tropical waters between latitudes 30° N and 30° S and to a depth of 30 m / 100 ft. Of the 600,000 km²/230,000 sq mls of reefs estimated to exist, about 1% occurs in the South Atlantic, about the same in the Eastern Pacific, and 14% in the Caribbean. Besides being one of the most diverse and biologically productive ecosystems on earth, they also provide habitat for many other organisms.

Reefs fall into two main categories: shelf reefs, which form on the continental shelves of large land masses; and oceanic reefs, which develop in deeper waters, often in association with oceanic islands. Within these two categories there are a number of different reef types: fringing reefs, which grow close to shore (i.e. most shelf reefs, although some develop around oceanic islands); patch reefs, which form on irregularities on shallow parts of the sea bed; bank reefs, which occur deeper than patch reefs, on continental shelves and in oceanic waters; barrier reefs, which develop along the edge of continental shelves and are separated from the mainland or island by a relatively deep, wide lagoon; and atolls, which are reefs more or less encircling a central lagoon and are found in oceanic waters, probably corresponding to the fringing reefs of submerged islands.

## Economic Importance

The IUCN *World Conservation Strategy* describes coral reefs as one of the 'essential support systems necessary for human survival and sustainable development'. Reefs protect coastlines against wave action and storm damage, and contribute to the formation of sandy beaches. Along with other coastal habitats, reefs have a crucial role as refuges and breeding grounds for numerous species of fish, molluscs and crustaceans on which many tropical countries generally, and island communities in particular, depend for food.

## Tourism

In recent years, and especially since the 1960s, coral reefs have found a new and rapidly expanding role as major tourist attractions of great economic importance in the Caribbean countries and elsewhere. Tourists are particularly drawn to islands such as the Bahamas, Puerto Rico and the US Virgin Islands, above all to their coastal zones, where a favourable climate, beaches, clear water and the presence of reefs are the major attractions, providing conditions that allow scuba diving, fishing, sailing and other marine sports. Tourism ranks as the principal source of foreign exchange in Kenya and the Seychelles and is becoming increasingly significant in countries such as Sri Lanka, Thailand, Indonesia and the Philippines.

## Fragility of Reefs

Although reefs may not be as fragile as at one time believed, their temperature, light, water clarity, oxygen and salinity requirements are nevertheless very specific. Reef-building corals are confined to depths of less than 30 m / 100 ft and are generally found in much shallower waters. The murkier the water the less light can penetrate, with a corresponding decrease in photosynthesis. In the course of evolution coral reefs have become tolerant of storm damage and changes in sea level, such as occur with abnormally high and low tides, but rain in monsoonal quantities deposited by hurricanes, cyclones and typhoons, can quickly reduce a reef to rubble, the shallower reefs being especially vulnerable. Recovery is a slow process – from 20 to 50 years for complex reefs, less for simpler reefs.

There is some concern at the apparent increase in irruptions of coral predators and other natural phenomena, possibly resulting from human activities.

## Climatic Influences

Excessively low or high temperatures can be damaging. The abnormally high seawater temperatures periodically associated with El Niño (see page 84) – as in 1982–83 – are thought to have been responsible for the widespread destruction of coral reefs which occurred in many parts of the tropical Pacific and western Atlantic. By the end of 1983, coral reefs in the eastern Pacific were estimated to have lost 70–95% of their living coral cover to depths of 15–18 m / 50–60 ft.

## Coral Predators

Reefs are highly vulnerable to predation. One of the most voracious coral predators, the crown-of-thorns starfish, has been responsible for the recent destruction of reefs over much of the Indo-Pacific region. Some 500 km / 300 mls of the Great Barrier Reef have been damaged and many others affected, among them reefs in New Guinea, Palau, Truk, Fiji, Tahiti, and the Tuamotos. All that prevents the crown-of-thorns starfish from invading the Caribbean is the Isthmus of Panamá. Major alterations to the Panamá Canal could allow the predator to spread through this narrow waterway, with potentially calamitous consequences for Caribbean reefs.

Although not themselves predatory, sea urchins can nonetheless cause substantial damage if present in large concentrations. By grazing on algal turf (which grows on coral rock) they can cause surface erosion which weakens the reef structure. Irruptions of sea urchins are believed to be attributable, at least in part, to a decline in predatory fish resulting from overfishing.

## The Impact of People

Like any other community of living things, reefs are undergoing a process of continual change. But whereas they can tolerate most natural change, there is mounting evidence to show that they are much less able to withstand human-induced change. This is becoming an increasingly important issue, especially in places where reefs are of economic importance. Most threats to coral reefs relate to the fact that tropical coastal areas are among the most densely populated regions on earth. The number of people living around Caribbean coasts, for example, has soared since the end of the Second World War: apart from Cuba and Dominica, the Antillean islands all have population densities of more than 100 per km²/260 per sq ml (Barbados has over 550 per km²/1425 sq ml). This, in combination with very low per capita income (as in Haiti, Honduras, St Vincent, Grenada and Dominica), exerts tre-

mendous pressure on marine resources, and it is no coincidence that most of these countries have lost significant portions of their reefs in the last decade.

## Run-off and Sedimentation

Lying close offshore, fringing reefs are particularly vulnerable to pollutants and sediments washed off the land. Deforestation – nearly 18,000 km²/6,950 sq mls of forest disappear from the Caribbean region each year – and slash-and-burn agriculture are the chief causes of soil run-off. Atoll and barrier reefs are not as vulnerable as fringing reefs but may be affected by marine pollutants from ships and carried on oceanic currents. About 20% of Indian Ocean reefs have been lost or seriously damaged.

Soil run-off is also responsible for sedimentation, one of the worst causes of damage to coral reefs, and one to which their static nature makes them particularly vulnerable. Reefs in Brazil, Colombia, Costa Rica, the French Antilles, Puerto Rico and Venezuela show very marked sedimentation. Fourteen countries in the Indian Ocean have been similarly affected, notably the coast of East Africa, parts of the Seychelles and Comoros, as well as Malaysia and Indonesia. The destruction or reduction of mangroves – which fulfil the important function of trapping sediment – serves to accelerate the process. In combination with the inflow of chemical pollutants – fertilizers, pesticides, etc. – sedimentation can cause extensive damage. Control is often difficult since the source of the damage frequently lies far from the point of impact and is often exercised by a number of different agencies and government departments.

Sewage, mostly untreated, is a widespread cause of reef pollution. Seventeen Caribbean countries are currently affected by it. Less than 10% of total domestic waste receives treatment in

● Whip coral and feather stars on a reef off Papua New Guinea. The beauty of coral reefs has, in many parts of the world, attracted excessive tourist pressure. Inadvertent damage to delicate corals by divers and snorkellers can sometimes cause almost as much harm as the deliberate removal of specimens. (Planet Earth Pictures / Gary Bell)

the region, causing eutrophication of coastal waters and ensuring accelerated algal growth which smothers coral. Effluents from the sugar industry and distilleries is a frequent cause of pollution in many Caribbean countries, contaminating a wide range of marine organisms.

As one of the world's largest oil producing regions, the Caribbean is particularly prone to oil pollution. Oil terminals tanker traffic, refineries and offshore oil reserves adjacent to reefs are all potentially hazardous. The huge number of offshore rigs makes periodic spills and blowouts inevitable. Tanker routes often pass close to reefs, pipelines are sometimes damaged, and spills have affected reefs in Panamá, the Florida Keys and Puerto Rico. While oiling may inflict greater damage to beaches than to reefs, the chemical detergents (often highly toxic) used in cleaning up the mess often cause greater damage than the oil spill itself, especially in confined waters such as the Red Sea and Persian Gulf.

## Coastal Development
Coastal developments – among them dredging, harbour and airport construction, housing and industrial development – have a marked impact on reefs. Malaysia, Singapore, Indonesia and the Philippines have concentrated considerable industrial development along their coasts, and in some countries (e.g. Thailand) mining is a coastal activity.

The accessibility of coral also makes it an obvious choice for building and road construction materials, particularly on islands where alternatives are limited; sometimes it is mined on a large scale.

## Over-exploitation
Ease of access also makes coral reefs a focus for recreational use. In places where tourism has become a dominant feature, a number of varieties of fish, crustaceans and molluscs – as well as the corals themselves – have been heavily over-exploited, with consequent reduction of the reef's productivity. Giant clams, mother-of-pearl and ornamental shells have been the chief objects of exploitation, and the sale of ornamental corals has resulted in localized reef damage. Corals and shells are exploited throughout much of southern Asia and eastern Africa, with the Philippines the centre of the trade. Shells are an important source of income and although many molluscs are able to tolerate reasonably heavy collecting, standards of management need to be improved and trade in ornamental coral and shells better controlled at both the national and international levels if damage to reefs and over-exploitation of the more popular species is to be avoided. Some scientists believe that over-collecting of the giant triton for its shell – 100,000 are estimated to have been taken from the Great Barrier Reef alone in the decade 1949–1959 – was the chief reason for the irruption of crown-of-thorns starfish (the giant triton's principal prey) that first became evident in 1963 and caused extensive damage to Pacific reefs. The use of dynamite and fish poisons – which despite being illegal is nevertheless still practised in many Caribbean countries and around the Indian Ocean – is not only wasteful of fish and other marine life but also destructive of the reef itself.

## Fisheries
For many Third World coastal communities reefs are the principal source of protein – in the form of fish, molluscs and crustaceans – for large numbers of people. Reefs also form the breeding grounds and provide the nutrients for many commercial and subsistence food species. With the development of tourism, reef fishing has also become an important recreational activity in these countries, even though it sometimes conflicts with subsistence fishing. The popularity of spear-fishing, which creams off the larger predator fish, creates ecological imbalance. The aquarium fish industry is also having a harmful effect.

## Protection and Conservation
If further damage to reefs and reef ecosystems is to be averted, and if the sustainable use of reef resources is to be maintained, it is essential that more effective means of protection should be introduced. The proper management of coastal regions is a universal need, but conservation programmes must be comprehensive in the sense of being drawn up within the context of the entire coastal zone, or even taking in the whole catchment area. The establishment of marine parks is an important step in that direction. In a few instances, the boundaries of terrestrial national parks can be extended to include marine habitats. Several countries such as the USA, the British Virgin Islands, the Netherlands Antilles, Kenya and the Seychelles now have comprehensive marine park programmes with a strong emphasis on the benefits accruing from tourism and recreation. Others, among them Saudi Arabia, Oman and Sri Lanka, are starting to introduce comprehensive marine park systems; but many – including Ethiopia, Somalia, Madagascar and both North and South Yemen – have done little to protect their reefs. Even national parks are only as effective as their level of management, and this frequently falls short of the requirement. Control over the levels of fishing and disturbance – at present very uneven – is essential. The international importance of some reefs – Aldabra, the Galápagos, and Australia's Great Barrier Reef – has been recognized by their designation as World Heritage sites.

# THE GREAT BARRIER REEF

LOCATION: Off the east coast of Queensland, Australia, extending for more than 2,000 km / 1,250 mls from a little south of the Tropic of Capricorn to the coastal waters of Papua New Guinea. 24°30′–10°41′ S, 145°00′–154°00′ E.

AREA: 348,700 km² / 134,633 sq mls.

ALTITUDE: Sea level to 40 m / 130 ft.

CLIMATE: Tropical. Wind patterns governed for much of the year by the Southeast Trades. Rainfall seasonally and geographically variable. Summer, the wettest season, is influenced by the monsoon and tropical cyclones and depressions. There may be heavy winter rain in the south. Temperatures average 24–30 °C / 75–86 °F in January; 18–23 °C / 64–73 °F in July.

The Great Barrier Reef is the world's largest system of coral reefs. A mosaic of reefs and islands, it consists of 2,900 separate reefs, ranging in size from less than 1 ha / 2½ acres to 10,000 ha / 25,000 acres, as well as about 300 individual cays (islets) and 210 continental islands. Biologically and scenically it is one of the natural wonders of the world.

The reef contains 400 species of coral, and there are extensive seagrass beds in inshore waters. Algae of many different kinds play an important part in the reef-building processes; they also constitute a major source of food for fish, turtles, molluscs and sea urchins.

In this complex marine ecosystem 1,500 species of fish, 4,000 species of molluscs, including important populations of giant clams, and a variety of sponges, anemones, marine worms and crustaceans have been identified, while 242 species of birds nest and breed on its coral cays. About one-third of the more important nesting sites of the entire region are to be found on the Capricorn Islands. Among them are nesting colonies of such species as the roseate tern and five other species of terns, as well as the white-bellied sea eagle, osprey, and reef heron.

Of the six species of marine turtles found in Queensland waters, three – the green turtle, loggerhead and hawksbill – live in close association with the reef. The far northern sector, the least disturbed part, contains some of the most important turtle nesting sites and seabird colonies in the western Pacific. Raine Island and Pandora Cay are among the green turtle's largest nesting grounds. The hawksbill also nests in the northern part of the reef and on islands in the Torres Strait .

The largest known populations of dugong are to be found in Australian waters: herds of over 100 have been sighted from the air. The seagrass beds and sheltered bays along the coast provide feeding and breeding grounds for many of these marine mammals.

Whales, notably the humpback, minke, and killer whale, are often seen off the Great Barrier Reef. Frequent sightings of females with newborn calves in the vicinity of the reef show it to be a humpback calving ground.

## NINGALOO

Although the Great Barrier Reef is in a class of its own, it is far from being Australia's only important reef. Impressive coral reefs, some in almost pristine condition, extend for 3,000 km / 1,860 mls along the coast of Western Australia. The Ningaloo Reef Tract, forming part of this coral complex, is the longest stretch of continental fringing reef in Australia. The only other reef of this type in the southern hemisphere to stand comparison with it is Grand Récif off the west coast of Madagascar.

Running along the coast between North West Cape in the north and Amherst Point in the south – about 260 km / 160 mls – the Ningaloo reef is separated from the coast by a shallow lagoon about 2–4 m / 6–13 ft deep, lying about 2.5 km / 1.5 mls offshore. Scattered throughout the lagoon are numerous coral pinnacles or 'bommies', some up to 6 m / 20 ft high and 10 m / 30 ft across. Surf breaking against the outer edge of the reef activates a circulatory system, causing water to move over the reef on a broad front before passing into the lagoon and ultimately flowing out through breaks in the reef.

Several green turtle rookeries are located inside the reef; the dugong occurs in the lagoon; and the migration route of the humpback whale passes close to the reef. Some 480 species of fish in 234 genera have been recorded.

Proximity to the coast and ready accessibility are making Ningaloo a popular tourist attraction. Until the 1980s, virtually no protection was accorded Western Australia's reefs. As a result, unregulated fishing both for recreational and commercial purposes – to meet the strong demand for aquaria species – as well as for shells and coral was creating pressures on the reef. But in 1987, the greater part of the reef and the nearby coast was designated the Ningaloo Marine Park (4,300 km² / 1,660 sq mls).

● An aerial view of part of the Great Barrier Reef, off the coast of Queensland. (WWF / Australian Information Service)

● A crown-of-thorns starfish feeding on coral. Population explosions of this starfish have in the past caused extensive damage to the Great Barrier Reef. (WWF / Rod Salm)

A heavy infestation of crown-of-thorns starfish, lasting from 1962 to 1976, brought a serious threat to the Great Barrier Reef, destroying a high proportion of coral in the central sector and causing coral reef communities to disintegrate. A further infestation followed in 1980-81. The cause of these outbreaks, and whether they are manmade or part of a natural cycle, has been the subject of much speculation. Countermeasures have ranged from collecting starfish by hand or injecting them with compressed air to legislation protecting the triton - a gastropod mollusc that preys on the starfish - and outlawing spearfishing, measures which have been limited in scope and only partly effective.

The growing popularity of the Great Barrier Reef as a tourist attraction brings difficult management problems. About two million people visit the reefs, islands and adjacent mainland each year, staying at island resorts or camping. Where the reef is close to population centres, such as Cairns, visitor pressure is intensified. Difficulties arise over the disposal of sewage and refuse from hotels - including floating hotels - and campsites. Boat propellers cause sedimentation, and the reef is damaged by fuel spillage, anchoring, coral and shell collecting, and even by walking. On some of the islands, introduced animals - dogs, cats and rats - have wreaked havoc on several turtle nesting grounds and bird colonies.

These problems are compounded by the rapid increase of industrial and agricultural (chiefly sugarcane) development on the nearby mainland. The use of pesticides and chemical fertilizers, the discharge of effluents and soil erosion, are among the factors that have contributed to pollution of the reef, affecting coral communities and the tropical fish associated with them.

Heavy poaching of giant clams by fishing boats from Taiwan was until recently a problem. The introduction of a 320 km / 200 mls Australian Fishing Zone, backed up with regular patrolling, has succeeded in bringing this form of poaching under control.

Most - 98.5% - of the reef has been protected by the creation, in 1975, of the Great Barrier Reef Marine National Park, supplemented by 13 Fisheries Habitat Reserves. Dugong and all sea turtles are fully protected within the park, except that Aborigines may continue to take them for their own consumption. Designated a World Heritage Site in 1981.

# THE GALÁPAGOS ARCHIPELAGO

LOCATION: East Pacific Ocean, 800–1,000 km / 500–620 mls from the nearest mainland; 1°40′ N–1°36′ S, 89°16′–92°01′ W. Located on the Equator, the Galápagos archipelago (administered by Ecuador) is a scattering of 15 principal islands, 42 islets, 26 rocks, and reefs. Over half the overall land area is taken up by the largest island, Isabela (4,588 km² / 1,771 sq mls). Three other islands – Fernandina (642 km² / 248 sq mls), Santiago (585 km² / 226 sq mls), and Santa Cruz (986 km² / 381 sq mls) – are within sight of Isabela. San Cristóbal lies a little farther out. Isabela is dominated by five still-active volcanoes rising from the depths of the ocean and towering up to 1,700 m / 5,500 ft above the sea. The continuing volcanic activity provides an underlying element of instability that greatly influences Isabela's ecology.

AREA: The archipelago extends 300 km / 186 mls from north to south and covers 7,800 km² / 3,012 sq mls of land, 96.6% (7,665 km² / 2,960 sq mls) of which has been declared a national park. The marine reserve extends to 79,900 km² / 30,850 sq mls.

ALTITUDE: Sea level to 1,707m / 5,600 ft (Volcán Wolf).

CLIMATE: Rainfall is spasmodic and brief, quickly disappearing into the volcanic soils. Mean annual precipitation at sea level reaches a maximum of 356 mm / 14 in, increasing at an elevation of 200 m / 656 ft to 1,092 mm / 43 in. Dry season temperature 17–22 °C / 62–71 °F; hot season 23–27 °C / 73–80 °F.

● The spectacular volcanic landscape of the Galápagos Islands. (WWF / Hartmut Jungius)

The larger islands of the Galápagos are characterized by dry coastal lowlands, with vegetation dominated by cacti and xerophytic scrub. Inland, the vegetation changes with altitude and the corresponding increase in precipitation in the form of mist and light drizzle. Beginning at heights of 200-500 m/650-1,600 ft is a belt of evergreen forest – the *Scalesia* zone – above which lies the more open fern-sedge zone reaching to the summits. The well-favoured highlands are of course the most desirable places for agriculture; they have been greatly altered as a consequence. Elsewhere, the land is too dry and unproductive for agriculture, and has therefore largely escaped ruination by man. In the short period that people have inhabited the archipelago, the forests have been felled to make way for pasturage and crops, and more than than 200 species of exotic plants, mostly herbaceous weeds, introduced. Many have become naturalized, and as weeds are generally quicker at adapting to prevailing conditions and able to endure heavy grazing, they may stand a better chance of survival in the long term than the indigenous plants.

As well as the giant tortoise, the reptiles include two types of iguanas: the Galápagos marine iguana, the world's only living marine lizard, which has evolved into seven distinct races, showing considerable variation from one island to another; and the Galápagos land iguana, which occurs on six islands. Both iguanas are peculiar to the Galápagos and both are vulnerable to predation: pigs, cats, dogs, and above all rats, devour their eggs, hawks kill their young, while, in the sea, sharks pose a constant threat.

The total number of birds breeding in the Galápagos is estimated at 750,000 pairs. They include what is believed to be the largest concentration of masked boobies in the world, as well as about one-third of all blue-footed boobies. No species of Galápagos seabird has yet become extinct; but one, the dark-rumped petrel, has been heavily reduced by introduced predators, mainly dogs and rats.

In addition to the seabirds, 28 species of land birds, most of them endemic, are known to breed in the archipelago. The mammals include three distinct groups of native rats, one of them a recently discovered giant rat, the size of a muskrat, known only from subfossil remains collected from owl pellets in caves. Most of the endemic Galápagos rodents are extinct; they are believed to have succumbed to some unidentified disease transmitted by the introduced black rat. But two species of native rats survive on Santa Fé and on Fernandina, both islands being free of black rats.

Two species of bats *Lasiurus* spp. are known from the Galápagos, as well as two seals: the Galápagos fur seal, which is nocturnal, and the Galápagos sea lion, which is active by day. The sea lion population stands at about 20-50,000, but numbers fluctuate widely, mainly as the result of seal pox epidemics.

The sole tropical representative of an essentially subantarctic genus, the Galápagos fur seal has adapted in a remarkable way to equatorial conditions. Its closest relative inhabits the Pacific coast of South America as far north as Peru. Unlike other southern fur seals, which are at sea for the greater part of the time, the Galápagos fur seal spends about one-third of its life ashore, and does not migrate. Many years of relentless exploitation ended only when numbers had been so severely reduced that further exploitation was considered uneconomic. By 1904 the Galápagos fur seal was close to extinction. Recovery was slow, but it gradually reoccupied beaches in various parts of the archipelago. The species now breeds throughout the Galápagos wherever conditions are suitable, and numbers have stabilized at about 30-40,000.

## EL NIÑO

The climate of the Galápagos does not conform with conditions generally associated with equatorial islands. The archipelago lies at the confluence of three oceanic currents, which strongly influence its climate: the westward-flowing Humboldt (or South Equatorial) Current, which brings cold water across the Pacific; the warmer North Equatorial Current bringing relatively unproductive waters from the north; and the cold Cromwell (or Equatorial Under-Current), which flows eastwards at considerable depth and upwells off Fernandina and Isabela, bringing nutrients to the surface. These three currents sustain a very diverse seabird fauna, ranging from cold-water species such as the endemic Galápagos penguin to tropical species such as the brown pelican. Some, such as the Galápagos flightless cormorant, never leave the island of their birth; others such as the waved albatross follow the Humboldt Current to waters off the coast of South America, returning to the Galápagos to breed.

For the greater part of the year the cold Humboldt Current flows through the archipelago, generating persistent cloud cover (garúa) and enveloping the uplands in mist and fine rain. Normally it meets warm waters to the north of the Galápagos, but from January to April or May the convergence moves south, surrounding the islands with warmer water, and giving rise to the hot season. Every fourth year or so is an El Niño year, when the cool waters are replaced with water 5–10 °C / 9–18 °F warmer than normal – with catastrophic ecological consequences. The rise in sea temperature is accompanied by torrential rain and followed by the disappearance of fish and other marine creatures. Seabirds and other animals dependent on the seas for food are suddenly deprived of it, and large numbers die.

The effects of an El Niño year are felt over the entire tropical Pacific region and exert a tremendous influence on global weather patterns, although other factors also play their part. Warmer than average water off the coast of Peru spreads across the Pacific, causing fluctuations in atmospheric pressure over the eastern Pacific from Australia to the Indian Ocean. The easterly trade winds in the equatorial Pacific reverse direction, drawing warm water eastwards and causing it to overlay the cool water that normally flows westwards along the equator. The significant rise in temperature delays the onset of the Indian monsoon and brings severe drought to Indonesia, much of Australia, southern Africa and northern South America, while the southern USA, western Pacific and coasts of Peru and Ecuador are saturated with rain. Between successive El Niños, the pattern is frequently reversed.

# DARWIN'S LABORATORY OF EVOLUTION

Although the various Galápagos islands have much in common, their differences are even greater, giving each island its distinctive character and vesting the archipelago as a whole with a scientific importance that is probably unequalled. Charles Darwin was the first to recognize this, and to be impressed by the realization that immigrant species arriving from the mainland within the relatively recent past had, in the course of establishing themselves in their new home, become changed. Darwin found the Galápagos a perfect natural laboratory of evolution, and although his visit to the islands in 1835 lasted little more than a month, his observations, and the conclusions he drew from them, enabled him eventually to put forward in *The Origin of Species*, his revolutionary, and at that time highly controversial, theory of evolution by natural selection.

The Galápagos land birds played a prominent part in the formulation of Darwin's theory of evolution, the best known being his 'most singular group of finches'. He was struck by the way finches, descended from common ancestral stock but dispersed on islands throughout the archipelago – and thus subjected to different ecological conditions – had adapted and evolved in differing ways, in the process becoming distinguished from one another in size, shape of beak, colour of plumage, song, and diet.

Several of the 13 distinctive species of finches developed highly specialized behaviour. The woodpecker finch, for example, uses a cactus spine or twig, held in its beak, to prise larvae from cavities in dead branches. Ground finches remove ticks from iguanas and tortoises in response to ritualized solicitation by the reptiles. An extension of this practice is the macabre habit of the sharp-billed ground finch, which pecks at the base of boobies' tails until it draws blood, which it then drinks.

After their discovery by the Spanish in 1535, the islands remained unoccupied for many years, except for adventurers and pirates drawn to the Pacific by the prospect of rich pickings from Spanish treasure ships. From the end of the 17th to the middle of the 18th centuries, the Galápagos proved a safe and convenient haven for English buccaneers, among them William Dampier, who wrote an interesting account of his exploits. Towards the end of the 18th century, both British and American whaling fleets used the islands as a base for whaling in the Pacific. The whalers were followed by sealers and, after Ecuador's annexation of the islands in 1832, the first settlers arrived on Floreana. The Ecuadorian government's decision to turn Floreana into a penal colony led to a period of anarchy, forcing the settlers to abandon the island. Not until 1929 was Floreana permanently reoccupied. Other islands followed a similar pattern. Altogether, four islands – Floreana, Isabela, Santa Cruz and San Cristóbal – were colonized; today they are inhabited by about 5–6,000 people.

In the last 30 years considerable strides have been made in conserving the Galápagos and its flora and fauna.

The turning point was the foundation in 1959 - appropriately enough the centenary of Darwin's *Origin of Species* - of the Charles Darwin Foundation for the Galápagos Isles. In that same year, the entire archipelago was declared a national park, with the exception of areas already occupied for other purposes. The Charles Darwin Foundation is an international organization responsible for coordinating scientific research and for advising the government of Ecuador on matters relating to the park. The Foundation's biological research station on the island of Santa Cruz provides a permanent scientific presence in the Galápagos. The close working relationship that has developed between Ecuador's National Parks Service and the Charles Darwin Research Station has been a key factor in transforming the outlook for the Galápagos.

Any management programme for the Galápagos must aim to meet two basic requirements: the avoidance of further loss or degradation of the native vegetation, combined with the restoration, where possible, of what remains; and the elimination, or at least the effective control, of alien predators. Useful progress has already been made in eliminating

alien animals. An intensive shooting campaign succeeded in eradicating goats from the islands of South Plaza, Santa Fé, Rábida, Española and Marchena. A similar eradication campaign needs to be directed against feral pigs.

Valuable measures have also been taken in establishing breeding programmes for giant tortoises and land iguanas, with the aim of replenishing depleted wild populations. Much more remains to be done, for the conservation of these islands is a challenging long-term enterprise.

The rise of tourism in the islands has become an important factor in the economy of the Galápagos, and has enabled the islanders to see for themselves the benefits of conserving their own natural assets. So far tourism has had little adverse effect on the native wildlife. As long as tourism continues to be based on cruise ships, the numbers of visitors remains limited, adequate controls continue to be exercised over tourist facilities, and adequate standards of authorized guiding are maintained the advantages of tourism will outweigh the disadvantages.

The growth of tourism has, however, led to an influx of entrepreneurs and workers from the mainland, increasing the population to nearly 15,000. Vigorous action needs to be taken to control the erection of Caribbean-style resorts, holiday flats and road building, and to restrict the number of tourists.

A high proportion of the native Galápagos fauna - the sea lions, fur seals, marine iguanas and seabirds in particular - is sustained by the seas surrounding the islands. Indeed, the extent to which the Galápagos fauna is dependent upon the marine environment is so all-embracing that adequate protection of the seas and the marine communities in the vicinity of the islands is of critical importance. This fundamental requirement has been met by the creation of a marine reserve embracing all the inner waters of the archipelago within a circumference of 15 nautical miles.

No other group of islands anywhere in the world is comparable with the Galápagos. Still retaining the aura of Darwin's matchless achievements, the archipelago continues to exert a profound influence on scientific thought, and on evolutionary theory in particular. The pursuit of scientific knowledge has been accompanied by effective practical advances in conserving the archipelago's unique flora and fauna and bringing it back from the brink of ecological disaster. These notable achievements have been complemented by substantial improvements both to the economic well-being of the islanders and to their physical environment. This has been made possible only by the close working relationship that has been established between the government of Ecuador on the one hand and the international scientific community on the other. To have attained such success against all the odds is a remarkable achievement, and one which by its example will, it is to be hoped, stimulate others to follow.

## THE INTRODUCED GOAT

Every one of the hundreds of whaling and sealing vessels to visit the Gálapagos exacted its toll of the giant tortoises with which the islands abounded. Tens of thousands were slaughtered, their meat being regarded as a delicacy. And when, after Ecuador's annexation of the islands in 1832, the first settlers arrived on Floreana, they continued the tradition, the giant tortoise becoming part of the islanders' staple diet.

After the giant tortoise was killed off, its place as principal grazing animal was taken by domestic livestock. Goats in particular had a profound impact on the indigenous vegetation, exemplified by the four goats released from an American warship in 1813 on the uninhabited island of Santiago. By the 1950s their numbers had risen to more than 100,000, and the native vegetation had almost entirely disappeared. On Española and

● Introduced goats have caused widespread damage to the native flora of the Galápagos. On one island, four goats released in 1813 had, by the 1950s, given rise to a population of more than 100,000. (WWF / Hirsch and Muller)

Santa Fé goats have even eliminated stands of *Opuntia* cactus. From the early 1960s attempts were made to eliminate goats from a number of small and medium-sized islands. On Pinta, for instance, some 40,000 goats – all descended from one male and two females – were shot, leaving only about 100. But goats remain on the larger islands, and are especially numerous on Santa Cruz and Santiago.

Altogether, 11 species of alien mammals have been introduced into the Galápagos; seven of them have become feral. As well as the goat, they include the black rat, house mouse, domestic cattle, horse, donkey, sheep, pig, dog, cat and guinea pig. Several islands have populations of feral pigs, which besides destroying vegetation also inflict great damage on bird colonies, eating the eggs not only of birds but also of tortoises, marine turtles, iguanas, lizards and snakes.

# MAURITIUS

LOCATION: Western Indian Ocean, about 840 km / 520 mls east of Madagascar and 1,900 km / 1,200 mls from the African mainland. 20°15′ S, 57°30′ E.

AREA: 1,865 km² / 720 sq mls.

ALTITUDE: Sea level to 800 m / 2,600 ft.

CLIMATE: Tropical oceanic. Rainfall and temperature differ widely according to altitude and location. Mean annual rainfall in the Macchabé Bel Ombre Nature Reserve 2,000–3,000 mm / 78–118 in, and mean annual temperature 14–26 °C / 57–79 °F.

Mauritius – ruled successively by the Dutch, French and British until independence in 1968 – is a pear-shaped volcanic island completely surrounded by coral reefs. Around its coast are several examples of elevated coral, some rising as much as 12 m / 40 ft above sea level. The interior plateau averages about 580 m / 1,900 ft in height, with higher ridges reaching a maximum elevation in the southwest of the island at the Piton de la Rivière Noire (828 m / 2,716 ft). The island is ringed by a series of solitary peaks – vestiges of the original volcanic dome – rising like gigantic fangs from the surrounding plateau.

● Mont Rempart and Les Trois Mamelles, Mauritius. (Planet Earth Pictures / Ivor Edmunds)

At the time of its discovery Mauritius was covered with dense tropical evergreen rainforest rich in valuable hardwoods. Heath and dwarf forest occurred at the higher elevations, and palm savannas in the drier eastern parts of the island. Except for a few relict patches, this indigenous vegetation has almost entirely vanished. The Mascarene palms, of which there are about a dozen species and are of particular interest, survive mainly in artificial cultivation; while many of the orchids, numbering more than 80 species, particularly the lowland species, are extinct. The only comparatively intact example of native lowland and coastal vegetation of any size that remains is on the Ile aux Aigrettes (25 ha / 62 acres) off the southeast coast of Mauritius. This island has been declared a nature reserve – though its protected status has done little to deter illegal woodcutters. The best remnant of upland forest is to be found around the Black River Gorges (in the southeast of the island). Such reafforestation as has occurred on some of the higher parts of the islands has been mainly confined to the planting of eucalypts, conifers and rubber trees.

The indigenous mammals include two species of flying fox, one – the Mauritian flying fox – endangered, and the other – the lesser Mascarene flying fox – extinct. Among the invertebrates are 130 species of land snails, about one-third of which are already extinct and another third seriously endangered, partly owing to habitat destruction and partly to the introduction of the carnivorous snail *Englandina rosea*.

Birds constitute the most remarkable feature of the Mascarene fauna, though few places have a worse record of extinctions. Taking the most conservative estimate, at least 22 species or subspecies of Mascarene birds – most famously the dodo and the solitaire – have become extinct since 1638, of which at least six are from Mauritius. Of the 11 surviving Mascarene forms currently listed as rare or endangered, seven are from Mauritius. As well as the Mauritius olive white-eye, they include the Mauritius pink pigeon, whose main feeding ground was destroyed in the early 1970s. Captive-bred birds released in the Pamplemousses Botanic Gardens adapted well, only to be killed by vandals with catapults. The Mauritius fody was also badly affected by habitat destruction in

## FORAGING FOR FOOD

As settlement expanded and wildlife became scarce on Mauritius and Réunion, foraging parties were obliged to go to Rodriguez, and even farther afield, to obtain food. The Vicomte de Souillac, Governor of Mauritius from 1779 to 1787, for example, was constantly worried about the supply of meat to the troops and hospitals in Mauritius, so he looked to the little island of Agalega (almost 1,000 km / 600 mls to the north) as an unfailing source of turtle meat. Exploitation for food was an important factor in the extinction of the Mascarene birds, made all the easier by their defencelessness and placid disposition. Indeed, the native birds were so foolishly tame that some scarcely bothered to move out of the way of intruders or merely flew into a nearby bush. Referring to the Mauritius black bulbul or merle, Richard Meinertzhagen the eminent ornithologist was '…told by an old forester that a man with a hair noose at the end of a stick could snare them as they sat singing on a branch'. He adds that '… when this bird was plentiful it was considered a great delicacy, and at shooting parties a dish of forty or fifty merles was no uncommon sight'.

the 1970s. The echo parakeet, a little known species, has been reduced to nine birds, including three pairs. The Mauritius cuckoo-shrike, a bird of furtive habit and thus seldom seen, is estimated to number no more than 40–50 pairs, while the Mauritius black bulbul or merle is only a little more numerous. The Mauritius kestrel had by 1973 been reduced to about five individual birds, but a captive breeding programme succeeded in raising numbers to 19–25 birds. The parakeet and pink pigeon are being similarly raised in captivity. A likely candidate for the endangered list is the cave swiftlet. During the 1970s, this species underwent a dramatic decline when its nesting caves were vandalized.

The Portuguese, who discovered the Mascarene Islands in 1512, made no attempt at settlement: they looked upon Mauritius chiefly as a convenient staging post for their ships, and as a natural larder – birds, tortoises, turtles and fish were to be had in abundance. With the aim of ensuring supplies of fresh pork, the Portuguese released a few pigs on the island. These bred so prolifically that, according to La Roques' account published in 1715 (by which time the island had come under French control), one *battue* resulted in killing more than 1,500 pigs in a single day.

The Portuguese also introduced the crab-eating macaque from Indonesia. (The Dutch followed suit in 1639 by bringing in the sambar deer from Java.) By the beginning of the 18th century this monkey had become so plentiful that La Roques reported seeing 'more than 4,000 … in a nearby garden'. The crude nests which Mascarene birds built upon the ground afforded scant protection from marauding pigs and monkeys.

But the animals responsible for causing the greatest havoc to both terrestrial and arboreal birds were the two species of introduced rats: the black rat, and the brown rat. They swarmed everywhere, even on the branches of trees. Cats and mongooses were turned loose in an attempt to control the hordes of rats, but themselves took to preying on the more easily caught native fauna.

## FINE HARDWOODS

Hardwoods, chief among them being ebony, of which Mauritius has at least eight species, were one of the first objects of exploitation – principally by the Dutch, who occupied the island from 1598 to 1710 – and have been largely destroyed. Mature black ebony trees of more than 50 cm / 20 in diameter (and perhaps 1,000 or so years old) are virtually extinct, although young trees may still be found scattered through the remnants of forest in regions of medium rainfall. The 'Bois de Canelle', belonging to the camphor family, is also now difficult to find. This tree was at one time greatly favoured by cabinet makers owing to its figuring, its beautiful rich brown colour, and above all its remarkable resistance to termite attack.

Eighteen nature reserves (many of them offshore islands and islets) and six fishing reserves have been established in Mauritius, but most are small, since only degraded remnants of the indigenous vegetation generally remain. An exception is the Macchabée/Bel Ombre Nature Reserve (3,611 ha/8,923 acres), parts of which contain comparatively undisturbed climax forest of the type originally covering much of the higher reaches of the island; but the lower elevations have been extensively overrun by exotic plants. Many species of orchids, ferns and mosses are to be found in the reserve; it also contains a small stand of *Hibiscus columnaris*, endemic to Mauritius and Réunion, of which only a scattering of individual plants remains. The reserve has been designated a World Heritage Site.

These nature reserves are all that stand between survival and extinction of the endemic Mauritian flora and fauna. The establishment of a more comprehensive system of reserves is inhibited by the inordinate pressure of people on the land. The eradication of malaria in the 1950s was followed by a population 'explosion': from 487,000 in 1950 numbers more than doubled to 1,100,000 in 1988. Mauritius has one of the highest population densities in the world, averaging 579 per km²/223 per sq ml – a microcosm of the Malthusian nightmare.

## ROUND ISLAND

Round Island (159 ha/393 acres), lying 24 km/15 mls northeast of Mauritius, is of unusual biological interest, as it possesses the last remnants of the primeval palm forest that once covered the northern plain of Mauritius.

Little of the native vegetation now remains. The palms seed freely, but most seedlings are destroyed by goats and rabbits, introduced in 1844 and 1869. Only one specimen of the endemic Round Island hurricane palm survives, very few of the bottle palm, and the once wide-spread fan palm has been greatly reduced: 80–90% of the island is now bare rock. The effects of wind and rain erosion are accentuated by the undermining of the few remaining trees by the burrowing habit of the wedge-tailed shearwater, of which there are 3,500–4,000 pairs.

Round island's palm forest sustains a distinctive reptilian fauna, including species that are extinct on Mauritius. Of the eight species of Round Island reptiles, four are endemic: two boas – the Round Island boa (which has not been seen for a number of years and may be extinct)

and the Round Island keel-scaled boa (the only survivors of a sub-family (Bolyerinae) of primitive snakes), as well as two lizards – Round Island day-gecko (of which only 100–150 remain) and Round Island skink. All four species are listed in the *Red Data Book*.

Among the seabirds to have breeding colonies on the island is the Round Island petrel (about 120 pairs), whose only other known breeding place is on the Trinidade/Martin Vaz islands southeast of Brazil.

The survival of Round Island's flora and fauna is clearly dependent on eradication of the alien species. In 1976 a determined effort succeeded in shooting all the goats. Ten years later, in 1986, the rabbits too were exterminated. Already there are encouraging signs of regeneration, but as palms are very slow-growing the process is bound to take time. Fortunately, Round Island has not had to contend with rats, and it is vital that they should never be allowed in. The danger of accidental introductions is always there, one of the most likely sources being fishermen landing on the island illegally to collect seabirds.

# RÉUNION

LOCATION: 780 km/485 mls east of Madagascar and 210 km/130 mls southwest of Mauritius. 20°51′–21°22′ S, 55°15′–55°54′ E.

AREA: 2,515 km²/971 sq mls.

ALTITUDE: Sea level to 3,000 m/10,000 ft.

CLIMATE: Tropical oceanic; the northeastern half of the island is exposed to the full force of the monsoon, when rainfall can be very heavy, with temperatures dropping to –5 °C/23 °F.

Réunion, a young oceanic and volcanic island, is an overseas *département* of France. Roughly circular in shape, the island rises steeply from the sea, most of it being formed of a high and steep-sided plateau dominated by mountain massifs. These massifs are topped by huge craters: the Piton des Neiges (3,069 m/10,069 ft) and Le Volcan (2,620 m/8,596 ft), the latter having two main craters on separate elevated cones, one extinct and the other – the Burning Crater or Piton de la Fournaise (2,631 m/8,632 ft) – still active.

At the time of its discovery by the Portuguese in the early 16th century the island was entirely covered with evergreen rainforest, the combination of fertile volcanic soil and humid climate nurturing a rich vegetation. A belt of forest still encircles the two pitons. Elsewhere, except in ravines and on the most inaccessible slopes – which still support an endemic flora of great interest – the

## THE SPICE TRADE

In the 18th century Réunion, by then a French colony, derived substantial prosperity from the cultivation of coffee (introduced from Arabia in 1715) and cloves. The clove tree was smuggled into the island from the Dutch East Indies (Indonesia), so breaking the jealously guarded Dutch monopoly of the lucrative spice trade. Indeed, the main reason for establishing the royal Botanic Gardens at Pamplemousses, Mauritius, in 1770, was to investigate the commercial potentialities of various exotic tropical plants. It was from Pamplemousses that cloves were introduced into Zanzibar in 1818, thus laying the foundations of that island's subsequent prosperity.

original forest cover has been cleared to make way for extensive sugarcane plantations, now the preponderant plantation crop. Soil erosion is very apparent.

The endemic land tortoise and 19 of Réunion's 33 native bird species are extinct, such as the Réunion solitaire. Of the surviving species, two are listed in the *Red Data Book*: the Mascarene black petrel, perhaps the rarest Indian Ocean tropical seabird; and the Réunion cuckoo-shrike, known only from the Plaine des Chicots and the Plaine d'Affouches. Other endemic birds include Barau's petrel, described only in 1964 and still little known.

Apart from human disturbance, the introduction of the Javan deer on to the Plaine des Chicots and its maintenance at artificially high densities for hunting purposes has resulted in extensive damage to the habitat, and ruled out any possibility of regeneration. Equally disastrous is

the official policy of clearing the native forest and replacing it with exotic plantations, notably the Japanese red cedar. Indeed, the Office National de la Forêt appears indifferent to the perpetuation of Réunion's native flora.

Important populations of marine turtles associated with some of the outlying islets, notably Europa and Tromelin, are protected at least to some extent by their inaccessibility. Several small nature reserves have been established but, as they remain unstaffed, their usefulness is necessarily limited. Proposals have been made to set aside some of the least disturbed areas of indigenous vegetation as strict nature reserves. Two of them – Hauts Bois de Nèfles (80 ha/198 acres) and Mazerin (2,000 ha/4,942 acres) – have been approved by the Ministry of the Environment, though the decision has yet to be implemented. This should be done while there is still some hope of safeguarding them.

# ALDABRA ATOLL

LOCATION: 420 km/260 mls northwest of Madagascar, 640 km/400 mls from the East African mainland, and 1,100 km/700 mls southwest of the Seychelles at 9°25′ S, 46°25′ E.

AREA: 350 km²/135 sq mls, of which 188 km²/72 sq mls are land, the balance saltwater lagoon and mangrove.

ALTITUDE: The island's elevated limestone surfaces are mainly 3.5–8 m/11.5–26 ft above mean low-tide level.

CLIMATE: Semi-arid with pronounced wet season (November to April). Rainfall variable, averaging 1,200 mm/50 in.

Aldabra is encircled by a coral rim cut by three entrances, two of them deep, which divide the atoll into four main islands, by far the largest being South Island. The large (150 km²/58 sq mls) shallow lagoon contains several islets and is surrounded by mangroves. The few sandy beaches are small and mostly located around the little inlets or *anses*, and there are sand dunes along the seaward side of the southern rim.

Oceanic islands are delicately balanced ecosystems that are highly vulnerable to external influences. Most tropical

islands have already been exploited by man and, in the process, irreparably ruined, their vegetation stripped, their soil eroded, and their native flora and fauna devastated by introduced species. Aldabra has the distinction of being not only the largest but the only elevated atoll (i.e. raised above sea level by geologic activity) in the Indian Ocean that has not been mined for guano, planted extensively with coconuts, or otherwise degraded. It thus retains an assemblage of plants and animals richer than on any other Indian Ocean atoll. But the scien-

tific importance of Aldabra lies not so much in the endemic plants and animals as in the totality of the biotic community; for Aldabra alone of all the islands in the Indian Ocean has preserved a relatively undisturbed ecosystem. This remote atoll can thus be regarded as a unique natural laboratory for the study of colonization and evolution, comparable with the Galápagos Islands in the Pacific. It should be kept intact and used only for purposes of scientific study, as originally proposed by Charles Darwin in 1874.

The terrain consists of two principal

● Aldabra is an important breeding ground for many oceanic birds, including frigate birds, boobies, and tropic birds. (WWF / Jack S. Grove / Eye on the World)

● The white-throated rail of Aldabra is the only surviving representative of a group of flightless birds that included the extinct dodo and solitaire. (WWF / Y. J. Rey-Millet)

● Giant tortoises were once found on several islands in the Indian Ocean, but all except the Aldabran species are now extinct. (WWF / Jack S. Grove / Eye on the World)

types – *champignon* and *platin*. Platin is mainly found at the eastern end of South Island, champignon almost everywhere else. Each supports a different kind of vegetation which governs the distribution of fauna. Champignon is covered with thickets of *Pemphis acidula* so dense as to be almost impenetrable. Soil is almost completely lacking except in the bottom of some potholes, while occasional tufts of coarse grass manage to cling to the bare rock surface. Platin is covered with a more open mixed scrub and trees, and an irregular turf, while conspicuous clumps of *Pandanus* have become established around the larger freshwater pools.

Aldabra's proximity to continental Africa and Madagascar, in combination with its comparatively wide range of habitats, has resulted in more favourable opportunities for colonization by plants and animals than on elevated limestone islands elsewhere. Furthermore, its isolation has enabled many of these plants and animals to develop special characteristics by adapting themselves to ecological niches which, in the absence of competitors, are in some instances quite different from those occupied by the stock from which they originated.

Thus, for a small coral island Aldabra has an exceptionally rich terrestrial flora, with 273 species of flowering plants and ferns, of which 19 are endemic. A further 22 species are shared only with neighbouring islands. Many of these plants are regarded as threatened.

Aldabra possesses an abundant insect fauna: about 1,000 species of insects have so far been recorded, including 127 species of butterflies and moths, of which 28% are endemic. In addition there are endemic crustaceans, a land mollusc, a fruit bat, the Seychelles flying fox, and even freshwater fish (living in rainwater pools) that occur nowhere else.

Because of its size, unspoiled vegetation and small human population – which is mainly confined to West Island and largely made up of itinerant turtle hunters and fishermen from the Seychelles – Aldabra is an important breeding ground for substantial numbers of seabirds. Most seabird colonies in the Indian Ocean have been severely reduced as the result of human interference and harassment. But because Aldabra has been spared such disturbance, it has become the principal breeding ground in the western Indian Ocean of both the greater and lesser frigate birds. The atoll also accommodates large breeding colonies of red-footed booby, masked booby (both members of the gannet family), fairy tern, noddy, and both red-tailed and white-tailed tropic birds.

Of particular interest is the Aldabra warbler, confined to a patch of coastal scrub 10 ha/25 acres in extent. Discovered in 1968, this species has been seen on only five occasions, making it one of the world's rarest birds.

The white-throated rail – sole survivor of a group of flightless birds that included the extinct dodo and solitaire – is endemic to Aldabra, where it appears to be restricted to Middle Island and Ile Polymnie. This species formerly inhabited South Island, but did not long survive the introduction of the domestic cat. Until the early part of the 20th century the genus was also represented on Astove, Cosmoledo and Assumption; but, by 1937, it had disappeared from these three islands. Aldabra is the white-throated rail's last remaining stronghold in the Indian Ocean region.

Though relatively unspoiled, Aldabra has not been altogether free from exploitation. Since 1888, the atoll has been leased for commercial turtling, fishing and mangrove extraction. In the process, cats, dogs and goats have been introduced. Contrary to the situation on some other tropical islands, however, these exotic animals do not appear to have greatly increased, though the reasons for this are obscure. Introductions have, moreover, been confined to South and West Islands: thus, the colonies of flightless rails and seabirds on Middle Island have not been affected.

Since 1967, Aldabra has been the object of intense scientific study. A research station established by the Royal Society of London in 1971 is now maintained by the Seychelles Islands Foundation. Present policy is to continue the research and monitoring programme, with particular emphasis on studying the ecology of exotic species as a basis for future management, while keeping human intrusion to the absolute minimum. Aldabra was designated a World Heritage Site in 1982.

# MADAGASCAR

LOCATION: Indian Ocean, separated from the southeastern mainland of Africa by the deep Mozambique Channel (425 km/250 mls wide). 11°57'–25°35' S, 43°14'–50°27' E.

AREA: 594,180 km²/229,413 sq mls; about 1,600 km/994 mls long, with a maximum width of 580 km/360 mls.

ALTITUDE: The central highlands average about 900–1,200 m/3–4,000 ft, with volcanic and granitic peaks reaching considerably higher.

CLIMATE: Temperatures vary greatly, from an average maximum of 34 °C/93 °F on the northwest coast to a minimum of –15 °C/5 °F on the Andringitra Massif (2,650 m/8,695 ft), the highest point of the inner escarpment.

The central highlands of Madagascar form a spine running almost the full length of the country and rising in the north to Tsaratanana (2,880 m/9,450 ft), the country's highest peak. This mountainous barrier lies across the path of the prevailing trade winds from the Indian Ocean. The northeastern part of the country thus receives a substantial rainfall, which progressively declines towards the southwest. The highest rainfall is on the Ile Sainte Marie (3612 mm/142 in); the lowest at Anakao (310 mm/12 in). Thus, while the east coast is hot and humid and the central highlands temperate, the west has a long dry season lasting six or seven months of the year; the southwestern region, lying beyond the

## MADAGASCAR'S ENDEMIC MAMMALS

Madagascar's prosimian fauna consists of five families, all of which are peculiar to the island: the Cheirogaleidae with seven species of mouse lemurs; the Megaladapidae with seven species of sportive lemurs; the Lemuridae with ten species of lemurs; the Indriidae with five species of woolly lemurs, sifakas and the indri; and the Daubentoniidae, containing only the aye-aye.

Other endemic families include the Tenrecidae, containing 30 species of insectivorous tenrecs, some of which have become adapted to a subterranean existence, some to climbing trees, and others to a semi-aquatic life; and the Myzopodidae, whose only member is the sucker-footed bat. In addition, there are three endemic subfamilies: the Nesomyinae, with ten species of Malagasy rats, including the giant rat *Hypogeomys antimena*; the Galidiinae, with eight species of Malagasy mongooses; and the Cryptoproctinae (a subfamily of the civets), accommodating the fossa, Madagascar's largest carnivore.

---

influence of the monsoon, is subdesert with low and uncertain rainfall.

Madagascar is the world's fourth largest island, exceeded in size only by Greenland, New Guinea and Borneo. It is virtually a miniature continent, differing so greatly from continental Africa that it qualifies as a distinct zoogeographic region – the Malagasy Region.

The antecedents of the Malagasy people are not known with certainty, but it is believed that some, including the predominant tribe, the Merina (who occupy the central plateau), are of Malaysian origin. Their forebears crossed the Indian Ocean in successive migrations commencing within the Christian era, but mainly between the 7th and 15th centuries AD. This theory receives strong support from similarities of language, the established use of outrigger canoes, and the long tradition of paddy rice cultivation in Madagascar. Most of the other Malagasy people came from Africa, many of them brought in as slaves. Arab influence, radiating from the coast of East Africa, reached Madagascar in the 12th century; but it was not until AD 1500 that the Portuguese captain Diego Diaz made known the existence of the island to the Western world.

Botanically, Madagascar is one of the richest countries in the world, possessing in the order of 11,000 species of flowering plants, 55–85% of them endemic and belonging to 8 endemic plant families in 238 endemic genera. They include more than 1,000 species of orchids, found mainly in the eastern rainforest. Vegetation ranges from dense tropical evergreen forest in the humid east, as well as in the extreme north and northeast, to open savanna in the west, grading into xerophytic scrub (plants adapted to dry conditions) in the extreme southwest. The degree of floral diversity is typified by the baobab, of which Madagascar has seven species compared with one on the entire mainland of Africa. Most of the indigenous forest has, however, been eradicated by indiscriminate felling and burning to open up new areas for agriculture. Nine-tenths of the original vegetation has already been destroyed. The central highlands, more than 1,100 km/700 mls long and 300 km/200 mls wide, and once covered with forest, have been devastated by fire and axe. The forest has been replaced with poor-quality grass on land impoverished by overstocking and burning and heavily eroded. Practically none of the original central plateau forest remains except for a few small isolated patches such as the Tsarafidy Forest and occasional narrow fingers of riverine growth.

A particularly interesting endemic family of succulents, the Didiereaceae, which superficially resembles cacti or cactiform euphorbias but has evolved entirely independently, is found in the southwestern part of the country. Adapted to withstand a high degree of aridity, it is confined to the coastal area between Morombe and Fort Dauphin. Madagascar's *Didierea* forests are unique. Entering this *forêt épineuse*, its long thorny branches forming fantastic shapes, is like stepping back into the Mesozoic era. In recent years, however, this extraordinary flora has been subjected to widespread despoilment: substantial areas have been cleared to make way

## AN ANCESTRAL PRIMATE

The aye-aye, the most primitive of all living primates, bears a superficial resemblance to a cat-sized black squirrel. Equipped with prominent rodent-like incisors, it has large eyes (as befits a nocturnal animal) and bushy tail. But its most striking anatomical features are its hands and fingers, in particular the elongated third finger – a remarkable instance of extreme specialization designed to assist the aye-aye in obtaining the larvae that form part of its diet. Searching for food, the aye-aye crawls slowly along a branch, gently rubbing the bark with the tip of its finger while using its nose and ears to detect scent and sound. Having located a larva, it tears away the bark with its incisors, using the pliant third finger to extract the larva from its cavity.

The female builds a spherical nest, some 60 cm/2 ft in diameter, in the fork of a tree, 10–15 m/30–50 ft off the ground. Constructed of twigs and leaves, the nest has a single opening at the side. In it she gives birth to a single young at intervals of two or three years.

Extensive deforestation has deprived the aye-aye of most of its preferred natural habitat. Further, the respect, even veneration, traditionally accorded the aye-aye by some Malagasy tribes (who regarded it as an ancestral reincarnation possessed of supernatural powers) has been replaced by indifference or downright hostility. In some places the aye-aye is thought to be a harbinger of death and killed on sight; in others it is killed because of the damage it is said to cause to coconut plantations and stands of sugarcane.

The continued existence of this extraordinary animal is dependent above all else upon adequate measures being taken to make the various national parks and reserves in which it occurs sanctuaries in fact as well as in name.

for sisal plantations; much of the remainder has been ruined by burning and by the intrusion of livestock.

From the zoological standpoint, Madagascar's exceptional interest stems principally from its position as the main stronghold of the prosimians, the most primitive of the primates. Some 50 million years ago, during the Eocene epoch, the ancestral prosimians were widespread in North America, Europe and elsewhere; but they disappeared in the face of competition from more advanced forms. A few species remain in parts of Africa (e.g. the bushbabies, potto and angwantibo) and in tropical Asia (the lorises and tarsiers), their survival no doubt attributable to their nocturnal habits, by means of which they avoid competition with the diurnal monkeys.

● Rice cultivation in Madagascar. Population pressure has led to most of the indigenous forest being cleared for agriculture. (Noel Simon)

● The magnificent emperor moth, with a wingspan of 15 cm / 6 in, is found only in Madagascar. (WWF / J. J. Petter)

● The unique *forêt épineuse* (spiny forest) of the arid southwest of Madagascar is dominated by a family of bizarre succulents, the Didiereaceae. (WWF / O. Langrand)

● A male fossa in the deciduous dry forest of western Madagascar. The fossa, a relative of the civets, is the island's largest carnivore. (WWF / O. Langrand)

## MADAGASCAR'S OTHER NATIVE FAUNA

Of Madagascar's 260 species of reptiles, the crocodile has been subjected to such intense hunting for its valuable skin that it is now uncommon. The last remnants owe their survival to living in places that are difficult of access – the subterranean caves of Ankarana being one – or in a few sacred lakes (e.g. Lake Anivorano, between Diego Suarez and Ambilobe) where crocodiles are still protected by *fady* (taboo). There are in addition four endemic species of tortoise, 60 snakes and 182 lizards (over 90% of them endemic), among which are two-thirds of the world's chameleon species, ranging in size from the tiny stump-tailed chameleons with a body length of 3.2 cm / 1.25 in, to the largest, Oustalet's chameleon, which attains a body length of up to 68 cm (over 2 ft). The leaf-tailed gecko has a flat body and a skin so well camouflaged to look like bark that when on a tree trunk it is virtually invisible even from close range. The reptile fauna also includes two species of boa, and two iguanas, whose nearest relatives are in South America. Some 26 species of lizard and four of snakes are currently endangered. There are moreover 144 species of amphibians – all frogs – of which all but two are endemic, and 22 of them threatened. Madagascar also has a rich invertebrate fauna: excluding 17 genera of skipper (of which seven are endemic) there are 260 species of butterfly (182 endemic), some 45–50 of which are endangered. Among them are such species as the emperor moth, which has a wingspan of 15 cm / 6 in. Other invertebrates include an interesting group of dragonflies, about 70% of them endemic, among them *Isomma hieroglyphicum*, known only from Nosy Be and Diego Suarez. Madagascar also possesses an important terrestrial mollusc fauna, with some 380 known species, of which all but 20 are endemic. There are 34 endemic species of fish, of which 14 are threatened.

But in Madagascar the prosimians remain supreme.

Madagascar is an evolutionary backwater where long isolation and minimal competition have enabled the prosimians to evolve into a greater variety of forms than anywhere else on earth. Widely contrasting climatic and ecological conditions, varying from the humid rain forests of the north and northwest to the savannas of the west and the xerophytic vegetation of the southwest, have given rise to a highly diversified group of lemurs and their allies that is one of the classic examples of adaptive radiation. The Malagasy prosimians – of which there are 30 species – include both nocturnal and diurnal forms, ranging from tiny mouse lemurs which can easily fit into the palm of one's hand, to the indri, largest of the living lemurs, which is the size of a gibbon, and the curious aye-aye.

Madagascar's flora and fauna are of inestimable scientific and cultural value. The island is a living laboratory containing a wealth of fundamental biological material not duplicated elsewhere, providing opportunities for the study of original evolutionary radiations and the complex problems associated with speciation. The Malagasy prosimians constitute the most important assemblage of primitive primates in the world. Their loss would be a zoological catastrophe; yet 24 species out of a total of 30 are already classified as rare or endangered. There are moreover remarkable insectivores, carnivores, birds, reptiles, insects and plants, most of which are found nowhere else and have been only superficially studied. It is difficult to think of another region of its size that is of comparable biological interest, or of one that is undergoing such precipitate degradation.

Madagascar's economy is based on agriculture, with 80% of the population engaged in farming. Measures taken since the end of the Second World War to eradicate malaria and to improve medical services were so successful that the rate of human increase soared to 2.9% annually, with the result that the population has doubled within the last thirty years from 5.4 million in 1960 to 11.2 million in 1989. This demographic eruption has led to more and more land, much of it marginal, being taken up for cultivation and grazing: more than 10 million cattle are kept primarily for prestige purposes; they are of little economic value. Unlike the African mainland, Madagascar has no tsetse fly to impose a limitation on livestock.

The continued existence of Madagascar's unique fauna is primarily dependent upon the retention of adequate tracts of unspoiled natural habitat. A relatively extensive network of protected areas has been established – two national parks, 11 nature reserves and 23 special reserves – but some rare species are not represented in any of them. Furthermore, although the system of reserves is basically sound, their general degradation by human intrusion, over-grazing, commercialized logging and administrative neglect has reduced some of them to irreparable ruin.

Most of the reserves resemble small oases standing out in stark contrast to the denuded and impoverished country around them. They are increasingly threatened by the local people, who are turning covetous eyes upon such unspoiled areas as have previously eluded them. Flagrant boundary infringements are commonplace, and there is mounting pressure from powerful vested interests for exploitation of timber and other natural resources within the reserves. These demands are frequently encouraged by local politicians, most of whom fail to see the importance of the reserves or the values that are at stake, and seek only short-term economic gain; they allude to the reserves as 'outmoded' and 'irreconcilable with proper development of the country'.

For these reasons the *Service des Eaux et Forêts*, the department responsible for administering the reserves, finds its authority constantly eroded, and has difficulty in obtaining the support that is essential for the proper performance of its responsibilities. In its efforts to protect the reserves under its control, the Service has to defend itself both against the peasant farmers and against the politicians, and is thus being forced into an increasingly isolated position.

It is vitally important that measures are taken to protect the few pristine areas that have survived, and that the system of national parks and reserves is strengthened and improved before the opportunity is lost for ever. Unless that is done, Madagascar's superlative wildlife will inevitably follow the giant lemur *Megaladapis* and the huge flightless 'elephant bird' *Aepyornis* into oblivion.

# TRISTAN DA CUNHA

LOCATION: South Atlantic, roughly midway between South America and Africa, 2,800 km/1,750 mls from Capetown and 3,200 km/2,000 mls from Brazil. 37°02'–37°24' S, 12°12'–12°42' W.

AREA: Tristan 104 km²/40 sq mls; Inaccessible 10 km²/4 sq mls; and Nightingale 2.6 km²/1 sq ml.

ALTITUDE: Sea level to 2,060 m/6,759 ft ('The Peak').

CLIMATE: A cool, humid oceanic climate. Mean annual temperature 15 °C/59 °F. Annual rainfall 1,700 mm/67 in. Prevalent westerly winds and frequent cloud cover.

The Tristan da Cunha group comprises three islands – Tristan, Inaccessible and Nightingale – all of volcanic origin, lying within little more than 30 km/20 mls of each other. A fourth island, Gough (65 km²/25 sq mls), situated about 350 km/220 mls SSE of Tristan (40°20' S, 10°00' W), is included within the group (a British dependency) for administrative purposes.

Tristan – the summit of a volcanic cone – emerges steeply from the sea about 450 km/280 mls to the east of the crest of the Mid-Atlantic Ridge, its cliffs reaching almost vertically to a height of about 600 m/2,000 ft before inclining more gently to the summit of Queen Mary's Peak (2,060 m/6,759 ft), known as 'The Peak'. The narrow, rocky foreshore affords no sheltered anchorage. Between the foot of the main precipices and the sea, however, are several narrow shelves of land, the largest being the Settlement plateau. Inaccessible, situated 40 km/25 mls to the southwest of Tristan, is surrounded by near-vertical cliffs rising to 490 m/1,608 ft; even under favourable conditions access is difficult. Nightingale, 38 km/23 mls SSW of Tristan, rises 400 m/1,300 ft above sea level.

Tristan's coastal region is dominated by tussock grass. In the lowlands and on cliffs are thickets of 'island tree' and tree fern. Wet heath vegetation occurs at about 750 m/2,500 ft, with moorland above 900 m/3,000 ft. Coastal waters contain extensive beds of giant kelp. Inaccessible and Nightingale are largely covered with tussock grass, forming a more or less dense cover up to 2 m/7 ft high. Nightingale's vegetation is mainly similar, but with scattered woodland patches of the juniper-like *Phylica nitida*.

The fur seal, at one time common but practically exterminated by sealers a century ago, is again breeding on Tristan, Inaccessible and Gough. Seal skins remain an article of commerce, but slaughter is now regulated. The elephant seal, once heavily exploited for its oil, is no longer of commercial interest: numbers are therefore steadily increasing.

The southern right whale, which is gradually recovering from the heavy slaughter of the 19th century, is a seasonal visitor. The right whales were so called because they were easy and profitable to catch, the females being particularly vulnerable owing to their habit of coming close inshore to calve and suckle their young. Right whales were pursued so remorselessly that they were almost wiped out. At least a hundred were recorded off Tristan in 1951. They spend the spring months in the waters around Tristan, arriving in July, and leaving again in December accompanied by small calves. Despite being a hazard to small boats, whales are not molested by the islanders.

Despite almost two centuries of human occupation, only two birds have become extinct on Tristan – what was probably a distinct race of the smaller bunting, and the Tristan subspecies of the gallinule or 'island hen', both believed to have been exterminated by rats accidentally introduced from a shipwreck in 1882. The Gough Island form of the gallinule remains quite abundant. The wandering albatross (with a wingspan of 3.5 m/11.5 ft or more) had ceased to breed on the main island by the turn of the century, but a few pairs survive on Inaccessible and there is a fairly large population on Gough. On Tristan the rockhopper penguin is now confined to two small colonies some distance from the settlement, but is still plentiful on Nightingale, Inaccessible and Gough.

Among the 31 bird species known to breed on these islands, the most numerous is the great shearwater. The entire

## COLONIZING TRISTAN

Tristan has been colonized since the end of the Napoleonic Wars, when Britain garrisoned the island to prevent it being used in any attempt to 'snatch' Napoleon from St Helena. After that scare had passed, some of the garrison elected to stay on. Long isolation resulted in the islanders continuing to speak with an early 19th-century accent not heard in England since Wellington's day.

The island has been occupied ever since, except briefly when earth tremors, first felt in August 1961, culminated in a volcanic eruption during the night of 9 October within 300 metres of Edinburgh, the only settlement. The lava flow engulfed the crayfish canning factory and the two main landing beaches. The following day, the entire population of 264 islanders was evacuated. Once volcanic activity had subsided, most of the islanders elected to return to Tristan, where they reverted to their customary, if somewhat modernized, way of life.

● The rockhopper penguin, once abundant on Tristan, is now reduced to two small colonies on the main island, although many more survive on the smaller neighbouring islands. (WWF / Fritz Polking)

---

## BIRD HARVEST

The islanders exploit some birds for food, notably the long-winged petrel, one of the few winter-breeding species of sea bird and therefore in considerable demand at a time of year when food is otherwise scarce. Dogs are used to locate them in their burrows, their numbers being much reduced as a consequence.

The great shearwater is another important source of food. Inaccessible and Nightingale are normally visited only two or three times a year for the egg and bird harvest. In the 1950s, the November 'egg trip' to Nightingale yielded about 15,000 eggs; and the March–April 'fat trip' accounted for about 15–20,000 young birds, which were either rendered down to fat or salted. This annual harvest, now somewhat reduced, has no effect on the overall numbers of shearwaters.

The yellow-nosed albatross or 'molly' is also eaten; over-exploitation caused a steady decline on Tristan and Nightingale, necessitating the introduction of a system of collection under permit to control the off-take. Over 1,000 breeding pairs nest on Inaccessible's summit plateau, which rises to 511 m / 1,676 ft.

---

world population breeds in the Tristan group; perhaps as many as 2 or 3 million breeding pairs inhabit Nightingale and its satellite islets, with a further 2 million pairs on Inaccessible. There is also a large and perhaps fairly recent population on Gough.

All of Tristan da Cunha's resident land birds derive from the Americas, but only one of them, the Gough Island bunting, is so recently established that it still closely resembles the parent stock in Tierra del Fuego. No doubt the prevailing westerlies helped the original colonizers to cross more than 4,000 km / 2,500 mls of ocean. Their feat is regularly repeated by the American purple gallinule, which turns up annually on migration; but none is known to have survived more than a few days owing to exhaustion and emaciation and, on the main island, to predation by cats and dogs.

The fauna of Inaccessible includes the world's smallest flightless rail, distinguished not only on account of its size but also by vestigial wings and degenerate hair-like plumage. Although confined to this one small oceanic island, the species is comparatively common, and is thought to number several thousand pairs. It lives in close association with tussock grass, with which much of the island is covered, and the tree fern *Blechnum pennamarina*. This rail's habit of moving through tunnels in the dense vegetation makes it difficult to see, and thus helps protect it from the marauding skua and the predatory 'starchy', the rail's only natural enemies.

The thrush or 'starchy' is represented by a distinctive race on each of the three islands of Tristan, Inaccessible and Nightingale. Predation by introduced cats and rats has reduced the Tristan race to a few hundred pairs, but the Inaccessible and Nightingale races remain abundant.

Tristan could benefit from a conservation programme, which should aim at enlisting the support of the islanders in conserving their own superb birdlife. Inaccessible (seldom visited owing to the difficulty of getting ashore) and Nightingale should become wildlife sanctuaries. Both islands are uninhabited, and have the further advantage of being free from the introduced rats, cats, dogs and other predators that have devastated so many island faunas. Provision would, of course, have to be made for continuing the islanders' traditional harvesting of eggs and young birds, and this should be carefully regulated to avoid over-exploitation.

---

## TWO TRISTAN GROSBEAKS

The only Tristan bird to give cause for anxiety is the large-billed bunting or Tristan grosbeak, which is restricted to Nightingale and Inaccessible islands. Of the two races, the larger, Wilkin's bunting, on Nightingale is estimated to number about 400 birds; the smaller race is extinct on Tristan but survives on Nightingale and Inaccessible. It is yellower in colour, with a harsher call, and possesses a smaller bill than Wilkin's bunting. The discrepant bills reflect the ecological differences between the two islands; both races eat the seed of the 'island tree' as their principal food, husking the tough seed with their bills before swallowing. Nightingale's woodland is comparatively extensive; Wilkin's bunting thus has adequate supplies of the hard seed that forms its almost exclusive diet, for which reason it has developed a large bill. Inaccessible's woodland, on the other hand, is sparse; the buntings there have therefore been obliged to supplement their diet with smaller seeds and insects, which account for the fact that their bills are markedly less developed. The main danger to the species stems from storm damage, which can temporarily reduce the supply of 'island tree' seed. But so long as adequate precautions are taken to prevent fires, woodcutting and, above all, the introduction of rats on to these two uninhabited islands there appears to be no reason why the Tristan grosbeak should not survive.

# THE HAWAIIAN ARCHIPELAGO

LOCATION: Consisting of 20 or so islands and numerous islets, the Hawaiian Archipelago reaches across 2,300 km / 1,430 mls of the Pacific Ocean, from Hawaii (19°35′ N, 155°30′ W) in the southeast to Kure Atoll (28°25′ N, 178°20′ W) in the northwest, making it the longest and most isolated chain of tropical islands in the world.

AREA: Of the eight main islands – the exposed summits of oceanic mountains – the largest, Hawaii, known as 'Big Island', covers 10,458 km² / 4,038 sq mls. The other seven, moving from east to west, are: Maui, Kahoolawe, Lanai, Molokai, Oahu, Kauai, and Niihau.

ALTITUDE: The smaller islands – all extinct volcanoes – are the oldest, having over a long period of time eroded almost to sea level. The youngest, Hawaii – still at the stage of up-building and active vulcanism – is distinguished by twin peaks, Mauna Loa (4,170 m / 13,680 ft) and Mauna Kea (4,206 m / 13,796 ft). Mauna Loa, a massive flat-domed shield volcano, rises 9,756 m / 32,009 ft from the ocean floor, more than half its height hidden beneath the surface of the sea.

CLIMATE: Generally mild in the lowlands, temperatures seldom falling below 21 °C / 70 °F or above 25 °C / 77 °F. The higher slopes are, of course, cooler: whereas annual temperatures average about 22 °C / 71 °F at sea level, they decrease to 7 °C / 44 °F at a height of 3,400 m / 11,155 ft. As rainfall is governed by the northeast trade winds, the northeastern slopes are wetter than the southwestern. Thus, while the windward side of Mount Waialeale on Kauai, for example, receives 11,400 mm / 450 in of rain a year, low-lying places in the lee of the wind may receive as little as 200 mm / 8 in. Mean annual rainfall at sea level on the windward side of the islands is 1,000 mm / 40 in, increasing to 3,810 mm / 150 in at medium altitude, whereas the leeward side receives only one-tenth of that amount.

● A fissure eruption with Puu Do in the background. The landscape of the Hawaiian Archipelago is constantly being reshaped by volcanic activity.

(Planet Earth Pictures / Dorian Weisel)

Ranking as the world's highest volcano, Mauna Loa is also among the most active. Volcanic activity, which has been very marked in the last three decades, causes the islands' features to constantly change: new lava flows have increased the size of the island by 81 ha / 200 acres over the last three decades, and erosion has left deep scars. In the last ten years, the East Rift Zone has produced extensive new lava flows and a cinder cone 300 m / 1,000 ft high. Coral reefs are well represented, being most highly developed on the sheltered south and southwestern coasts and in bays that are not subject to wave action.

The islands have an extremely diverse flora in a very confined space. On Big Island, five major ecological zones have been described, from persistently wet rain forest to dry desert scrub, and from coastal beaches to alpine regions. The native flora includes a great variety of ferns. Indeed, tree fern-dominant rain forest reaches its highest level of development in Hawaii. Geographical isolation has resulted in an unusually high degree of endemism; but about half the plant species constituting the archipelago's native flora are believed to have been exterminated.

Most of Hawaii's endemic birds are either extinct, rare or endangered: of the 140 species of birds native to the archipelago, at least 57 are already extinct and 30 endangered, 12 of them reduced to such low numbers that their recovery is considered unlikely. The *IUCN Red List* includes 14 species of Hawaiian

honeycreepers (family Drepanididae), among them akepa, o'u and akiapola'au. The endemic birds also include the Hawaiian thrush or omao, the non-migratory Hawaiian goose or néné and the Hawaiian hawk or i'o. A single bat, the Hawaiian hoary bat, is the only native terrestrial mammal.

The original Polynesian colonists settled in the archipelago in about AD 400. They brought with them domestic pigs, rats, dogs and jungle fowl and cleared and burned about 80% of the lowland forest to grow crops. Further introductions – ranging from goats, sheep, cattle and horses, to several species of exotic deer, mongoose and Argentine ant – followed Cook's discovery of the islands in 1778. In the absence of native predators, they increased rapidly, some becoming serious pests. By the 1930s, feral sheep on Mauna Kea were estimated to number 40,000. Strenuous efforts were subsequently made to destroy them: by 1950, only 200 remained, while goats (in the national parks at least) were completely eradicated. But they left a legacy of environmental damage that will be almost impossible to put right.

Exotic plant species, of which 4,500 have been recorded – though only about 100 of them are a serious threat – have also wrought havoc on native plant communities, and have become much more numerous than the native species. Responsibility for modifying or eradicating much of the indigenous forest (particularly at the middle elevations) is attributable to land clearance for the ranching

of beef cattle and the establishment of plantation crops – chiefly sugarcane, cotton, pineapple and other tropical fruits. The feral pig is widespread; its wallowing habit creates shallow depressions holding standing water that make ideal breeding places for mosquitoes, giving rise to avian malaria, which has had a devastating impact on endemic birds.

Since the end of the Second World War, Hawaii's climate and beaches, especially famed Waikiki, have attracted many people. Big Island's popularity has led to its intensive development as a holiday resort. Another of Hawaii's economic mainstays comes from the islands' strategic position in mid-Pacific. US military installations and associated service industries are a major source of income. The quickening impact of modern development has altered the landscape of Hawaii almost beyond recognition.

The southeastern part of Big Island, including the summit calderas of Mauna Loa and Kilauea, has been designated the Hawaii Volcanoes National Park (929 km² / 358 sq mls). For administrative purposes the park is divided into two parts: the summits and part of the sides of Mauna Loa and Kilauea Volcanoes covering 840 km² / 324 sq mls; and the Ola'a Forest Tract (39 km² / 15 sq mls), separated from the main park by privately owned land. The aim of management is to effect the maximum reduction of exotic plants and animals, particularly in areas that are still relatively intact, priority being given to controlling pigs and goats by fencing. Local residents are permitted, indeed encouraged, to participate in control work. The Hawaiian Volcano Observatory, on the rim of the Kilauea Caldera, has trained most of the world's volcanic scientists and developed research techniques that have been adopted as standard international procedure. In the process Mauna Loa and Kilauea have been more intensively studied than any other volcanoes. The Hawaii Volcanoes and Haleakala national parks were jointly designated the Hawaii Island Biosphere Reserve in 1983 and a World Heritage Site in 1987.

● The endemic Hawaiian goose, or néné, reduced to about 50 in the late 1940s, has been replenished by the reintroduction of birds raised in captivity outside the archipelago. (WWF / Philippa Scott)

# DESERTS

*Failure to understand the insidious effects of grazing on desert vegetation has without question led to the collapse of more civilizations than all other factors together, including war, conquest and pestilence.*
A. Starker Leopold, 1969

Civilization had its roots in the arid regions of the Middle East. It was there that Neolithic man first developed from hunter-gatherer into pastoralist. The establishment of settled communities which followed not only met the need for greater security and stability than was possible under pastoral nomadism. It also provided a food surplus – a fundamental prerequisite for human progress, and one that can be achieved only by settled agriculture.

## The Ancient World

Cereal growing began in the Levant (Syria, Lebanon, Israel and Jordan), perhaps 12,000 years ago, when wild wheat and barley, plants indigenous to the Near and Middle East, were brought into cultivation. This momentous human advance bridged the gulf between the simple human groupings of prehistory and the emergence of more complex settled societies by a gradual evolutionary process extending over many centuries, and was centred on the 'Fertile Crescent', the region stretching from the Euphrates Valley to the eastern Mediterranean littoral, in present-day Syria, Iraq and Iran.

The Sumerians are generally credited with construction of the world's earliest cities and thus with being the people who created one of the first true civilizations. The key to the Sumerians' success lay in water control through the development of irrigation technology, which harnessed the waters of the Tigris and Euphrates, enabling crops to be grown intensively. Salt played a decisive part in Sumer's decline. Having risen to power by successfully converting the alluvial Mesopotamian plain into a renowned granary, excessive irrigation eventually caused the salt level in the soil to rise – a threat that exists wherever semi-arid, poorly drained land is placed under irrigation. As crop yields declined, the cultivation of wheat had to be abandoned and barley – a more salt-tolerant cereal – grown in its place. The decline of its agricultural power base left Sumer so weakened that it gradually collapsed.

The Mediterranean basin, the fountainhead of Western civilization and once a region of plenty, is a classic example of people's misuse of the land. The succession of cultures and empires that sprang up around the Mediterranean owed much of their wealth and power to the exploitation of natural resources. These resources were used increasingly wastefully; indeed, dissipation of natural resources at home was one of the underlying reasons for expansion overseas.

As early as the 4th century BC, Plato in his *Crito* attributed the desiccation of Athenian land to deforestation. He describes how the forests had been felled, the once-rich soil washed away leaving

---

## THE MA'RIB DAM

The Semitic peoples of the Fertile Crescent had close ties with the Kingdom of Saba (biblical Sheba) in the southwestern uplands of Arabia. Their ruler, the almost legendary Queen of Sheba, married King Solomon in the 10th century BC, and it is from this union that the Ethiopian royal house traces its descent.

Little is known about the Sabeans, and the chronology is uncertain; but their commanding position at the entrance to the Red Sea enabled them to control the trade routes – both by sea and overland camel caravan – from the Orient to the Mediterranean. They have left the ruins of substantial buildings and irrigation works, including the great dam of Ma'rib, known as the Queen of Sheba's Dam, though not in fact built by her. Constructed about 850 BC, of massive, interlocking blocks of stone, it was probably the largest dam in the ancient world. For 1,300 years, in conjunction with an extensive system of rock-walled terraces, it helped make Sabean agriculture the most productive in Arabia.

The downfall of the Sabean civilization is generally attributed to the bursting of the Ma'rib Dam in AD 450. Its loss is said to have come about through the breakdown of the *hema* system – a centuries-old way of protecting land through the prohibition or restriction of grazing, grass-cutting, or tree-felling. *Hemas* were found throughout the Arabian Peninsula and used for a variety of purposes, including, as in this instance, protection of the catchment area surrounding the dam for purposes of watershed management.

Abandonment of the *hema* system led to destruction of the vegetation, resulting in erosion and heavy siltation that eventually destroyed the dam. Within a century of the destruction of the Ma'rib Dam, the coastal irrigation systems broke down, peasants deserted the land and trade collapsed. The resulting crisis created pressures that, in AD 636, finally impelled the Arabs to overrun the Levant and overthrow the second Persian Empire. Thereafter, the centre of the Arab world shifted farther north, followed by the Arab conquests of the 7th century which were instrumental in establishing an Islamic Arab empire that at the height of its power stretched from India through Asia Minor and North Africa as far west as Spain.

● Women searching for water near Tahoua, Niger, a region where the Sahara is spreading inexorably into the Sahel. (WWF / Werner Gartung / Wings)

● The Sahara was not always desert. Saharan rock paintings dating from around 3500 BC depict giraffes, ostriches, and buffalo, and even riverine species such as hippopotamus and crocodile, which were able to live in the region at a time of higher rainfall. (WWF / Wlodzimierz Lapinski)

the land littered with stones, and how springs had ceased to flow. That his account of the destruction of Arcadia is no exaggeration is testified by the still visible ruins of ancient temples standing amidst stone-strewn wastes that at one time were green and fertile land, and by shrines marking the sites of former springs that no longer flow. Athens' growing reliance on imports to counter her own declining agricultural productivity weakened her position and undoubtedly contributed to her decline.

Greece, Rome and other countries bordering the Mediterranean met the problem of erosion by extensive terracing. Thousands of kilometres of terraces at one time held the soil in place on hillsides and mountains. Terrace building is a Herculean task, spread over suc-

ceeding generations, and maintenance is an ongoing process which, if neglected, quickly leads to ruin. Over the centuries the ravages of war created conditions that led once-productive terracing to fall into disrepair and be abandoned. Pastoralism replaced cultivation. Sheep and goats overgrazed the terraces, retaining walls broke down and the soil washed away, thus completing the process of destruction. The deciduous open-canopy forest with dense undergrowth that once fringed the northern Mediterranean, and was originally the habitat for an abundant native fauna, has largely gone. Trampled by the hooves of sheep and goats, the exposed soil was blown away by the wind or washed into rivers and carried out to sea. Maquis or scrub grassland is now the characteristic vegetation

over much of this region. Estimates suggest that Greece has retained no more than 2% of its original topsoil. The mountains, once covered with woodland, are now so eroded that reafforestation is often impossible because no soil remains. Yet this has not deterred the Greek government, as recently as 1988, from passing a law opening 61% of the country's remaining woodland to grazing.

## Spanish Desertification

Nowhere in the Mediterranean region has the land been more grievously maltreated than in the Iberian Peninsula. Forest clearance was a more or less continuous process from pre-Roman times to the Moorish occupation; but thereafter the pace of destruction quickened as the settled agricultural communities were replaced by nomadic pastoralism and livestock raising. The devastation of once-fertile Spain seems to have reached its climax towards the end of the Middle Ages. In the 14th century the forests of Castile were still extensive; but, by the mid-16th century, Castile was being described as a treeless waste.

This dramatic change came about through pastoralism based on Merino sheep, introduced into Spain by the Moors. Spain's typically dry Mediterranean summers, influenced by the proximity of the Sahara Desert, encouraged the adoption of a system of transhumant stock-raising – the seasonal migration of pastoralists and their livestock from the lowlands to the uplands. After wintering on lowland pastures in Estremadura and Andalusia the flocks spent the dry summer in the hills of northern Spain. During the course of these migrations they followed well-marked routes or sheepwalks. By the end of the 15th century, the tax on transhumant sheep and the export of wool were the Spanish Crown's most lucrative sources of revenue – until gold and silver started to arrive from the New World.

The *Mesta* – an organization representing the owners of all transhumant sheep – became increasingly influential and despotic. By the early 16th century, the number of transhumant sheep had soared to nearly 3.5 million – and the *Mesta* almost ruled Spain. A century later, the *Mesta* pronounced the death sentence on Spain's forests by obtaining royal consent to feed their sheep on tree shoots during the dry season – the very time when they were most vulnerable. The Spanish forests were ruined and never recovered. The high price paid for the *Mesta*'s land-use practices can be seen from the derelict landscape covering much of the Iberian Peninsula, now almost beyond hope of reclamation. Many Spaniards took advantage of the discovery of the New World to abandon their land, the parched province of Estremadura in particular, and emigrate to America. The combination of drought, forest destruction and inefficient farming methods is currently estimated to be losing Spain about one billion tonnes of topsoil every year. Seventeen per cent of the country's surface area is officially

● Deserts can be remarkably fecund. When rain does fall, seeds that have remained dormant for months or even years germinate, and the desert is briefly carpeted with flowers. Here Californian poppies have sprung up in the Mojave Desert of California.
(Jacana / François Gohier)

## THE *FOGGARA* SYSTEM

Under the Romans North Africa had been a highly productive region: its prosperity derived from water-control systems – dams, canals, aqueducts, terraces and methods of flood control – that were models of ingenuity. One extremely effective device was the *foggara* – or *qunat* – a series of vertical shafts reaching below the water table and linked together by an almost horizontal subterranean connecting channel that compounded the trickle from each shaft into a moderate flow. After the departure of the Romans the *foggara* system was abandoned, and the carefully tended crops which caused North Africa to be dubbed the 'granary of Rome' were replaced by flocks of sheep and goats: desertification quickly followed. The ruins of the Roman cities of Sabratha and Leptis Magna and the Greek cities of Appolonia and Cyrene bear witness to the prosperity of North Africa in ancient times.

designated as desert, and a further 90,000 km²/34,750 sq mls is threatened by desertification.

### Destruction in the Levant

The Levant, in bygone days a prosperous and productive region, is now a wilderness and a burial ground for cities and civilizations of the ancient world. The former prosperity of the Beka' district of Lebanon, for example, can be seen from the remains of its principal town, Baalbek (ancient Heliopolis), with its impressive temples dedicated to Bacchus and Jupiter.

Much of the Levant was at one time well forested. Despite constant exploitation, the forests survived into the Byzan-

tine period; but after the 7th-century Arab invasion they were progressively reduced by the encroachment of Bedouin nomads intent on opening up more land on which to pasture their flocks.

The Levant nevertheless retained a degree of agricultural prosperity until 1258, when Baghdad was sacked and the region ravaged by Genghis Khan's grandson, Hulagu (c. 1217–1265). The irrigation system that had for generations sustained the great Mesopotamian civilizations was destroyed. The land reverted to desert and never recovered. Towns lie buried under mounds of sand that was originally topsoil on nearby fields. Raqqa, for instance, situated on the Euphrates in the fertile Al Jazirah region of Syria – which Harun al-Rashid (766–809), ruler of the Arab empire at the height of its power, considered the loveliest place on earth – was transformed by the Mongol invaders into a desolate wasteland.

A century and a half later – while the Islamic world was still reeling from the effects not only of Genghis Khan's malevolence but of two outbreaks of the Black Death (1348 and 1381) and the accompanying famine – came Tamerlane's (Timur's) equally pitiless visitation. At the end of the 14th century he invaded the Levant, sacked Aleppo, defeated the Mameluke army, occupied Damascus, destroyed Baghdad and depopulated entire regions, his passage marked by heaps of heads piled into grisly cairns.

Lands farther east, in what is now Uzbekistan, were subjected to the same treatment. Midway between the Syr Darya and the Amu Darya flows the Zeravshan, smaller in volume but no less renowned than the other two giant rivers, for on its banks arose two of Asia's most historic cities: Bukhara and Samarkand. While Bukhara lies on the open plain, Samarkand stands among the foothills where the river emerges from the mountains. At its back the mountains reach to the Zeravshan's source among the glaciers of the Pamir Range; before it lies the open steppe, the Mongol hordes' highway to world conquest. Samarkand (ancient Macaranda), the oldest city in Central Asia, was once described as 'embedded in a veritable Garden of Eden, in a land flowing with milk and honey, where all the fruits of the earth seem to grow in profusion and to perfection, where man finds life easy and the climate

## THE WORLD'S MAJOR DESERTS

Arid lands – that is to say, lands receiving an average of less than 250 mm / 10 in of rain a year – occupy about 20 million km² / 8 million sq mls (14%) of the earth's land surface. Semi-arid lands, receiving between 250 and 500 mm / 10–12 in, occupy a further 14%. More than one quarter of the earth's land surface is thus desert or semi-desert. The hot deserts include much of Africa, the greater part of the Middle East, large expanses of Central Asia, substantial parts of both North and South America, and almost half of Australia. The principal hot deserts are:

**The North American Desert** (1,300,000 km² / 500,000 sq mls), extending from the southwestern US into northwestern Mexico.

**The Atacama-Peruvian Desert** (360,000 km² / 140,000 sq mls), occupying a long narrow strip on the Pacific coast of Peru and Chile, the most arid and barren region of all.

**The Patagonian Desert** (675,000 km² / 260,000 sq mls) on the Argentinian side of the Andes, reaching 600 km / 375 mls into the interior plains and running the full length of Argentina.

**The Sahara** (9,000,000 km² / 3,500,000 sq mls), by far the largest, covering nearly one-third of the African continent and almost the size of the United States.

**The Kalahari-Namib-Karroo Desert** (1,100,000 km² / 425,000 sq mls) in southwestern Africa.

**The Somali-Chalbi Desert** (1,300,000 km² / 500,000 sq mls) in the Horn of Africa, extending into the lowlands of Eritrea and Ethiopia, and northern Kenya.

**The Arabian Desert** (3,500,000 km² / 1,350,000 sq mls), covering the Arabian Peninsula and extending northwest into Israel, Jordan, Iraq and Syria.

**The Iranian Desert** (390,000 km² / 150,000 sq mls), lying largely within the mountainous periphery of Iran's great central plateau, reaching into Afghanistan and Pakistan.

**The Thar, or Great Indian, Desert** (300,000 km² / 115,000 sq mls), covering the arid parts of western India and eastern Pakistan, and stretching from the Indus eastwards across the Indo / Pakistan frontier.

**The Turkestan Desert** (1,800,000 km² / 700,000 sq mls) in the south-central part of the former Soviet Union.

**The Takla Makan-Gobi Desert** (1,300,000 km² / 500,000 sq mls) in Sichuan (China) and Mongolia, the coldest of the world's hot deserts.

**The Australian Desert** (3,300,000 km² / 1,300,000 sq mls), occupying about 44% of the Australian continent.

There are in addition the cold deserts, among them not only Antarctica but the tundra regions of northern Eurasia and North America, as well as the highlands of Central Asia and the Tibetan Plateau.

is perfect . . . a beautiful city in beautiful surroundings, a pleasure city as it were, created for some superman . . .'. Although located at the intersection of the northernmost of the three Silk Routes from China and India to Baghdad and Damascus, its importance derived not from commerce (for that was the prerogative of Bukhara) but from its unrivalled position as a centre of learning, art and religion. Such distinction did not deter Genghis Khan from razing Samarkand to the ground in 1220.

The initial thrust of the Mongol invasion was directed against Transoxania, which had long been a flourishing province of the Islamic world. The many great cities of Khurasan (which included

much of present-day Afghanistan) lying beyond the Oxus held a special attraction for the destroyers; it was a region very different from the sparsely populated land of today. What is now the Sistan Desert, straddling the Iran / Afghanistan border, was then the 'Granary of the East', where large estates flourished on fertile land. What are now heaps of rubble were then busy cities. Genghis destroyed them all. Balkh, the great 'Mother of Cities', is now no more than a name on the map. Merv, once a centre of learning and culture that earned it the title 'Queen of the World', never recovered from Genghis' visit. Herat used to be the embodiment of luxury. Genghis destroyed it utterly, as he did Bamian (where one of his grandsons was killed), razing that great city to the ground so thoroughly that no trace of it remains. Nishapur suffered the same fate.

## China

Genghis Khan was not alone in transforming once populous lands into desert. On the eastern flank of what was the Mongol empire, the immense industry of the Chinese peasantry has been directed towards acquiring land and fuel. The ineluctable pressures of the Chinese masses on the land over the centuries have brought all the low-lying regions of their country under cultivation; in the process virtually the whole of central and eastern China and the greater part of northern China (excepting Manchuria) have been stripped of forest cover. In travelling for hundreds of kilometres over northern China, Arthur de Carle Sowerby (1885-1954), Director of the Shanghai Museum in the 1920s, saw nothing larger than a stunted willow or herbage thicker than the scantiest yellow grass or thorn scrub. The only extensive areas of forest he could find were in the higher and more remote parts of Sichuan and Yunnan, and the southeastern provinces of Zhejiang and Fujian.

Having destroyed the forests, the Chinese peasant had to look elsewhere for fuel: roots, grasses and dung were diligently collected and burned, with the inevitable result that unimpeded erosion has long been commonplace, and siltation a major problem. As the rivers become choked with silt and thus unable to carry off excess water, they overflow their banks and inundate huge areas, causing untold damage to crops and property and heavy loss of life. Nowhere is this more evident than along the Huang He (Yellow River), described as 'the world's greatest earthmover'. During the rainy season or at the time of the snow-melt in Tibet and Mongolia, the silt-laden waters of this great river sometimes burst its levees, bringing catastrophic flooding to its lower course across the Great Plain of China. Depositions of silt are currently raising the bed of the river at the rate of 75-150 mm/3-6 in a year. The likelihood of flooding is increased by the widespread reclamation of lakes and wetlands for agricultural purposes, which reduces floodwater storage capacity.

The worst floods in China this century occurred in June 1991, affecting 18 of the country's 30 provinces, Anhui and Jiangsu being particularly severely hit. Six million people were made homeless, tens of thousands injured, and many killed, while diseases such as dysentery and malaria reached epidemic proportions: in places the incidence was as high as 25% of the population. Thousands of houses, factories, hospitals and schools were destroyed or damaged, and road and rail facilities disrupted. In Anhui, 60% of the grain crop was ruined. Losses were estimated at US$5 billion.

Flooding is part of a broader problem of land-use malpractices and is exacerbated by the relentless encroachment of the deserts of Central Asia upon China's northern and northwestern provinces. Sowerby, who was well acquainted with the region between the two world wars, described the effect of this encroachment when he visited the town of Yü-lin Fu ('the City of Elm Woods'), lying in the great loop of the Huang He. The elm woods have disappeared under a sea of sand from the Ordos Desert, which has surrounded the city walls and swept on deeper into China. Within living memory the northern bank of the Huang He (Yellow River) has changed from a fertile area to its present absolutely desiccated condition. Even the south bank and the once fertile valley and hills to the south are rapidly losing their fertility. Not only has North China been affected, but the influence of these new conditions has become increasingly apparent as far south as the lower Chang Jiang (Yangtze). Residents of Chen-Chiang and Nanjing told Sowerby that 25 years previously such a thing as a dust storm had never been heard of, but had subsequently become comparatively frequent, and the sand they brought down was Gobi sand.

In an attempt to halt the widespread degradation of land, the Chinese government has established numerous nature reserves, as well as instituting reafforestation and erosion control programmes. Among these is a scheme for establishing the 'Great Green Wall of China' - a network of tree plantations reaching more than 5,000 km/3,000 mls across the northern grasslands and deserts.

From China, one of the world's oldest civilizations, to Australia, one of the newest, is only a step; but land degradation is equally evident there. About three-quarters of Australia's land is arid - a proportion that is exceeded only in North Africa and the Middle East. Although the land is of relatively low biological value, the native flora and fauna - together with the continent's Aboriginal people - had successfully adapted to living under these particular ecological conditions. But European settlement and the introduction of exotic plants and animals - above all the Merino sheep and the rabbit - were accompanied by farming practices which though perfectly appropriate for temperate Europe were unsuited to the very different conditions in Australia. In under two centuries more than half Australia's farmland has been reduced to desert, with some of it so badly degraded that the damage is irreversible.

Nowhere have the deserts advanced more rapidly than in the Horn of Africa. In the 1880s and 1890s, Somalia was a sportsman's paradise, and for a time it was fashionable to hunt there. Elephants were widespread from the highlands to the coast, lions and rhinos were abundant, wild ass plentiful, and hartebeest so tame that shooting them was considered unsporting. Swayne's hartebeest, elephant and rhino have now entirely disappeared from Somalia, and the wild ass is close to extinction. Early visitors were struck by the delightful park-like country between Berbera on the coast and Hargeisha 177 km/110 mls inland. Today, it is stark desert. Indeed, Somalia is the most glaring contemporary example of a man-made desert.

## SAMARKAND AND PAPER-MAKING

Samarkand's most profound contribution to Western civilization was aided by an accident of history: in AD 751, the city – then under the control of the Arabs – repulsed an invasion by the Chinese. Among the captives were several Chinese versed in the art of paper-making. Samarkand soon established itself as a paper-making centre. Knowledge of the manufacturing process gradually spread through the Arab world, although four centuries elapsed before the Moors introduced this craft into western Europe.

# THE HORN OF AFRICA

LOCATION: Northeastern Africa, at the entrance to the Red Sea. Extending from Somalia's narrow coastal plain (the Guban), inland to the foothills of the Ethiopian Highlands, merging into the Danakil Desert in the north and the Northern Frontier Region of Kenya southwards to the Tana River.

AREA: 1,300,000 km² / 500,000 sq mls, of which 637,000 km² / 246,000 sq mls lie within the political boundaries of Somalia.

ALTITUDE: From sea level to about 1,220 m / 4,000 ft.

CLIMATE: Tropical. Hot and humid on the coast; hot and dry inland. Under monsoonal influence, the long rains are in April-June and the short rains in October-December. Mean annual rainfall ranges from 100 mm / 4 in in the northwest, 200–300 mm / 8–12 in on the central plateau, increasing to 500–600 mm / 20–24 in in the northwestern and southwestern parts of Somalia. Instead of being evenly spread, the little rain that falls is concentrated in a few downpours of high intensity.

Few countries have had a more turbulent history than Somalia. Its present difficulties can be said to have started with the opening of the Suez Canal in 1869, which transformed the Horn of Africa from a backwater into a region of vital strategic and economic concern to the Great Powers. France had occupied Djibouti when, in 1885, Britain assumed control of northern Somaliland. Four years later, Italy took possession of southern Somaliland and Eritrea.

At the end of the 19th century a series of natural disasters – drought, famine and disease – struck the Horn of Africa in quick succession. Smallpox killed thousands of people, while a disastrous series of epidemics of rinderpest, a disease of cattle, starting in the Eritrean port of Massawa in 1888, swept through Africa south of the Sahara, leaving a trail of desolation in its wake. The pastoral people lost 90% of their cattle; the effect on them was calamitous.

The impact on wildlife was equally catastrophic. During the 1897 outbreak,

● The gerenuk's ability to browse on thorny vegetation and to go without water allows it to live under conditions that are too arid for most animals. The degradations of land in the Horn of Africa and southwards into Kenya has actually enabled this species to extend its range. (WWF / Mauri Rautkari)

the Somalis on foot and with their bare hands pulled down sick animals for their hides. The Commissioner for the Somaliland Protectorate around the turn of the century, E.J.E. Swayne, witnessed the dramatic effects of rinderpest. Referring to his first visit in 1891, he describes entering a vast plain covered with immense herds of hartebeest, oryx, Soemmerring's and Speke's gazelles. On firing a shot, the mass stampeded, one herd communicating its fears to another, until right up to the horizon there was a crowd of galloping animals. Swayne counted 400 oryx in one herd, and estimated the total number of animals at not less than 10,000. Returning in 1905, he saw fewer than a dozen animals at a time; hartebeest had practically disappeared, and oryx were reduced to a few scattered herds. He found the contrast with his earlier visit distressing.

Swayne's distress would be even greater could he see Somalia today. Branches have been lopped to provide fodder for goats, and trees felled for fuel until almost none remains, while pasturage is now so scarce that a passing shower results in sheep and goats being loaded into trucks and rushed wherever the rain has fallen to take advantage of the flush of fresh grazing.

Under the conditions of the Horn, where life is a perpetual struggle for survival, pasturage and water are the most elemental requirements and thus the traditional causes of feuding. They are potential flash points and, where people are living permanently on the knife-edge of existence, the willingness to fight for survival is often the only alternative to dying.

Since independence, in 1960, Somalia has plumbed the depths of political, economic and environmental indigence. It has been subjected not only to drought but to inept and corrupt government, torn by two disastrous wars with Ethiopia, besides suffering more than a decade of civil war. Some 4.5 million people - about half the population - are refugees in their own country. Administrative collapse has left the country ungovernable. With the economy in ruins, the currency worthless, the banking system no longer functioning, foreign aid cut off, development projects suspended, malnutrition, starvation and disease stalking the country, and banditry a way of life, the conditions necessary for exist-

## GRÉVY'S ZEBRA

Largest of the zebras, Grévy's zebra is characterized by narrow stripes extending to the hooves, white belly and large, well-rounded ears, as well as by its call: whereas the common zebra utters a bark, Grévy's makes a deep braying sound akin to the wild ass.

A species of the subdesert steppe and arid bushland, and primarily a grazer, Grévy's zebra will browse when fodder is scarce, and even dig for roots and tubers. Although it can go several days without drinking, water is essential to its existence. During the dry season the herds undertake lengthy migrations in search of water.

Until the 1960s, it enjoyed a wide distribution in Ethiopia, western Somalia and the Northern Frontier region of Kenya. Now confined to the western part of northern Kenya, and a small part of southern Ethiopia; in Somalia, it has been extinct since 1973.

Reduced from abundance to scarcity in 20 years, the species declined by 90% between 1960 and 1980; from 1977–1988 numbers fell still further. Warfare, civil unrest, and casual shooting by military personnel all took their toll, as did the practice adopted by pastoral tribes, such as the Samburu, of fencing waterholes to exclude wildlife. But the principal reason for this zebra's decline was to satisfy the demands of the curio trade. Many thousands were slaughtered for their distinctively marked skins. Single skins were sold in Nairobi at US$360, coats at US$600 and handbags at US$85. Kenya has now imposed a total ban on trade in wildlife products.

The Ethiopian population is mainly concentrated in the Chew Bar Wildlife Reserve (4,212 km$^2$/1,626 sq mls) in the vicinity of Lake Stephanie, close to the border with Kenya. Smaller groups occur in the Alledeghi Wildlife Reserve (1,832 km$^2$/707 sq mls) and the Awash National Park (756 km$^2$/292 sq mls). In Kenya the species is seasonally represented in five protected areas: Marsabit National Reserve (1,132 km$^2$/437 sq mls), Sibiloi National Park (1,570 km$^2$/606 sq mls) east of Lake Turkana, Buffalo Springs National Reserve (131 km$^2$/50 sq mls), Samburu National Reserve (165 km$^2$/63 sq mls) and Shaba Game Reserve (239 km$^2$/92 sq mls). In 1964 and again in 1977, Grévy's zebra were introduced into Tsavo National Park (20,812 km$^2$/8,035 sq mls) but, although the animals survived, their numbers did not increase. In planning this species' conservation, Kenya's Buffalo Springs, Samburu and Shaba reserves are of critical importance. Buffalo Springs is a vital source of water in an arid region, and a traditional zebra foaling ground.

● Grévy's zebra in the Samburu National Reserve, Kenya. Confined to northern Kenya and Somalia, where it was once widespread, it is now much restricted in its range, having been hunted to the verge of extinction for its skin. (WWF / Mark Boulton)

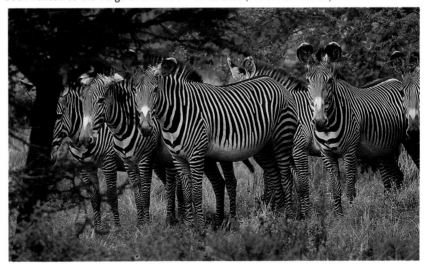

ence have largely broken down. The gun has become the one recognizable authority. Arms dealing is said to be the only commercial activity to prosper. A generation born into anarchy has learned to live with it, knowing nothing else, understanding nothing else, and caring for nothing else.

Of all the factors responsible for insecurity and conflict in the Horn of Africa, none is more fundamental than the question of land. It is the basic resource, and its degradation has been on a scale and at a level of intensity that are cataclysmic.

When peace returns, Somalia will be in desperate need of rehabilitation. This will inevitably be a lengthy and expensive process, and some parts of the country will probably be beyond recovery. The tortured land is there for all to see, but all too often the wrong conclusions are drawn and the wrong solutions proposed. Any rehabilitation programme needs to ensure that the chosen ways forward identify the underlying causes of instability and take them into account rather than dismissing them as irrelevancies. First World solutions are not necessarily right for Third World problems; too often their introduction has failed to raise the standard of peasant living, indeed has even reduced it.

In striving for policies that are ecologically sound it would be wrong to ignore or belittle the traditional systems of land use that have withstood the test of time. Systems which westerners decry as backward or outmoded are often well adapted to prevailing conditions as well as being compatible with the interests and capabilities of the people most directly concerned. The importance of satisfying the legitimate aspirations of the local people and of gaining their confidence and collaboration suggests that indigenous land-use strategies should, where possible, be adapted to present-day requirements.

Traditional land-use systems are based on self-sufficiency – on producing enough to meet immediate needs, but little more. Western systems, on the other hand, are grounded on a market economy involving the production of cash crops that generate the money to buy food and other necessities, and to meet such obligations as taxes. Foreign aid is of course geared to the monetary system; hence its support for large-scale development schemes that maximize foreign exchange

## THE LONG-NECKED ANTELOPES

The Horn of Africa is the home of two species of long-necked antelope, both somewhat similar in appearance: the dibatag and the gerenuk. These graceful animals are distinguished by long slender limbs and necks, and small narrow heads. Only the malés carry horns.

Deriving from the Somali words *dabu* (tail) and *tag* (erect), the dibatag's name aptly describes the animal's habit of holding its neck and tail erect when alarmed or disturbed and running away. The gerenuk ('giraffe-necked' in the Somali tongue), on the other hand, plunges through thorn bush with head and neck thrust out in line with the back, and body hunched low on the legs, in a rather awkward manner.

Both species are specialized browsers. To reach the higher branches of the forbidding thorny vegetation on which they feed, they are given to standing on their hind legs with one forefoot against a branch, often using the other to pull down higher branches. Their independence of water makes it possible for them to live in waterless regions.

Restricted to parts of Somalia and eastern Ethiopia, the dibatag was at one time fairly common in the Haud and central Somalia, but severe poaching, competition from livestock, drought and, above all, the war in the Horn of Africa have greatly reduced its range and numbers to the extent that it is now regarded as endangered.

The gerenuk enjoys a far wider distribution, extending through the dry thornbush country from the Horn of Africa into Kenya's Northern Frontier Region, southwards almost to the foot of Kilimanjaro. Indeed, degeneration of habitat, as in the Kajiado district of Kenya, has actually increased its range.

earnings. But with armaments and prestige projects draining the country's resources, the peasant agriculturalist and pastoralist derive little benefit. Between 1977 and 1982, for example, about 60% of Ethiopia's economic resources were consumed by military expenditure.

Returns that were satisfactory under a subsistence economy, founded on self-sufficiency and barter, are no longer adequate under a cash economy. In the circumstances of the Horn, money is useless and food the only item of value. The pastoralist able to subsist on his livestock is incomparably better off, even under drought conditions, than if he were at the whim of global market forces. This is particularly true when drought conditions oblige him to sell his chief asset, his cattle, at the bottom of the market – the very time when cereal prices are at their peak – leaving him increasingly dependent on state-run organizations and foreign aid handouts.

The most efficient way of utilizing the Horn's sparse and fragile grasslands in a way that can be sustained is through traditional nomadic pastoralism. But the decline, and sometimes the displacement, of the nomadic pastoralist has been a feature of the post-colonial period. Move-

ment is the very essence of nomadism, and where mobility has been abandoned and replaced by a static system the death knell has been sounded to a way of life that, however precarious, is as culturally satisfying as it is ecologically successful.

Traditional pastoralism developed survival techniques that, in times of drought or war, involve migration to dry-season grazing areas – the Haud, for instance – reserved for just such emergencies. Land enclosure effectively cuts off these traditional famine escape routes. Nomad self-sufficiency and self-esteem have been replaced by reliance on food aid, a form of close dependence that will not only continue but will create an Alice in Wonderland situation whereby the more that is given the more will have to be given just to stay in the same place.

Even though complete reversion to the traditional system may not be a practical possibility, it is nevertheless essential to recognize its inherent advantages and to devise ways in which the old system could with advantage be integrated into the new.

With more than one-third of the Horn's inhabitants engaged in pastoralism, the need and the challenge is to devise policies that are ecologically sound and at the same time compatible with the requirements of the pastoralist. Nothing less will win the confidence and support of the people most directly concerned – and whose attitude will ultimately determine the success or failure of any scheme that is introduced.

If success in the use of arid lands is measured in terms of sustaining the greatest biomass, the indigenous fauna should be given the chance of making its own contribution through tourism, game farming or whatever way is most suitable for the prevailing conditions. Failure to do so represents the waste of a prime asset. In the context of the Horn, the need is not for national parks and game reserves from which the nomad is excluded – thus at once alienating local support by appearing to be more concerned with animals than with people struggling to stay alive – but for a comprehensive and fully integrated land use strategy covering the whole region, with the dry lands carrying both wildlife and livestock, their numbers regulated to keep them within the carrying capacity of the land.

# THE NAMIB DESERT

LOCATION: Centred on Namibia, extending for 2,800 km / 1,750 mls from south of the Orange River to the Kunene River and into Angola, varying in width from about 25 km / 15 mls to 140 km / 85 mls.

AREA: 135,000 km$^2$ / 52,000 sq mls.

ALTITUDE: Rising from sea level to about 900 m / 3,000 ft on the inland escarpment, with occasional isolated massifs such as the Brandberg (2,579 m / 8,462 ft).

CLIMATE: Oceanic. Mean annual rainfall 23 mm / 0.9 in, varying from 5 mm / 0.25 in or less along the coastal strip (where precipitation is more than doubled by condensed fog water), increasing to 100 mm / 4 in inland. Mean daily temperature 16–18 °C / 60–64 °F; maximum summer temperature 35 °C / 95 °F.

Unlike the Kalahari and the Karroo, the Namib is true desert. It forms a narrow strip of arid country, made up of bare gravel plains in the north and sand dunes in the south, sandwiched between the Atlantic coast and the inland escarpment marking the edge of the subdesert plateau. Between 18° and 29° S the country is extremely arid, becoming progressively less arid as it merges into the Kalahari and the Karroo. The central sector, from Walvis Bay to Lüderitz, is a region of huge sand dunes that are among the largest known, some attaining a height of almost 300 m / 1,000 ft.

These golden dunes, which at first sight appear so lifeless, in fact support an astonishingly rich invertebrate fauna, including spiders, scorpions, ants, crickets and several hundred species of tenebrionid beetles. They form the staple food for many birds, reptiles and a few mammals, notably bats. These creatures have adapted themselves in a variety of ingenious ways to the very specialized conditions of life in the dunes. The huntsman spider, one of Africa's largest, discovered as recently as the 1950s, makes a tunnel with a silk lining to support the sand above and with a trap door at the entrance to keep out wind-blown sand. By day it remains in its tunnel, emerging at night to wander the dunes in search of the beetles, crickets and other insects – sometimes other spiders – on which it feeds.

The reptiles include various kinds of lizard, the most characteristic being the web-footed gecko, a nocturnal creature with webbed toes that enable it to move more easily over the loose sand of the

## FISH RIVER CANYON

To the east of the Namib, in a region of arid upland plateaux, the Fish River has cut a cleft 60 km / 40 mls long through sedimentary rock; in places the gorge is 750 m / 2,000 ft deep. The green thread of acacias and other vegetation marking the course of the river stands out in sharp relief against the grey background of its harsh surroundings. This strip of riverine vegetation is a haven for a broad spectrum of mammals, birds and insects, some of which live in close association with the desert fauna although not themselves true desert species.

A few scattered pools at the bottom of the gorge – among them the hot springs at Ai-Ais, renowned for their healing properties – contain the only surface water that remains throughout the dry season; their popularity is attested by the density of spoor covering the approaches to the gorge. Well stocked with fish, freshwater crabs and other aquatic life, these pools attract occasional visits from the Cape clawless otter, a species normally residing in the permanent waters of the Orange River, but in the habit of travelling overland.

dunes. The diurnal lizards include the snouted lizard, whose shovel-like snout, long hindlimbs and fringed toes allow it to race over loose sand at high speed. To escape from a pursuer, it dives head first into the sand, disappearing from sight in a moment. This highly effective method of avoiding predators is adopted by many of the dune dwellers, including the beetles. The individual grains of sand are so loosely packed that the surface of the dune is rather like liquid, enabling small animals to 'swim' through it almost as though through water.

The semi-stable sand dunes south of the Kuiseb River are the home of the Namib golden mole, the only mammal endemic to the region. Wind-blown sand provides this inoffensive insectivore with fortuitous protection by quickly obliterating all trace of its tracks, mounds and

● In contrast to the extreme aridity of the Namib, the seas off the Skeleton Coast teem with life, providing an abundance of food for this colony of Cape fur seals at Cape Cross. (WWF / Gerald Cubitt)

● Webbed toes are a special adaptation that enables this gecko to move easily over the loose sand of the Namib. (Jacana / CNRS / A.R. Devez)

● The primitive *Welwitschia mirabilis* is one of only a handful of surviving members of the division Gnetophyta, a group of plants that first arose in the Triassic period, 250–205 million years ago. Endemic to the Namib, *Welwitschia* obtains most of its water from condensed fog. (Noel Simon)

## AN INOFFENSIVE HYENA

The Namib Desert is one of the principal strongholds of the brown hyena. Notwithstanding its heavy build and powerful jaws, the brown hyena differs from other hyenas in being purely a scavenger as distinct from an opportunist predator. Content to feed on the remains of other animals' kills, it is also adept at poaching from more powerful predators such as leopard and cheetah. It locates kills by watching the sky for vultures or listening for the calls of other predators announcing the successful conclusion of a hunt. In the dry season, when food is scarce, the brown hyena supple- ments its diet with rodents, birds and other small prey, not scorning insects, termites and other small creatures, or even tubers and wild fruits.

Births take place underground, usually in an abandoned aardvark's or bat-eared fox's den. To start with, the female leaves her cubs while she forages for food, returning periodically to suckle. When the cubs are 3 or 4 months old, their mother carries them one at a time to a large 'creche', where they join other cubs of her clan. From then on the cubs are raised communally, all the adults helping to bring food, while nursing females suckle cubs other than their own. Such cooperative living has the advantage both of improving security for the cubs and of increasing the supply of food. And should one of the females be killed, her orphaned cubs will be adopted by the others.

Intensive poisoning and trapping campaigns had in the past largely eradicated this shy, nocturnal animal from the closely settled areas. More recently, however, it has successfully adapted to living close to man, partly because of its secretive habits and partly because it is now known to pose little threat to domestic livestock, and so its presence is tolerated.

---

burrow openings. At night, however, when it emerges from beneath the sand to wander across the surface of the dunes, it becomes subject to owl predation, as pellet analysis has shown.

Despite its barren appearance, the Namib has a rich and varied flora, estimated at 3,500 species of flowering plants, ranging from true desert species to savanna and woodland types. The Brandberg alone has almost 500 species of plants, including 15 that are endemic. Best known are the aloes, euphorbias and other succulents, such as *Pachypodium namaquanum*, known as the 'halfmens' or 'elephant's trunk'. The wild prune growing in Damaraland is of great ecological interest as it requires very little moisture to revive from a state of dormancy. Its winged seeds form an important item of food for many animals.

In such a fragile environment, growing pressure from excessive numbers of people and livestock, over-grazing and veld burning, lead very quickly to a decrease in palatable types of grasses – resulting in bush encroachment, damage to plant communities and lowering of productivity. Rare plants and animals have become the object of a lucrative trade that not even the threat of severe penalties has prevented. Some plants have become scarce through excessive collection by tourists.

Climatic and ecological conditions along the coast of Namibia are strongly influenced by the Benguela Current, which flows from the Southern Ocean northwards along the coast towards the Equator. Cold, nutrient-rich water, upwelling from the depths of the ocean, forms the base of a highly productive food chain. Immense shoals of fish sustain large numbers of seabirds, among them many guano-producing birds, notably the Cape cormorant and the Cape gannet. The Benguela Current also enables the Cape fur seal to maintain itself on various islands along the southwestern coast of Africa. The most northerly point of this seal's range, Cape Cross – named after the cross erected in 1485 by the Portuguese navigator, Diogo Cão, the first European to visit Namibia – is the only seal breeding colony on the African mainland south of the Sahara.

The faunal diversity of the Namib is exemplified by the presence off the coast of the Antarctic jackass penguin in close proximity to such tropical species as the lesser flamingo. The latter feeds in the saline pans and brackish lagoons that form along the coast at the mouths of seasonal rivers. Namibia's coastal lagoons are one of the few sources of permanent water for desert wildlife and both resident and migratory waterfowl. During the winter months, these coastal wetlands are thronged with European waders and other Palaearctic migrants. The Walvis Bay wetlands – one of the five most important coastal wetlands in Africa – sustain more shore birds than anywhere else in southern Africa.

A further effect of the Benguela Current is to generate a bank of sea fog which, at night, rolls inland, blanketing the desert for a distance of more than 30 km / 20 mls before being gradually dispersed by the morning sun. As the fog condenses it deposits a layer of moisture over the desert sufficient to support a scant vegetation. Near the coast, where precipitation is heaviest, the desert is carpeted with orange-coloured lichen, principally *Telochystes flavicans*, a plant that plays an important part in colonizing bare surfaces, as well as being a basic constituent of desert food chains and thus of fundamental importance in the ecology of the Namib seaboard.

Sea fog governs the distribution of *Welwitschia mirabilis*, a plant that obtains most of its water requirements from fog condensing on its two (sometimes three) tentacle-like leaves. Unique to the Namib, this plant, which has been described as a living fossil, is among the longest-lived organisms on earth: estimates based on carbon dating show that it can live for as much as 2,000 years.

Fog is an important source of water for plants and for many animals. The sand-diving Namib lizard emerges from beneath the sand to drink up to 12% of its body weight in droplets of condensed fog water. Stored in the bladder, this water lasts the animal for several weeks.

Rain seldom falls on the Namib; but showers and even storms are occasionally known. When this happens the desert becomes temporarily transformed. Grass seed which has lain dormant in the sand, perhaps for years, germinates, converting the normally parched land into green pasture. For a short while, pools of water lie scattered across the floor of the desert, attracting animals from far around; at such times clusters of springbok, mountain zebra, and other animals indulge the rare luxury of grazing hock-deep in grass.

The Namib's only perennial rivers are the Orange in the south and the Kunene in the north; but a number of sand rivers flow intermittently. Usually their banks are lined with trees standing out like a green thread against their barren surroundings. Heavy rain, often at some distance, causes flash floods that transform these sand rivers into raging torrents. For a few hours or days, the floodwaters sweep along the normally dry river courses, before subsiding as suddenly as they began. Within a short while the only evidence of their occurrence are streamers of vegetation suspended from overhanging branches and a trail of uprooted trees and other debris partly buried under still-damp sand.

Though seemingly dead, these sand rivers are in reality lifelines for the wild animals of the Namib: after the floods have subsided water remains concealed beneath their surface, held in the sand as though in a sponge. For most of the year, no other water is to be had in the desert. But, as it lies underground, it is available only to those animals that know how to tap it. The gemsbok and Hartmann's mountain zebra are among the few able to do so. Using their hooves, they scrape out holes in the dry river bed and wait for the water to seep slowly in. When enough has collected they drink, moving off when they have had their fill, leaving the waterholes for other animals that are unable to dig holes for themselves. In this way the gemsbok and the zebra perform a service of inestimable value, which elsewhere in Africa is undertaken by the elephant.

The aardvark or antbear also performs an important ecological function by excavating numerous burrows that are taken over by warthogs, hyenas, porcupines, snakes and many other animals, and adapted for their own purposes.

The confluence of the Fish and Orange rivers marks the southern limit of Hartmann's mountain zebra, a much larger animal than the Cape race, equalling Grévy's zebra in size. Its main habitat is the high country on the eastern flank of the Namib, but it frequently ranges deep into the desert, sometimes as far as the Atlantic shore.

Some of Namibia's finest game country is in the northwestern part of the country, in the Kaokoveld. In years of good rains, large concentrations of both mountain zebra and plains zebra, springbok and gemsbok move from the hills to the low-lying plains. Most remarkably, the Kaokoveld is the only place in the world where elephant and black rhino have successfully adapted to desert conditions. Rock engravings at Sossus indicate that elephants have lived in the northern Namib for a very long time. The Kaokoveld is also the home of the black-faced impala, a distinctive subspecies, numbering no more than about 750–1,000 animals. Permanent water is essential to its existence, therefore it seldom moves far from the Kunene River.

From 1907 until 1970, the northern Kaokoveld was a game reserve, while the southern part was included in the Etosha National Park (at that time reaching to the Atlantic coast). The Skeleton Coast Park lying between the Kunene and Ugab rivers protected the coastal sector. But, in 1970, the western part of the Etosha National Park was deproclaimed to become the 'homeland' of Kaokoland and Damaraland.

Despite assurances by the South African minister responsible that a huge new game reserve would be established in the region, this pledge was never honoured. By 1977, it became known that high-ranking civil servants and South African military personnel were hunting in the area. One minister admitted to shooting elephant and black-faced impala from a military helicopter.

In the late 1970s and early 1980s SWAPO guerrillas started to operate in Kaokoland, whereupon the government issued the local tribesmen with between 2,000 and 3,000 rifles and over 200,000 rounds of ammunition for their own protection. As this was a time of soaring ivory prices, with poaching at its peak, these weapons were, needless to say, used for hunting. By the end of 1981, both elephant and rhino had been virtually wiped out in central and western Kaokoland. The focus of poaching activity then shifted to northern Damaraland.

Early in 1982, by which time no more than 300 elephants and 60 rhinos were left in the region, a group of concerned individuals formed the Namibia Wildlife Trust, a privately funded, non-governmental organization, which aimed to assist in bringing poaching under control.

The Trust persuaded both the Herero and Damara tribal leaders to help in conserving their own natural resources by imposing a complete ban on hunting, at the same time introducing a system of auxiliary game guards recruited from among the local people. These measures succeeded in suppressing poaching, and the elephant population appears to have been stabilized. The wildlife of Kaokoland has nevertheless been badly mauled and remains in a precarious state.

Substantial parts of the Namib have been set aside as national parks and reserves, the best known being the Etosha National Park. The largest, the Namib-Naukluft Park (49,768 km$^2$/19,215 sq mls) lies between the central Atlantic coast and the Naukluft Mountains 170 km/105 mls inland. The Skeleton Coast Park (16,000 km$^2$/6,177 sq mls) protects Namibia's northern coast from the Ugab River to the Kunene River on the Angolan border. This is the only locality south of the Sahara to have true barkhan dunes – crescent-shaped with the horns pointing downwind. Among the most effective wildlife preserves are the diamond areas of southern Namibia, which are entirely closed to the general public.

More effective measures are nonetheless still required to protect the northern Namib and its remarkable fauna. A reserve should be established in the Kaokoveld, preferably linked to the Etosha National Park. Besides benefiting the elephant and rhino, it would protect the migratory routes of the two zebra species. Through the development of tourism, it would also provide an alternative form of employment to the subsistence farming currently practised in an area that is entirely unsuited to any type of conventional agriculture.

Despite many setbacks, the wildlife of Namibia still stands comparison with any in the world. It is to be hoped that the newly independent government will have the wisdom to recognize the indigenous fauna as one of Namibia's greatest assets, and the determination to maintain what is unquestionably one of the finest wildlife regions remaining in Africa.

## ETOSHA

Nowhere can Namibia's fauna be seen to better advantage than in Etosha National Park (22,270 km² / 8,598 sq mls). The focal point of the park is the Etosha Pan, a 6,130 km² / 2,367 sq mls basin (or, more accurately, *playa*) lying at the bottom of a shallow natural depression, and nourished by sporadic rain falling over a large catchment area, but with no outlet. For the greater part of the year the pan remains dry. The resultant salinity inhibits most plant growth, leaving the salt-encrusted pan almost devoid of vegetation. Its shimmering surface creates the illusion of water, but this becomes reality only in years of exceptional rainfall – apart from occasional open stretches of saline water frequented by thousands of flamingos (which breed on the pan), pelicans and other waterfowl. Water is nevertheless stored beneath the surface, as shown by the numerous springs and natural pools fringing the pan.

During the dry season, these waterholes attract large numbers of plains game into the area. At such times Etosha offers an incomparable spectacle: herds of plains zebra, wildebeest, hartebeest, gemsbok, kudu, springbok, giraffe, and many other species, together with their attendant predators, congregate within easy reach of the pan. Patches of thorn scrub scattered across the open grasslands are inhabited by the dainty damara dikdik, while herds of elephant frequent the mopane woodlands to the south and east of the pan.

Over 300 species of birds have been recorded from Etosha. Prominent among them are the ostrich; the secretary bird, which spends much of its time striding across the short grass plains in search of the snakes, lizards, chameleons and locusts and other creatures on which it feeds; and the wattle-nosed guineafowl, found in only a few parts of Namibia and Angola. Perhaps the most endearing of Etosha's terrestrial birds is the male black korhaan, whose white ear coverts stand out against its black face as, with head out-thrust, it hurls defiance at all comers.

The original Etosha Game Reserve covered 99,526 km² / 38,427 sq mls, and was thus more than four times the size of the present park. It ranked with Botswana's Okavango Delta, Tanzania's Serengeti Plains and Sudan's Ilemi Triangle as among the few places where exceptional concentrations of game could be seen. Despite its reduction in size, Etosha remains one of the most important national parks in southern Africa.

● A rare storm over the Etosha Plains. (WWF / Michel Gunther / BIOS)

● A gemsbok among the huge sand dunes of the central Namib, some of which are nearly 300 m / 1,000 ft high. (Jacana / Pat Wild)

# THE TURKESTAN DESERT

LOCATION: Central Asia, extending from the Caspian Sea in the west to Dzungaria in the east, and from about 48° N to the borders of Iran and Afghanistan in the south, reaching to the foothills of the mountain ranges of Kyrgyzstan and Tajikistan in the southeast.

AREA: 1,800,000 km² / 700,000 sq mls.

ALTITUDE: From 26 m / 85 ft below sea level in the vicinity of the Caspian to about 360 m / 1,200 ft in the east.

CLIMATE: Continental. Summers torrid; winters vary from cold with severe frosts in the north to moist and comparatively warm in the south. Mean annual precipitation 160–170 mm / 6–7 in at Kyzylkum, and 75 mm / 3 in along the lower reaches of the Amu Darya. Turkestan is the hottest part of the former Soviet Union. Mean monthly temperature for July at Repetek 32 °C / 89 °F. Winter temperatures in the south may fall to −26 °C / −14.8 °F (Ashkabad) and to −42 °C / −43.6 °F (northern Aral). A characteristic feature of these deserts is the exceptionally wide variation between maximum and minimum daily temperatures, which can fluctuate as much as 21 °C / 70 °F during a 24-hour period. In places exposed to the sun, temperatures may reach 79.4 °C / 175 °F by day and drop to 21 °C / 70 °F at night, a variation of about 38 °C / 100 °F in a few hours.

Arid lands occupy almost the whole of Turkestan, from the Caspian Sea to the borders of Dzungaria, among them the Karakum and Kyzylkum – their names meaning 'black sands' and 'red sands' respectively – the Bet Pak Dala, or 'steppe of hunger', in Kazakhstan, and the Ust Urt, between the Caspian and Aral seas, one of the most desolate places in Central Asia. They vary from extensive areas of sand or gravel to pebble or rock. Three-quarters of the Karakum is sand, either loose or in the form of sandhills and shifting dunes. *Takyrs* – expanses of bare clay with a hardened surface polished smooth by wind-borne sand, and sometimes covering hundreds of square kilometres – are a characteristic feature of the Karakum. The region is poorly drained: most of the rivers originating in the mountains of Kyrgyzstan and Tajikistan, including rivers as substantial as the Murghab, Tedzhen and Zeravshan, seep into the desert and disappear – though not before nourishing several important oases, Samarkand and Bukhara among them. Only the Amu Darya (the Oxus of antiquity) and Syr Darya (the ancient Jaxartes), the largest rivers in the region and among the largest in Central Asia, succeed in traversing the desert. After covering 2,500 km / 1,560 mls from its source in the Hindu Kush, the Amu Darya discharges into the southern end

of the Aral Sea, while the Syr Darya enters it from the north. Twice a year the Amu Darya is in spate: in spring this is caused by thawing snow, and in summer by melt-water from icefields and glaciers in the high mountains. In some years the two periods follow so quickly that they merge into a continuous flood lasting for up to five months. When swollen with floodwater, this mighty river carries a heavier silt load than the Nile and, when it overflows its banks, the silt is deposited over the land.

The banks of the Amu Darya are lined with strips of *tugai* vegetation 2-3 km / 1-2 mls wide. *Tugai* is formed of a tangle of reeds, trees and shrubs, sometimes so dense as to be almost impenetrable, growing on the seasonally inundated floodplain. It is a perfect habitat for wildlife, including innumerable birds and such large mammals as the tiger and its principal prey, the wild boar. Until the end of the 19th century, the *tugai* was regarded as superlative tiger country; this applied in particular to the Amu Darya's headwaters and its delta on the Aral Sea. The Caspian tiger was common until the First World War, but had vanished by 1947.

Away from the rivers, surface water is very scarce. Vegetation is equally scant, with shrubs and annuals predominating, and occasional trees growing along dry watercourses. Sagebrush and saltbush

grow where the soil is not too poor. Where there is sand, white saxaul can be found; the black saxaul prefers saline soils. Saxauls are hardwood shrubs that grow in almost impenetrable thickets. They store water in their bark, and conserve moisture by having exceptionally small leaves; the black variety has no leaves whatever. To reduce evaporation to the minimum, both shed their branches in winter, growing new ones in the spring.

Feather grass is widespread, and innumerable annuals and ephemerals are found throughout the desert. Annuals complete their life cycles during the brief rainy season, before spending the remainder of the year as seeds lying dormant in the soil. Ephemeral plants limit their activity to periods when moisture is available. Ephemerals are of particular importance in the deserts of Central Asia. With the coming of the rains, they develop extraordinarily quickly. The deserts, which for the greater part of the year have only a thin scattering of plants, suddenly bloom like a garden – a regime that exerts a great influence on the lives of desert animals.

Central Asia is the home of a remarkable array of wild animals. The combination of sparse grazing, low and erratic rainfall, torrid summers and bitter winters requires the large grazing animals to be highly mobile. One of the most economically important animals in the region is the saiga antelope. Preferring the subdesert steppe, this species originally had a very extensive range stretching across Eurasia to Dzungaria and even, according to some accounts, to the Gobi. Gregarious and highly nomadic, the saiga migrates southwards in winter, driven by cold and snow. In particularly severe winters, blizzards sometimes bury entire herds under snowdrifts. When this happened in the past, they fell an easy prey to Kalmuk and Kyrgyz hunters, who slaughtered them in hundreds.

The saiga was hunted not only for its meat but also for its amber-coloured horns, which fetched a high price in Chinese medicine shops. Small armies of nomads spent the summer months hunting them. Despite a prohibition on hunting imposed shortly after the Russian Revolution, numbers continued to decline. By 1930, only a few hundred remained in Kazakhstan and Kalmykia, and stringent measures had to be taken to protect them. Numbers then soared, and within thirty years exceeded two million. The saiga is now the most abundant hooved animal in Central Asia. More than 250,000 are culled every year, each yielding 27 kg / 60 lbs of high-quality meat

and a valuable skin. The males alone have horns. Saiga horn finds a ready market in the Far East where, as an increasingly acceptable substitute for rhino horn, it currently fetches about US$150 a kilo.

At one time, the Persian, or goitred, gazelle was even more common in Turkestan than the saiga. Like so much of the fauna of Central Asia, it was hunted almost to extinction: by 1960, only 130 remained. The species owes its survival to the creation of the Shirvan Nature Reserve (177 km² / 68 sq mls) in Azerbaijan, and the Badkhyz Reserve (877 km² / 338 sq mls) in southernmost Turkmenistan, close to the border with Afghanistan. By 1976, numbers had risen to more than 20,000. The Badkhyz Reserve, at the confluence of the Tedzhen and Murghab rivers, protects a wide range of desert fauna, including one of the few remaining herds of Turkmenian wild ass or kulan. Both the wild ass and the Persian gazelle were introduced into the Barsa-Kel'mes ('the place of no return') Reserve (183 km² / 71 sq mls), located on an island of the same name in the Aral Sea. Representative samples of sand desert and its associated fauna have been set aside as the Kyzylkum Nature Re-

serve (101 km² / 39 sq mls) and the Repetek Reserve (346 km² / 134 sq mls) in the southeastern part of the Karakum, while the Kaplanky Reserve (5,700 km² / 2,200 sq mls), in Turkmenistan, protects an area of gravel, stony and clay desert.

Nature reserves also play a leading part in protecting rare birds. The Krasnovodsk and North-Cheleken Bays Ramsar Wetland (1,887 km² / 728 sq mls) on the shore of the Caspian, for example, is of great importance for wintering aquatic birds. More than a million ducks and geese pass through during their spring and autumn migrations. It is, moreover, the principal wintering ground for about 15,000 flamingos, *Phoenicopterus ruber*. The Kurgal'dzhino Nature Reserve (2,371 km² / 915 sq mls) in Kazakhstan, which was established to protect a flamingo nesting site, also benefits such waterfowl as the grey goose and the mute swan.

The vast size of the Turkestan Desert might give the impression that no special protective measures were necessary. But the pace of change in the deserts of Central Asia is so rapid that the indigenous fauna of this region is likely to survive only in national parks and re-

## MARCO POLO'S SHEEP

The size of a small pony, the Marco Polo sheep *Ovis ammon polii* stands 1.2 m / 4 ft at the shoulder and weighs up to 140 kg / 300 lbs. This splendid animal, the most prized of all living sheep, is renowned for its massive spiral horns which, in the older rams, describe a double convolution and are immensely thick at the base. A pair of horns presented to Lord Roberts of Kandahar by the Maharajah of Kashmir, measured 1.9 m / 6.25 ft from base to tip, 1.38 m / 4 ft 6.5 in from tip to tip, and had a basal circumference of 40 cm / 16 in.

The argali inhabits open country with an uninterrupted field of view, for it relies on its astonishingly sharp eyesight – said to be comparable to a man using eight-power binoculars – to give warning of its chief enemies, man and wolf. Its sense of smell is so acute that it can detect a human a kilometre away.

Instinctively gregarious, argali congregate in winter herds of up to 100 led by

a mature female. With the coming of spring the herds break up; the females and young stay together – immature males remaining with the females until they are about five years of age – while the mature rams move to the higher elevations to live apart. The autumn mating season is marked by fighting. Rival rams charge each other head on, colliding with an ear-splitting crash. As well as butting, the rams sometimes run side by side, striking at each other with their horns.

Contrary to most accounts, argali are by nature creatures not of the high mountains but of the downlands. They have, however, been driven into the most inaccessible and difficult country by the increasing herds of domestic livestock. But in winter, deep snow forces them to descend to lower elevations in search of food, where they are subjected to severe predation by both man and wolf. Human predation includes the practice, tradi-

tional among shepherds in eastern Asia, of slaughtering wild sheep and goats for food in preference to animals from their own domestic herds; and as wild goats are more difficult to hunt on account of their habit of frequenting rocky cliffs and sheer slopes, the wild sheep are the principal target. Domestic livestock has also been responsible for the transmission of disease: the 1897–1899 rinderpest epidemic, for example, almost wiped out the wild sheep in the Pamir Mountains of Tajikistan.

The decline of the argali in the lowlands needs to be viewed against the species' successful adaptation to higher elevations, where large herds are currently maintaining themselves satisfactorily in areas unfrequented by man. In the eastern Pamirs there is a regular interchange of animals between herds on both the Russian and Chinese sides of the frontier.

# SNOW LEOPARD

The snow leopard occupies an immense range in the mountainous regions of Central Asia, extending from Lake Baikal in the east to the Pamir and Karakoram ranges in the west, thence across the Himalayas and Tibetan Plateau into China.

Within this vast region of mountain and desert the snow leopard occurs wherever terrain is suitable and prey sufficient for its needs. It spends the summer just below the snow line (c. 4,000 m/13,000 ft) where the open alpine meadows carry the herbivores on which it preys – the *bharal* or blue sheep, the ibex, and many others. As summer advances, the snow leopard follows its prey higher into the mountains (up to 5,500 m/18,000 ft). With the onset of winter it descends to the lower valleys and foothills. The total population is estimated at 3,400–4,100, with the highest density believed to be in the western Tien Shan.

Both summer and winter ranges are being increasingly occupied by domestic livestock, thus inevitably bringing the snow

● The snow leopard, native to the mountains of Central Asia, has been heavily hunted for its valuable fur.
(Jacana / Michel Denis-Huot)

leopard into conflict with people. Its pelt is also among the most valuable of all spotted cat furs, so it is heavily hunted, usually in winter when it descends to lower levels and can the more easily be snared, trapped or caught in pitfalls.

The Great Leap Forward of 1958 had disastrous consequences for Xinjiang's once-abundant wildlife: the drop in agricultural productivity led people to slaughter wild animals for food. The Cultural Revolution of 1966–1976 had similar results: one report describes 'trainloads of ibex, argali and other carcasses being shipped east from the Tien Shan'. More recently, the Chinese government has introduced measures to conserve the native fauna, but law enforcement is difficult and hunting continues, with snow

leopard skins sold openly. The snow leopard's habit of preying on domestic livestock naturally incenses herdsmen who retaliate by systematically destroying it. (Even marmots have in places been exterminated because they are said to be depriving domestic animals of grazing.)

The species is now accorded full legal protection in the former USSR, China, Pakistan, India and Nepal, and it occurs in a number of protected areas. Mongolia has established the Great Gobi National Park (53,000 km$^2$/20,463 sq mls), and Tibet the Chang Tang Wildlife Reserve (240,000 km$^2$/92,400 sq mls), both containing substantial snow leopard populations. But, vitally important as these sanctuaries unquestionably are, they can achieve their purpose of ensuring the snow leopard's survival only if the game laws are more strictly enforced, and there is more effective control of the trade in spotted cat skins. Proper management of prey species is a further important requirement.

serves that have been set aside for the purpose. And as the larger desert mammals range widely, sanctuaries need to be large to accommodate them.

After the Russian Revolution, and particularly since the end of the Second World War, development of the virgin lands became a central feature of Soviet agricultural policy. After transforming the Eurasian steppe by colonization and the plough, the communists next turned to the parched lands farther south. Traditional nomadism was superseded by sedentary agriculture on state farms and collectives, founded on exploiting the desert's latent fertility through irrigation. Nowhere has desert reclamation and development taken place on a larger scale. Kazakhstan and adjacent regions have a long tradition of irrigated agriculture in such places as the Fergana Valley – where fine vineyards, orchards and cornfields are widespread – Samarkand, Tashkent, Bukhara and elsewhere. But,

in the last four or five decades, the scale and pace of irrigation have enormously expanded. Every river in the region has been tapped; many have been dammed to create large reservoirs for storing water and for generating hydroelectric power, while a vast network of canals carries water across the desert. The harnessing of the Syr Darya included the construction of the Kyzyl-Orda Dam and the huge Charara Reservoir, to feed the main Kyzylkum Canal, which irrigates extensive rice-growing schemes, dairy farms and karakul sheep-raising projects.

The lower reaches of the Vaksh, Pianj and Kafirnigan valleys in Tajikistan are the most important cotton-growing region in the former Soviet Union. Nine dams along a 300 km / 186 mls stretch of the Vaksh River irrigate 60,000 ha / 150,000 acres of fine-fibred cotton in the valley itself, and more than 500,000 ha / 1,235,000 acres in Uzbekistan and Turkmenistan. The Takhiatash hydroelec-

● The saiga of the Eurasian steppe and the arid grasslands of Turkestan. This antelope's curious downward-pointing nostrils are believed to be an adaptation to winter cold, warming and moisturizing air prior to inhalation. (Jacana / Frederic)

● The bleak landscape of the Alaiski Mountains, Uzbekistan. (Explorer / P. Roy)

tric development in the Amu Darya basin has brought a further 1.7 million ha / 4 million acres under irrigation. The Ural-Kushum scheme in Kazakhstan irrigates up to 1 million ha / 2.5 million acres in the Urals, and there are other similar projects.

To meet such an ambitious programme, virtually all the waters of the Amu Darya and Syr Darya have been diverted into irrigation, instead of flowing 'uselessly into the Aral Sea' – the phrase used by the Soviet planners of the past. Land reclamation included extensive destruction of the belt of *tugai* vegetation, which serves the important purpose of anchoring the river banks and thus helping to keep the river on course. Clearance of the *tugai* by fire and plough, especially along the middle reaches of the Amu Darya, was accompanied by parties of soldiers deputed to act as 'tiger extermination squads', as a prelude to close settlement and intensive cultivation. Removal of the *tugai* has increased the incidence of flooding. To boost the economy the muskrat, native to North America, was introduced, an aquatic animal whose habit of tunnelling causes extensive damage to river banks. Commercialized cropping of the muskrat has developed into an important industry employing thousands of full-time trappers.

Unspoiled stretches of *tugai* are now rare. The confluence of the Vaksh and Pyandzh rivers, on the upper reaches of the Amu Darya, is one of the few places where sizeable stands remain. Part of this area has been set aside as the Tigrovaya-Balka ('Tiger Valley') Reserve (497 km² / 192 sq mls) in Tajikistan and the Aral-Paygambar Reserve (40 km² / 15 sq mls) in Uzbekistan. These two reserves help protect the endangered Bactrian deer, an animal that lives in close association with *tugai* and whose survival is dependent on it.

The former Soviet Ministry of Water Resources envisaged extending the area under irrigation. This would have involved developing the land around oases as well as irrigating tens of millions of hectares of desert pasturage to increase livestock production. Under this plan the Karakum Canal, carrying water from the Amu Darya into the deserts of Turkmenistan, was to be extended to the Caspian. This formed part of a scheme to maximize the use of all the water resources of Soviet Central Asia – rivers, subterranean waters, and atmospheric precipitation – to bring into cultivation about 20 million hectares / 50 million acres of virgin lands, amounting to about 10% of the entire region. The remaining land, which could not be irrigated, was estimated to be capable of supporting about 30 million head of livestock, mainly karakul sheep, and including about 250,000 camels.

Implementation of these ambitious schemes exacted a heavy ecological price. When this became evident, a halt was called to further development; but not until all the water of the Amu Darya and Syr Darya had already been extracted for irrigation. None now reaches the Aral Sea, which is thus drying up, while lakes in the Fergana Valley and the Zeravshan floodplain have been reduced to swamps.

In such a fragile environment the margin of tolerance is so low that any error may have far-reaching ecological consequences. At first, irrigation increases productivity many fold. But when too much water enters the soil, the land becomes waterlogged, and as the water table rises, dissolved minerals are brought to the surface by capillary action: evaporation adds to the problem by concentrating a layer of toxic mineral salts just beneath the surface, creating a dependence on chemical fertilizers which in turn gradually saturate the land with poisons. Monoculture compounds the difficulty by encouraging weeds and insect pests – which have to be treated with ever-larger applications of herbicides and insecticides, thus adding to the mélange of chemicals in the soil. This process of salinization – which for all practical purposes may be irreversible – has brought ruination to millions of hectares of land – an attempt at harnessing nature that has clearly failed. FAO estimates that salinization has been the cause of some reduction of crop yields over half the world's irrigated land. About 10,000 km² / 3,860 sq mls of irrigated land has to be abandoned each year.

As well as extensive agricultural development, strenuous efforts have been made to exploit other resources. Rich deposits of copper have been discovered in Kazakhstan, and large quantities of oil and gas in Uzbekistan and Turkmenistan. Indeed, almost half the proven reserves of industrial gas in the former Soviet Union are in the deserts of Central Asia. Exploitation of these deposits necessitated the construction of a network of oil and gas pipelines, one of which – the Bukhara-Ural gas line – runs for hundreds of kilometres before branching into Western Europe and the Far East. Pipelines are also used to bring water to some of the new townships and settlements, drawing on the immense reserves of fossil water lying beneath the desert.

The Commonwealth of Independent States – the former Soviet Union – has inherited a great many social and economic problems, not least those relating to some of the Central Asian republics. The changes currently taking place in Tajikistan, for example, are almost revolutionary in their extent. Thousands of displaced people, removed from their homelands under Stalin's draconian labour policies and compelled to labour on state collectives and cooperatives, are now returning and re-establishing themselves. They are coming back to an economy based predominantly on pastoralism, but with the effects of overgrazing everywhere apparent, and with about one-third of the cultivated land, mainly in the Vaksh and Fergana valleys, ruined by salinization and saturated with residual pesticides. Against a background of environmental degradation, and with limited resources being exploited at a level that is unsustainable, the country is having to wrestle with such fundamental issues as the changeover to a free market economy and the reintroduction of private land ownership. This necessitates a change of attitude towards the land. By doing away with individual responsibility, collectivization destroyed the traditional feeling of respect for the land. Restoration of that attitude is an essential first step towards rehabilitation.

Difficulties such as these are being exacerbated by the revival of long-suppressed ethnic and religious tensions, and by one of the highest birth rates in the world. Since the beginning of the 20th century, Tajikistan's population has increased six-fold – from under one million to over 5 million today – and, at the present rate of 3.5%, is expected to double again by 2005. Whether this essentially pastoral country can attain economic viability, without which it will have difficulty in retaining its new-found independence, is a question that only time can answer.

# THE AUSTRALIAN DESERT

LOCATION: Australian interior.

AREA: 3,300,000 km²/1,300,000 sq mls. Semi-arid lands cover an additional 2,500,000 km²/1,000,000 sq mls.

ALTITUDE: Varies from 12 m/40 ft below sea level at Lake Eyre to more than 1,220 m/4,000 ft in the Macdonnell Range (Northern Territory), averaging up to about 300–450 m/1,000–1,500 ft.

CLIMATE: Dry continental and subtropical, characterized by warm or hot summers and mild winters. Rainfall is influenced by two separate systems: the Antarctic System from the south in the winter, and the Monsoonal System from the north in the summer, and varies from less than 125 mm/5 in a year at Lake Eyre in South Australia to more than 760 mm/30 in in the Kimberley region of northern Western Australia. Daily temperatures at Alice Springs in January (the hottest month) range from a maximum of 35 °C/95 °F to a minimum of 20 °C/68 °F. The absolute extremes are 47 °C/117 °F and –5 °C/23 °F. Fluctuations between day and night temperatures are very wide.

The Australian Continent is only a little smaller than the United States, yet almost half of it (44%) is desert; a further 37% is semi-arid. The rainfall, despite being erratic and in some years negligible, is slightly higher than in most other deserts. Thus, although the climate is harsh, there is vegetation of a kind almost everywhere.

The western part of the interior consists of an extensive plateau, the Great Western Plateau or Shield, which is less than 600 m/2,000 ft above sea level. The uniform flatness is relieved by a few east-west ranges, notably the Macdonnell and Musgrave ranges, which lie in the 'red heart' of the continent.

The Great Western Plateau, now covered with dunes and sandy desert, was laid down long before there was life on earth. Its archaic rock formations, similar to those in parts of Antarctica, eastern South America, Africa and peninsular India, provide evidence of the existence of an ancient southern continent that was later split by continental drift.

The northwestern part of the Great Western Plateau is taken up by the Great Sandy Desert; the central and southern parts by the Gibson and Great Victoria Deserts, much of these areas being stony or 'gibber' desert. They merge into the pebble-sand desert of the Nullarbor Plain, which stretches for about 1,000 km/600 mls along the southern coast of Australia. Except for a scattering of bluebush and saltbush, little grows on the Nullarbor's thin soil. To the east, in the Central Eastern Lowlands, the Simpson (or Arunta) Desert forms an extensive lowland region centred on Lake Eyre. A large saline playa or salt flat, about 240 km/150 mls long and lying 15 m/50 ft below sea level, Lake Eyre is flooded only occasionally. It is the largest of the dry lakes that are a feature of this arid land. The Simpson consists mainly of sand dunes forming parallel ridges up to 30 m/100 ft high, 250–500 m/800–1,600 ft apart, and extending for 250 km/150 mls, and is regarded as the most extreme desert in Australia although, paradoxically, it has more ground water than any other.

Although small in relation to the immensity of the Australian continent, the Macdonnells are nonetheless the largest of the central ranges. Extending for 650 km/400 mls, they form a series of up-tilted ridges of red rock, generally no

## SURVIVING DROUGHT

Shrimps have been discovered 250 m/800 ft up the sides of Ayers Rock, in a series of shallow depressions which sometimes contain water. Living in these transitory pools are several species of shrimps, ranging from the size of a dewdrop to about 5 cm/2 in long. The principal types are the shield shrimp, so named because of its broad, flattened carapace and long, segmented tail; fairy shrimp, with long narrow body and multiple legs and gills; and ostracods, tiny creatures that resemble bivalve shellfish. These desert crustaceans have drought-resistant eggs that are carried on the wind, and are thus constantly appearing in places where they have never previously been known. Once the eggs have colonized a rock hole they can wait, sometimes for long periods, for the rain that stimulates them to hatch.

In the ranges, similar rock holes contain species of freshwater fish, crayfish, and a great many aquatic insects not found in the surrounding sand desert. When drought afflicts the plains causing widespread loss of life, these mountain populations serve as reservoirs from which the surrounding country is re-colonized. Many bird species are permanently resident in the ranges, moving to the surrounding plains only when conditions are favourable. The ranges are also the haven of the euro or wallaroo (a large stocky hill kangaroo), the little rock wallaby, and the echidna or spiny anteater. And, when all the desert water holes have dried up, the Aborigines too fall back on the ranges.

more than 500 m/1,500 ft above the surrounding country, topped by a number of peaks, the highest being Mount Ziel (1,509 m/4,950 ft).

The action of wind and water over long periods of time has reduced the greater part of the Great Western Plateau to sand, the monotony occasionally relieved by huge rocky outcrops which, being of harder material, have better resisted the effects of weathering. The most spectacular of these protuberances – Ayers Rock, Mount Olga and Mount Connor – lie between the Macdonnells and the Musgraves within a comparatively short distance of one another. Ayers Rock, resembling a gigantic red hummock 3 km/2 mls in diameter and rising more than 300 m/1,000 ft above the plain, is particularly impressive.

At one time much of the interior was well watered, as is shown by deep canyons and gorges cut by ancient rivers, by dead lakes and dry rivers that characterize the Australian deserts, by the fossil record, and by the occurrence of certain relict plants and animals. A stand of palms *Livistona mariae* thrives in a gorge in the Macdonnell Range, almost 1,000 km/620 mls from its nearest relatives on the coast. Together with the cycad *Macrozamia macdonnelli* and a variety of other plants, they are survivors from the remote past when they were widespread. Today, rainfall is light and erratic; but rare downpours cause the deserts to bloom and dry rivers to run again, occasionally even filling the lakes.

The principal vegetation over much of the interior is 'mulga', a spindly scrub dominated by acacias and pockets of porcupine grass, a spiny xerophytic (adapted to dry conditions) grass that grows in large deep-rooted tussocks. Each tussock forms a miniature refuge for a host of insects, lizards, birds and small

● The Little Sandy Desert of Western Australia. Plants are often spaced in very regular patterns in deserts, and new colonists excluded by intense root competition, or sometimes by chemical exudates from other plants. (Explorer / Jean-Paul Ferrero)

● Ayers Rock, in Australia's Northern Territory, has been shaped by the action of wind and water. Shallow depressions in the rock are home to tiny shrimps, which hatch from drought-resistant eggs following rainfall. (WWF / Australian Information Service)

mammals. One bird, the rufous-crowned emu wren, lives in very close association with porcupine grass. Like all grass wrens, it seldom flies unless hard pressed: when disturbed it makes for the security of the nearest clump of this grass.

Porcupine grass is also associated with Australia's rarest bird, the night parrot. Until the end of the last century, this species was comparatively common in the Macdonnell and Musgrave ranges; but there have been no certain records of it since 1935 and it may be extinct. Little is known about this rare nocturnal bird, but its ground-dwelling habits and near-flightlessness would make it particularly vulnerable to introduced predators.

To the south, the mulga is replaced by a taller, more open type of scrub, known as 'mallee', formed of dwarf eucalypts, which stand out above the surrounding shrubs and porcupine grass.

Acacias are widely distributed throughout the desert, but eucalypts are mainly confined to the ranges. An exception is the river red gum, which grows in low-rainfall areas where few other trees can live, particularly along the edges of desert watercourses. This large, widespreading tree is a haven for birds and insects, providing them with shelter from the sun by day and a safe refuge by night. And when in bloom, its white flowers attract nectar-eating birds, such as the white-plumed honeyeater, from far afield.

The birds of the Australian Desert are of immense variety and exceptional beauty. They range from the wedge-tailed eagle, which preys on rabbits and other small mammals, to a great variety of parrots, parakeets and cockatoos, including the budgerigar. The budgerigar still appears in flocks of many thousands; it is the most common of the so-called grass parakeets, a group of exclusively Australian birds that rank among the most colourful in the world.

The rare pink-coloured Bourke's grass parakeet, well adapted to conditions in the central desert, is reported to be increasing its range. Similarly, the 'mulga parrot', also an inhabitant of the central desert, appears to be in no danger. This is in refreshing contrast to the status of some of the other grass parakeets and their allies: before the introduction of effective protective legislation, the survival of several species was threatened by the demands of the avicultural trade.

The Apterygiformes, an order of flightless running birds exclusive to Australasia, is represented by three living species: the kiwi in New Zealand and the cassowary in New Guinea and northeastern Australia are both forest dwellers, while the emu has been driven out of the more closely settled areas in the southeast of Australia and now lives on the interior plains. Stockmen see the emu (and the red kangaroo) as pests on account of the damage they cause by trampling crops and knocking down fencing, and because they are alleged to compete with sheep for the sparse grazing.

Reptiles are common in the Australian desert. Snakes, goannas, skinks, geckos and lizards of various kinds are widely distributed. Some of the lizards occupy ecological niches that in other continents are filled by mammals. One of the most bizarre is the moloch or spiny devil lizard whose multi-coloured body is a mass of spiky protuberances. The frilled lizard has the ability to erect a flap of skin around its neck resembling an open parasol; the animal's ferocious aspect, enhanced by open-mouthed hissing and feigned attacks, is a guise that suffices to deter most enemies.

The most abundant large marsupial is the red or plains kangaroo. The favourable parts of its range have been taken up for stock farming, but it remains widely distributed over most of the open central plains. The rust-coloured adult males when standing erect may be over 1.8 m / 6 ft tall. Massive hind limbs propel them along in gigantic bounds at speeds approaching 50 kph / 30 mph.

For many millions of years the Australian flora and fauna remained largely unaffected by extraneous influences, leaving them free to evolve in isolation and to retain their distinctive character. The two centuries since the arrival of the first Europeans is but a moment in comparison. But that relatively short time has been a period of profound change.

Despite the importance of minerals to the economy, Australia is predominantly an agricultural country. Subsistence nomadism, traditional in other deserts, has never been practised in Australia. The system of stockfarming adopted by settlers from Europe is founded on huge cattle and sheep stations, some covering thousands of square kilometres. Graziers follow a pattern of rotational grazing based on dry-season concentration and wet-season dispersal;

and, like pastoralists the world over, try to maximize production by increasing the size of their flocks and herds. Sheep thrive on the dry open plains that produce one-third of the world's wool clip. Approximately one-third of all the sheep in Australia – totalling 129,894,000 in 1991 – are carried on arid and semi-arid land. But numbers vary widely, the flocks incurring heavy losses during droughts. To guard against such losses, dams have been built, wells sunk, artesian waters tapped, and fodder crops such as lucerne (alfalfa) grown under irrigation.

Many graziers look upon overstocking as a necessity. Some are of the opinion that the stocking rate of marginal lands can be increased by the installation of water and fencing. But provision of permanent water in previously waterless areas, and a lowering of the incidence of aboriginal (i.e. natural) fire as the result of more effective fire control measures (encouraged by the need to protect buildings and fences) results in the more edible native shrubs and plants being progressively eliminated.

Artificial watering points cause cattle to concentrate within a five-mile radius. Only in wet periods, when temporary surface water becomes available, is livestock able to use distant pasturage. The concentration of livestock forces it to graze grass to the roots, damaging it beyond recovery. Stripped of vegetation, the land lies exposed to wind erosion and every passing storm – the ultimate phase in the transformation of Australia's low-rainfall areas into desert.

Habitat degeneration has been accelerated by a series of misguided introductions of exotic plants and animals, one of them the prickly pear. Without its natural predators and competitors, this spiny cactus flourished, quickly taking over large expanses of grassland. Conventional methods of control proved useless. The decision was therefore taken to use biological controls. In 1925, the moth *Cactoblastis cactorum* was introduced from Argentina. Within eight years, the moth had cleared 24 million hectares / 60 million acres of cactus thicket and restored it to pastureland.

The most calamitous introduction was of course the rabbit in 1880: within 20 years it had spread across all southern Australia, consuming grazing and sparse water, and reaching plague proportions. No countermeasures were effective until 1950, when the introduction of myxomatosis succeeded in killing several hundred million rabbits. Even so, it failed to exterminate the animal entirely. The survivors developed a resistance to the disease and have become as big a pest as before. In 1988, 200 million starving rabbits roamed the country searching desperately for food. One South Australian cattle station alone was overrun by 24 million rabbits. A dozen full-time hunters made little impact on overall numbers.

The damage to grasslands has been catastrophic. In places, pastures have been left bare of any vegetation. Changes in the flora have naturally affected the native fauna. Remorseless competition from rabbits has reduced a number of indigenous species to rarity. But there are also instances where changes in the vegetation have been to the advantage of the native fauna: the red kangaroo and the euro or wallaroo are among the species to have benefited.

When rabbit numbers were first seen to be getting out of hand, several natural predators – ferrets, stoats, and weasels – were introduced in the hope of containing the horde; but these exotic carnivores turned instead on the native mammals and birds, as did the European

red fox, brought in for sporting purposes. These factors, in combination with the cumulative effects of long-standing climatic change, have seriously affected the marsupial fauna and brought a number of species to the brink of extinction. The rat kangaroo or burrowing bettong, for example, at one time abundant throughout Australia, now survives only on two islands off the coast of Western Australia. The disappearance of this burrowing species is believed to be attributable to predation by the introduced fox and feral cat.

The arrival of the Europeans had a devastating impact on the Aborigines also. In the early days they were hunted like vermin. Indeed, the native Tasmanians were wiped out completely; others died of diseases, against which they had no natural immunity. The Aboriginal population was reduced from an estimated 300,000 to only 40,000. Many of the survivors, particularly in the more fertile areas, were dispossessed of their land. For a people who regard the land as a fundamental natural resource of which they themselves are an integral part, loss of tribal habitat and the way of life associated with it, together with exposure to an alien culture, was devastating.

The Australian Aborigine is a nomadic subsistence hunter-gatherer adapted to living under exceptionally harsh conditions. Aboriginal culture is based on the necessity of living in harmony with an essentially fragile environment and of maintaining a delicate ecological balance. An important part of the process involves ensuring that their own numbers do not exceed the carrying capacity of the land or outstrip available food resources, an aim they achieved by a series of natural checks and through self-imposed social and religious taboos.

The contrast between Australia's two cultures, at least insofar as their approach to the land is concerned, could hardly be greater: on the one hand the indigenous Aborigine, who has successfully come to terms with his environment; and, on the other, the grazier whose methods of stock raising caused far greater environmental degradation in 200 years than occurred during the preceding 20–30,000 years of Aboriginal supremacy.

The impact on the environment has never been more vividly described than by James Cotton, an Australian stock inspector, giving evidence before a Royal Commission at the turn of the century. Referring to an area of semi-arid steppe

## MONOTREMES AND MARSUPIALS

Australia's monotremes and marsupials are the most primitive mammals on earth; for at least 75 million years they have been isolated from the mainstream of evolution, and have thus remained unaffected by the emergence of the more advanced placental mammals, with which they would have been unable to compete. They have followed the law of adaptive radiation by exploiting most of the available ecological niches. Of the five possible lines of evolutionary development – aquatic (swimming), fossorial (burrowing), cursorial (running), scansorial (climbing) and volant (flying) – the aquatic niche is taken by the duck-billed platypus; the fossorial by a host of marsupial mice and their allies; the cursorial by the kangaroos and wallabies; the scansorial by the phalangers, tree kangaroos, and koala; and the volant by the flying phalangers.

The Australian marsupials provide

some notable examples of 'evolutionary convergence': i.e., the development of superficially similar characteristics and habits by unrelated species living in isolation from each other. They include the jerboa-marsupials, which bear a close superficial resemblance to the placental jerboas of Africa and Asia and the kangaroo rat of America; the marsupial mice and several related species which superficially look like rats; the hare-wallabies, which show marked convergence with true hares; the numbat or banded anteater, a specialist anteater equipped with prominent foreclaws for ripping open the nests of the termites and ants on which it feeds, and an elongated muzzle and long sticky tongue for lapping them up; the wombat, a powerful badger-like burrower with rodent-like dentition; and the koala, whose cheek pouches and exclusive diet of leaves recall some of the leaf-eating

lemurs and monkeys. Particularly striking examples of convergence are found among the marsupial predators, ranging from the small dasyures, aptly known as 'native cats', to the wolf-like thylacine, largest of the carnivorous marsupials. Perhaps the most remarkable of all is the marsupial mole, which in both appearance and habit is barely distinguishable from the African golden moles.

Like the placentals of other continents, the marsupials evolved giant forms. The largest, *Diprotodon*, was the size of a modern rhinoceros. The remains of diprotodons have been found in many parts of Australia, including a group of more than 500 discovered in 1953 in an ancient lake bed. Carbon dating has shown that this gigantic herbivorous marsupial survived until well into the period of Australia's colonization by the Aborigines 20–30,000 years ago.

with rainfall in excess of 250 mm / 10 in, he told the Commission how 'in the years 1880 and 1881 ... the Cobar district was covered with a heavy growth of natural grasses – kangaroo grass, star grass, blue grass, mulga and other grasses. The Western half of the district abounded with salt and cotton bush.... The ground was soft, spongey and very absorbent. One inch of rain then in spring or autumn produced a luxurious growth of fresh green grass.... The country abounded also with numerous edible shrubs and bushes, and pine scrubs and other noxious scrubs were not noticeable.... There has been a gradual deterioration of the country caused by stock which has transformed the land from its original soft, spongey and absorbent nature to a hard, clayey, smooth surface, ... which instead of absorbing the rain, runs it off in a sheet as fast as it falls carrying with it the surface mould, seeds

● A kangaroo killed by a car in central Australia. Although parts of its range are now taken over by ranching, the red kangaroo is still widely distributed in this part of the continent. (WWF / G. Martin / BIOS)

● Nambung's Pinnacles, an extraordinary limestone formation in Western Australia. (WWF / Fredy Mercay)

● The short-beaked echidna, one of only three surviving genera of monotremes (egg-laying mammals). (Matthew Hillier SWLA)

# NUMBAT

An inhabitant of open eucalypt woodland – principally *Eucalyptus wandoo* – with a shrub understorey, the numbat's essential needs include a plentiful supply of hollow logs and an abundance of termites. These insects not only form the numbat's primary diet, they also consume the eucalypt heartwood, and so provide the hollow-log shelters that are fundamental to the numbat's mode of life, and from which it seldom strays.

Even in suitable habitat numbers are low. The species is believed to have declined dramatically since the 1970s. The reasons are not fully understood, but are believed to be mainly due to modification of habitat arising from changes in the fire regime: occasional high-intensity wild fires arising spontaneously from natural causes have been replaced by frequent low-intensity burning by people, resulting in a much more open understorey, which is unsuitable for the numbat. Introduced predators, such as the fox and cat, may have been a contributory factor, but opinion differs as to their role in this little marsupial's decline.

Of the two subspecies, the eastern form, the rusty numbat, which differs only in the colour of its fur, last recorded from the Warburton Ranges, Western Australia, in 1950, is believed extinct. The surviving subspecies is now the subject of a research programme aimed at finding out how to conserve it.

● A numbat wearing a collar as part of a research programme designed to ascertain how this rare marsupial may best be conserved. (WWF / Fredy Mercay)

of all kinds of plants, sheep manure, sand, etc....'

James Cotton's first-hand account is even more valid today. Unsuitable farming practices have brought Australia to the brink of a manmade ecological and economic disaster comparable with America's great Dust Bowl of the early 1930s. And it has come about for broadly similar reasons – inappropriate methods of land use causing extensive erosion and permanent loss of topsoil and soil nutrients. Some Australian farms are no longer capable of supporting either crops or cattle, and are being abandoned. More than half Australia's farmland is in need of rehabilitation, of which at least 4 million ha / 10 million acres have already been so severely degraded as to become permanent desert. Two-thirds of Australia's tree cover has been destroyed, and many river systems, wetlands and subterranean water sources degraded.

Australian scientists have been aware of the problem for years, but only recently have their warnings been taken seriously. It is now seen that the land use methods of the past 200 years are ecologically unsustainable, and that systems of management based on sound ecological principles need to be introduced if widespread desertification is to be averted.

# MOUNTAINS

*In nature's infinite book of secrecy*
*A little I am read.*
Shakespeare, *Antony and Cleopatra*

Mountains are among the last natural frontiers, nature's ultimate strongholds. Cold, remote, and difficult of access, they have lain until recently at the fringes of human existence. Like most wild places, mountains have long been feared and abhorred. To the ancients they were often sacrosanct, the abode of pantheons of gods and deities exerting control over human destinies. To the frontiersman they were a dismal howling wilderness, the haunt of savage beasts and equally wild men, an adversary to be either avoided or subjugated and bent to man's will.

Even today mountains retain their primordial mystique. Like works of art, people find mountains aesthetically enriching, providing the spiritual dimension that is so integral a part of our being. Despite an erosion of spiritual values, mountains still possess the power of moral uplift for those who seek tranquillity and beauty among the peaks, and a test for climbers bent on stretching themselves to the limit.

The inaccessibility which long discouraged people from going to the mountains gradually vanished as cars and air travel became available to large numbers of people. From being a world apart, mountains have become more closely integrated into life in the lowlands. But the process of integration has not been accomplished without exploitation.

In Europe at least exploitation largely centres on the tourist industry. Tourism's very success is creating its own problems. Wherever large numbers of people congregate there are sure to be environmental pressures: questions of soil, water and air pollution, for example, and the disposal of waste matter. A factor affecting the more popular sites arises from the effects of trampling. The damage caused by thousands of hikers' feet is so serious that in several parts of the world paved footpaths and flights of steps have been constructed to mountain summits.

Mountains cover one-fifth of the earth's surface and are inhabited by about 10% of the world's population – almost 500 million people, ranging from the relative affluence of those living in northern industrial countries to those living at subsistence level in the tropics – Amerindians, hill tribes of Southeast Asia, southwest China, and elsewhere. In Indonesia and the Philippines, for instance, waves of new colonists from over-populated regions, desperate for land, are encroaching onto relatively sparsely occupied territory, displacing the original inhabitants and forcing them to move ever higher into the mountains.

Besides being sanctuaries for people pursuing a way of life based on hunter-gathering, mountains are also natural refuges for endemic species and plant-animal communities, the wide variety of environments giving rise to a high degree of biological diversity. Between the foothills and the summits lies a whole range of habitats – from low-level wetland and woodland to forest and alpine meadow – each with its specialized plant-animal associations. Mount Kinabalu (4,101 m/13,455 ft) in Sabah (North Borneo), for example, is estimated to have over 4,000 species of plants – more than a quarter of the total found in the whole of the United States.

Mountains also serve as genetic reservoirs: the Caucasus, Carpathians, and a few other parts of southern Europe harbour some of the last remaining examples of wild fruit and nut trees – wild pear, apple, plum, pistachio, pomegranate, fig – as well as wild wheat and other relict food plants that no longer survive on the plain, while the Sierra de Manantlan (2,880 m/9,449 ft), in Mexico, contains the only known stands of wild maize. Such gene pools are of vital importance in plant breeding.

Mountains are storehouses for natural resources – water, minerals, timber. Montane forest serves the important role of protecting the land from flood, avalanche, mudslide and erosion. Vastly improved technology coupled with ceaseless demand for timber has led to widespread exploitation. More than half the forests of the Andean uplands have already been cut down, with seemingly little thought given to their replacement. The archa (juniper) forests that once covered the Tien Shan and Pamir-Alai, in Kyrgyzstan and Kazakhstan, have been similarly eradicated. Of the 20,000 km²/ 7,700 sq mls of archa forest in the mountains of Soviet Central Asia at the beginning of the present century, only 4,600 km²/1,775 sq mls remain. The practice in those areas of turning livestock loose in the forest to graze has destroyed all possibility of natural regeneration.

Over the centuries many hill peoples have protected their land by terracing. But terrace maintenance is labour-intensive. Warfare and political and economic instability have caused much terraced land to be abandoned. In our own day, the migration of young people to towns and cities has removed the labourers who previously maintained the system. The modern machinery brought in to take their place is mostly too large to be operated in the confined space of a terrace.

## The Domino Effect
The combination of steep slopes, thin fragile soils, and sparse vegetation leaves

mountains highly vulnerable to neglect or mismanagement. Lack of terracing in conjunction with deforestation is a certain recipe for soil erosion, as can be seen from the highlands of Ethiopia, where soil losses are estimated at 16 billion tonnes a year. As the Ethiopian highlands contain the headwaters of the Blue Nile, Ethiopia's land use practices are a matter of legitimate concern to riparian countries far downstream. Nowhere is this process more keenly watched and monitored than in Egypt and the Sudan where, since time immemorial, the Sudd has acted as a huge natural reservoir, the annual inundation spreading life-giving silt over the land. Until little more than a century ago, this fortuitous renewal was regarded as the gift of the gods, for none knew whence it came. Egypt's future will be increasingly influenced by systems of land use adopted by a distant country over which it has no control. While the countries bordering the Nile have, it is true, subscribed to the Nile Waters Agreement, it conspicuously fails to make provision for misuse of the land by any of the signatories.

The knock-on effect of mountain land use malpractices can be seen in Nepal where emigration was until recently the

● Montane vegetation below Mount Ossa in the Cradle Mountain–Lake St Clair National Park, Tasmania. (Auscape International / Dennis Harding)

traditional means of relieving population pressure. Now deprived of that possibility, Nepalese subsistence farmers are moving higher and higher into their mountains and cultivating slopes too steep for agriculture. The consequent erosion has been compounded by the introduction of plantation forestry which, unlike natural forest, lacks understorey. People who previously relied on the understorey for woodfuel are driven to cut down the trees themselves, thus depriving the land of its protective cover. Such quantities of soil are washed downriver by monsoonal rains that the Ganges delta is silting up and its drainage channels are being choked, increasing the spread of floodwaters over the delta, with disastrous consequences for the people of Bangladesh.

Of all montane resources none is more important than water. By collecting water in the mountains and holding it in dams and reservoirs for later use and to generate hydroelectric power, those living in the lowlands benefit. But although industry is mainly concentrated in the lowlands, its emissions nevertheless affect the mountains. Atmospheric pollution carries great distances, as can be seen from the heavy loss of forest in the northern hemisphere attributable to acid rain. Forest die-back, caused by airborne chemicals produced by industrial emissions and fumes from vehicle exhausts, has already destroyed 15% of Europe's alpine forests, and contaminated

● Cerro Paine Grande (3050 m / 10,004 ft) in the Torres del Paine National Park, Patagonia, southern Chile. (Planet Earth Pictures / David Horwell)

over 50%. Apart from the loss of trees, acidification also lowers the quality of water, sometimes rendering mountain lakes uninhabitable for fish and other aquatic animals.

Litter pollution may be more localized – but is no less offensive for that. Even Mount Everest is not exempt. Since it was first climbed in 1953, about 250 expeditions have scaled the mountain, each requiring 5–10 tonnes of equipment, of which about one-third is abandoned during the descent. Sir Edmund Hillary describes Everest as having become a 'junk heap' and suggests closing the mountain for five years to allow the mess to be cleaned up.

Mountains should be harnessed to contribute to human welfare – dams for power, reservoirs for water storage and irrigation, trees for timber and watershed protection, minerals for extraction. The aim of conservation lies not in preventing utilization of these natural resources but in ensuring that whatever form development takes, it recognizes the essential fragility of montane ecosystems and adheres to the principles of sustainable development.

# THE ANDES

LOCATION: Extending through about 66° of latitude from Tierra del Fuego at approximately 55° S to Colombia and Venezuela at 11° N, the Andes cover a distance of about 7,250 km / 4,500 mls. This great mountain range has an almost unbroken connection with the Rocky Mountains of North America, thus forming part of a mountain system extending the full length of the Americas, from subantarctic latitudes to the threshold of the Arctic.

AREA: The Andes cover an area of more than 2 million km² / 772,000 sq mls.

ALTITUDE: Numerous peaks over 6,000 m / 20,000 ft culminate in Aconcagua, in Argentina, at 6,960 m / 22,834 ft.

CLIMATE: The Andes mark the watershed between the Atlantic and Pacific oceans. On the west they plunge steeply into the Pacific, while on the east they slope more gently to the Atlantic. The two sides are equally dissimilar in climate. The southern Andes is a region of violent winds. Practically all the rain beating in from the Pacific is deposited on the Pacific side of the cordillera, leaving the eastern part of southern Patagonia in a rain shadow that makes it as impoverished as any desert.

Like an immense spine running down the western flank of the South American continent, the Andes form a mountain system which for sheer scale and scenic grandeur is unquestionably the most spectacular in the world.

The Andes are divided into two parallel cordilleras – the eastern, only a little less high than the western – separated by a complex of plateaus that becomes wider in northern Chile, Peru and Bolivia. In this central region the Cordillera Blanca, 160 km / 100 mls long and overlooking the source of the Amazon, includes among its peaks Huascarán (6,768 m / 22,205 ft), the highest point in Peru. The western cordillera contains a number of lofty peaks which rise like gigantic castellations and are shrouded in almost perpetual cloud and mist. Over the whole of their length the cordilleras are studded with volcanos. In the Ecuadorean Andes, between Mount Sangay (5,410 m / 17,750 ft) and the frontier with Colombia, are no fewer than 20 still-active volcanos capped with glaciers and eternal snow, among them Chimborazo (6,267 m / 20,561 ft), Ecuador's highest mountain. Earthquakes are of frequent occurrence and there is hardly an Andean city that has not at one time or another been devastated by them.

Where they terminate in Colombia the Andes divide into three distinct massifs – the Cordillera Occidental, Cordillera Central and Cordillera Oriental – extending from end to end of the country and separated by broad valleys. This has given rise to a mosaic of biological zones: cold, temperate uplands alternate with hot lowlands resulting in such anomalies as, for instance, tropical forests adjacent to snow-covered mountains.

The topography of the southern end of the Andes is infinitely complex, and the landscape among the most extraordinary in the world. In places the cordillera has sunk beneath the sea, leaving a great many steep-cliffed islands. From Chiloé in the north to Tierra del Fuego in the south, these islands protrude from a labyrinth of channels, the largest being the Strait of Magellan separating Tierra del Fuego from the mainland. The coast is heavily indented by fiords – glacial valleys flooded by the sea – many of them overlooked by active volcanos. Towering above tranquil lakes as beautiful as those of the European Alps but on an altogether grander scale, huge icefields spawn glaciers that move coastwards to spill their burden into the sea.

Differences in elevation give rise to a succession of plant communities, with polar and tropical species frequently juxtaposed. Lowland rainforests of unexampled luxuriance around the headwaters of the Amazon merge with the temperate forests (predominantly Ant-

## THE PUYA TREE

One of the few trees associated with the altiplano is the puya. Resembling a palm tree about 4–9 m / 15–30 ft high, this giant bromeliad has a cluster of long, hairy, lanceolate leaves sprouting from the top of its trunk, each individual leaf lined with sharp thorns and tipped with hooks. Despite providing so few opportunities for perching, puya trees, despite their formidable defences, are extremely popular with birds. As many as 30 different species have been seen perching together on a single puya.

Some of the native plants of the páramos have been likened to the giant grounsels of the East African mountains which grow under somewhat similar environmental conditions – an interesting example of parallel development. The most typical plants of the páramos are the frailejohns. Their leaves, initially forming a tight cluster on the ground, are raised higher as the plant grows; and when individual leaves die they droop down to hang from the trunk like a grass skirt.

# MANU NATIONAL PARK

Located on the eastern side of the Peruvian Andes, at an altitude of 365–4,000 m/1,200–13,000 ft, the Manu National Park covers 15,328 km²/5,918 sq mls. It is one of the last great unspoiled wilderness areas remaining on the eastern side of the Peruvian Andes, and the only part of the Peruvian Amazon region to have escaped exploitation. Manu's pristine state, diverse habitats and wealth of wildlife make it biologically richer than any other part of Latin America.

Manu lies in the *selva*, the great tropical forest region of the Amazon basin. The park contains virgin forest with emergent trees reaching through the canopy to heights of 50 m/160 ft or more. While most of Peru's *selva* animals have been heavily hunted for meat and skins, 200 mammal species have been recorded in Manu, representing more than half the mammalian fauna of Peru. Of particular interest is the intermingling of species of North American origin, such as deer and bears, with ancient South American forms – marsupials such as opossums (the only marsupials to be found outside Australasia), and edentates (armadillos, sloths and anteaters). Manu's five species of opossums include the rare black-shouldered opossum, known from only six specimens. The edentates are represented by three species of anteaters: the giant anteater, the tamandua, or lesser anteater, and the dwarf, or silky, anteater; two species of sloths – the three-toed and the two-toed sloths – and at least two species of armadillo, the rare giant armadillo and the common nine-banded armadillo. The richness of the ecosystem is also reflected in 13 species of primates, among them the woolly monkey, and Emperor tamarin. Other endangered species include the giant otter, jaguar, ocelot, bush dog and North Andean huemul. Manu's birds are equally diverse: of Peru's 1,500 species of birds, 800 have been recorded in the park. They range from the Andean condor, harpy eagle and hoatzin to no fewer than 18 species of macaws and parrots. Two species of endangered reptiles also occur in Manu: the black caiman and spectacled caiman, both of which are elsewhere vanishing at the hands of hide hunters.

Small groups of nomadic forest Indians live in the park, following their traditional way of life, based on shifting cultivation, hunting, fishing and collecting turtle eggs. It is important that they should be left undisturbed. But oil exploration poses a threat to their stability: detribalized Indians employed in oil company concessions to the north of the park are procuring the forest Indians to poach animals and fell timber. Other problems arise from the encroachment of several thousand cattle into the *puna* zone in the higher reaches of the park, accompanied by burning of the grasslands to bring on fresh grazing. Logging – both licensed and unlicensed – is taking place along the park's eastern and southern boundaries, more particularly in the upland forests. A North American company has acquired the right to mine gold along the Palatoa River on the park's eastern boundary. If significant amounts of gold are found, large numbers of people could be drawn into the area. Proposals have been made for constructing a road along the Manu River between Urubamba and the Madre de Dios. This would inevitably encourage people to settle along it. Attempts are being made to have the road realigned outside the park. Problems such as these are unlikely to be resolved unless the park administration is strengthened. Manu was declared a biosphere reserve in 1977.

arctic beech) of central Chile, while the low-lying Atacama Desert reaches to the dry *puna* zone of the Peruvian altiplano before grading into the slightly cooler and damper *páramos* of Ecuador, Colombia and Venezuela.

Contrasting with the sparse vegetation of the High Andes are the lush wetlands. They range from seasonally inundated depressions, whose shallow waters are wintering grounds for immense numbers of migrant waders from North America, to large permanent lakes such as Titicaca, Junín, Saracocha and Poopó. Biologically rich, they form an important ecological component of the High Andean plateauland.

Partly because of its prehistoric links with the continent of Gondwanaland and partly as a result of the two Americas being alternately separated and rejoined by a land bridge across the isthmus of Panamá, South America evolved a highly distinctive fauna, which includes such diverse animals as marsupials, edentates (anteaters, sloths, armadillos) and primates. Birds are particularly numerous: at least 25% of all known bird species are to be found in South America, including some with no Old World counterpart. They range from Antarctic species – penguins, albatrosses and the like – to tropical species such as flamingos, and from tiny hummingbirds like jewelled moths to the huge Andean condor, the world's largest living bird.

The continent's greatest and most varied concentration of birds is in Ecuador and Colombia: Colombia has 1,556 species, representing more than half of all the species of birds in South America. The hummingbirds alone number more than 200 species: endemic to the New World, they are distributed from Tierra del Fuego to Alaska. It is perhaps surprising to find such delicate looking birds, normally associated with tropical conditions, in so cold and hostile an environment; but several species of hummingbirds live in the High Andes. They have even been recorded nesting in a ravine less than 200 m/650 ft below the snowline on Ecuador's Mount Sangay.

Harsh climatic conditions and an absence of trees have impelled a number of birds to develop subterranean habits. The rock flicker, a gregarious woodpecker, has evolved special adaptations to underground living: its long conical beak and extensile tongue tipped with horny projections are used not only to dig for the worms, insects and larvae on which it feeds but also to excavate a burrow in which to nest, loosening the soil with its beak and scraping away the spoil with its claws. Congregating in colonies on a cliff overlooking a river, their burrows protect them not only from the climate but also from foxes and other predators.

The lakes of the Andean plateauland are a haven for birds. Of particular interest is the giant coot, a goose-sized bird that builds its nest on a raft of floating vegetation about 3.5 m / 12 ft in diameter and capable of supporting a man. Its relative, the horned coot, assembles a pile of stones on which to build its nest. The profusion of ducks is typified by such species as the puna teal, yellow-billed teal, Andean pintail, and crested duck. Fast-flowing Andean streams are the habitat of the torrent duck. This remarkable bird forages in swirling rapids for the caddis fly larvae on which it feeds. Most other animals would be swept to certain destruction, but the torrent duck negotiates the white waters with ease; even waterfalls do not deter it.

Some of the Andean lakes are encrusted with mineral salts which leave them too saline for most animals; but they suit the flamingo, three species of which inhabit the High Andes: the Chilean flamingo, the puna flamingo of the Bolivian Andes, and the Andean flamingo, rarest of the flamingos, which lives under conditions so extreme as to be at the very limits of existence.

The fish of the High Andes include a number of endemic forms which, during the course of evolution from limited ancestral stock, have come to provide interesting examples of radiation and speciation. The introduction of exotics such as the salmon has resulted in the elimination of a number of indigenous fish species.

The larger Andean mammals include the spectacled bear, the only representative of the bear family in South America, found from western Venezuela to Bolivia. The mountain or Andean tapir, covered with a long, black, woolly pelt, lives at high altitude, migrating up and down the mountain at different seasons to obtain the bamboo shoots, ferns and *páramos* grasses on which it mainly feeds. The Andean deer include the huemul and the northern pudu, the latter restricted to the Andes of Ecuador and Colombia. The viscacha, a gregarious rodent living colonially in rock clefts

● Vicuña in the Pampa Galeras National Park, Peru, one of the animal's main refuges. Thirty years ago there were only 6,000 vicuñas left, but now, with the establishment of reserves in several Andean countries, numbers have risen to 80–85,000. (WWF / Hartmut Jungius)

● The torrent duck forages for caddis-fly larvae in the swirling rapids of Andean streams. (Noel Simon)

● The Moreno Glacier entering Lake Argentino in the Los Glaciares National Park, southern Argentina. (WWF / J.W. Thorsell)

## VICUÑA

The smallest New World representative of the camel family, the vicuña is the paramount animal of the *puna*, the high-altitude rangeland of the Central Andes.

The vicuña possesses the lightest and finest quality fleece obtainable. Under the Incas, robes and garments made from vicuña wool were reserved for the exclusive use of the royal family, beautification of the temples and sacrificial offerings to the sun. The spinning and weaving of vicuña wool was the privilege and responsibility of the Virgins of the Sun.

Hunting was the prerogative of the Inca himself. In a royal hunt, or *chaco*, several thousand beaters encircled a huge area and gradually drove the animals towards the centre. Deer and other animals hemmed in by the phalanx of beaters were clubbed or speared, but the vicuña were captured alive, shorn and released. A *chaco* was not held in any district more frequently than every three to five years, thus giving time for the vicuña fleece to grow and stocks of other animals to recover. This was in marked contrast with the methods adopted by the Spanish Conquistadors, under whom the carefully regulated system of conservation adopted by the Incas gave way to indiscriminate hunting.

In Inca times the vicuña is believed to have numbered several million. By 1965 the population had been reduced to 6,000 and the species was heading for extinction. Peru's establishment of the Pampa Galeras National Park (500 km² / 193 sq mls), in 1967, enabled measures to be taken to protect the herds and build up numbers. This programme proved so successful that, by 1980, the vicuña population had risen to 48,000. Before long, there were signs – notably deaths from starvation – that numbers were exceeding the carrying capacity of the reserve. With a natural increment of 19% a year, the herds had risen to the point at which culling had become necessary, indeed essential, if large-scale starvation was to be avoided. Some of the surplus animals were used to re-stock other areas. Today, in addition to 48,000 in Peru, Chile has more than 8,000 vicuñas, mostly in Lauca National Park; Bolivia 4,500, mainly concentrated in Ulla-Ulla Nature Reserve and Huancaroma National Park; and Argentina 8–10,000 – an overall total of 80–85,000.

In every country in which they occur, vicuña are now protected by law. Trade in vicuña wool, hides and other products is regulated by the 1979 Convention on the Conservation and Management of Vicuña, to which the four countries possessing vicuñas – Peru, Chile, Bolivia and Argentina – are signatories.

and crevices, is a characteristic animal of the High Andes. At one time the chinchilla was equally abundant, but demands for its luxurious pelt led to trapping of such intensity that the species was virtually wiped out. Although numerous in captivity, there is no certainty that it still exists in the wild. The high country is also the domain of the guinea pig, extensively domesticated for food.

Cultivation in the High Andes is limited, but the *puna* – the high open plateauland covered with coarse, tufted grasses interspersed with a variety of herbaceous plants – is suited to grazing. The Indians have long utilized the Andean grasslands for their herds of alpaca and llama. These New World representatives of the camel family are used as beasts of burden; they are also a valuable source of meat and wool. Their wild relative, the vicuña, restricted to the altiplano of Peru, produces a soft downy fleece of exceptionally high quality wool. The guanaco occurs farther south and at lower elevations, in Patagonia.

Among the principal threats to the Andes are widespread deforestation, soil erosion caused by cultivation of steep hillsides and overgrazing. Tourist pressures are damaging some of the more popular sites such as the old Inca Way from Wayllamba to Macchu Picchu; and fires started either accidentally by careless tourists or deliberately by ranchers wishing to improve their pastures are a perpetual hazard, one example being the destruction of 3,000 km² / 1,150 sq mls of the Macchu Picchu National Forest in 1988. Inadequate financial support for conservation is inevitably reflected in lack of control over hunting.

A comprehensive conservation plan covering the entire range, sanctioned by all the countries concerned, would be an invaluable aid in avoiding further degradation of the Andes.

## ANDEAN TAPIR

Restricted to the upper reaches of the Andes in Colombia and Ecuador, where it lives at a height of 2,000–4,400 m / 6,500–14,500 ft, with minor extensions of range into northern Peru and western Venezuela's Sierra de Mérida. This is a region of low temperatures and almost perpetual mist. The mist forest, characterized by stunted tree growth, gives way at the higher elevations to bush and scrub, almost impenetrable except by way of the labyrinth of tunnel-like tapir trails criss-crossing it in all directions.

A shy and retiring species, the mountain tapir is unable to tolerate disturbance and harassment; nor can it adapt to changing environmental conditions. Undisturbed natural forest is essential for its existence, and extensive deforestation the principal cause of decline. Starting with large-scale charcoal burning at the end of the 19th century, the pace of environmental degradation has quickened. Roads penetrating deep into the Andes facilitate the settlement of previously inaccessible areas; fingers of cultivation reach ever higher into the cordilleras, while pastoralism expands into the uplands, progressively destroying woodland and scrub.

The only hope for this species' continued existence lies in the establishment of well-protected national parks or similar sanctuaries in some of the few unspoiled parts of the forest that remain, and in giving effect to the protective legislation that already exists but is seldom enforced.

# THE ROCKIES

LOCATION: The western limits of the flat prairielands of North America are marked by an immense mountain barrier towering 3,500 m / 11,500 ft above the plain. Extending down the western edge of the continent and reaching almost its full length, the Rockies form North America's main watershed.

AREA: North America's dominant physical feature and one of the most extensive mountain systems in the world, the Rocky Mountains reach 5,150 km / 3,200 mls from the Brooks Range in Alaska to Mexico's Sierra Madre Oriental. Varying in width from 110–160 km / 70–100 mls, boundaries – particularly in Alaska – are largely arbitrary.

ALTITUDE: The Rockies are at their highest in Alaska. The St Elias Range includes Mount Logan (5,951 m / 19,524 ft), Canada's loftiest peak, and only a little less tall than Mount McKinley (6,194 m / 20,320 ft) in the Alaskan Range, the highest point in North America. The Canadian Rockies comprise two parallel ranges: the main (eastern) range, an area of scattered mountains interspersed with plateaus, has almost 700 peaks over 3,050 m / 10,000 ft, the highest being Mount Robson (3,954 m / 12,973 ft) and Mount Columbia (3,747 m / 12,294 ft). West of the main range, and 600–900 m / 2,000–3,000 ft above sea level, lies the Rocky Mountain Trench. From 3–16 km / 2–10 miles wide, it extends for 1,600 km / 1,000 mls from the Yukon to the United States. Beyond and to the west of this great trough are the Columbian Mountains, separated from the Coast Mountains by an undulating plateau cut by deep valleys and studded with lakes. The Coast Mountains, much dissected by glacial action, rise in a series of serrated peaks to tower above the islands and fiords with which Canada's western coastline is strewn. Innumerable ice-carved fiords reach far inland, their sheer sides cut by hanging valleys and dramatic waterfalls. Offshore and partly submerged is yet another range, its peaks forming Vancouver Island, the Queen Charlotte Islands and others.

South of the Canadian border the Rockies extend in a succession of discontinuous ranges separated by river valleys. In the north, the Cascade Ranges reach into Washington, Oregon and California, while in the centre, from the Yellowstone River to the Wyoming basin, are five principal ranges: the Beartooth Mountains, Absaroka Range, Bighorn Mountains, Teton Mountains dominated by Grand Teton (4,190 m / 13,747 ft), and Windriver Mountains topped by Gannett Peak (4,202 m / 13,787 ft). Farther south, the Colorado massif has several dozen peaks close to 4,250 m / 14,000 ft high. The snows and glaciers of the Colorado mountains give rise to six great rivers, the Missouri, Snake, Colorado, Arkansas, Platte and Rio Grande. The next line of peaks, the Wasatch Mountains, lies beyond the Wyoming basin and the Colorado Plateau, close to the Great Salt Lake, a wilderness of dry plateaus, excessively hot in summer and bitterly cold in winter.

CLIMATE: So vast a natural barrier forms a climatic divide. Winter fogs, a product of the warm humid oceanic climate, affect a narrow strip of Pacific coast, allowing exotic plants to thrive in the winter warmth of Vancouver. But a short distance inland the mountains draw off almost all the rain, depositing it on the Pacific side, leaving little to reach the interior plateau which as a consequence is arid, in places almost desert – a parched land of grass steppe, sage bush and cactus. Thus, forest and semi-desert, humidity and aridity occur side by side. In the more northerly latitudes snowfalls and blizzards are a frequent occurrence: along the border between Canada and the United States annual snowfalls in the order of 2,000–2,500 mm / 80–100 in are normal.

The entire Rocky Mountain region has been shaped by glacial action, leaving characteristic U-shaped valleys, deep lakes and fiords. Glaciers in the range are vast: the Malaspina Glacier, part of the St. Elias Mountains glacier system, covers 3,885 km² / 1,500 sq mls. The Columbia Icefield (375 km² / 145 sq mls), the largest icefield in North America outside the Arctic, spans the Continental Divide. Its glaciers feed the headwaters of three major river systems – Columbia, Athabasca and North Saskatchewan – which flow into three oceans: the Pacific, Arctic and (via Hudson Bay) the Atlantic. Glacial meltwater also fills many of the lakes with which the Canadian Rockies are so richly endowed.

Between the icefields of Alaska and the arid lands of Utah and Nevada lies an exceptionally diverse range of habitats. In the Canadian Rockies, lodgepole pine and white spruce are dominant in the valleys, with aspen and poplar along the

● A mountain goat in winter coat in the Banff National Park. (WWF / Eric Dragesco)

● The Grand Canyon, between the Sierra Nevada and the Colorado Mountains, is one of the natural wonders of the world. (Explorer / François Gohier)

## GRAND CANYON

Between the Colorado Mountains and the Sierra Nevada, the Grand Canyon forms one of the most remarkable geological spectacles on earth. The almost fathomless immensity of this great cleft – 445 km / 276 mls long – lies outside the realm of normal experience. The scale is set by the Colorado River: from the canyon rim, 1,500 m / 5,000 ft above the floor, this mighty river resembles a slender silver thread, and the roar of torrents coursing through rapids and plunging over waterfalls is dissipated in the full-ness of space. The pageant of colour, constantly changing according to season and hour, is especially vivid at dawn and dusk when buttes and promontories, bathed in vermilion and gold, stand out in contrast against mauve-blue shadows lurking in the recesses of the gorge.

The Grand Canyon provides the most comprehensive geological time chart that exists, a stratified cross-section of the past reaching back across more than 2,000 million years of geological history long before there was life on earth.

rivers. Douglas fir is the principal tree at medium elevations, and Engelmann spruce the main constituent of the subalpine forest community. Spruce-fir forest (including alpine fir) grows close to the timber line, grading into alpine meadow flora which, besides grasses, heaths and sedges, includes such plants as dwarf birch and alpine bearberry.

Along the Pacific coast from southern Alaska to the Olympic Mountains in Washington State magnificent forests arise from the very edge of the ocean, creating a strip of rich temperate rainforest frequented from late spring to autumn by hummingbirds. Here, too, are some of the mightiest trees in the world, the Douglas firs, which attain such huge size – reaching heights of over 76 m / 250 ft – that they rank among the world's tallest living organisms. Farther south, warm mists drifting inland from the Pacific sustain gigantic redwoods and sequoias. Still farther south, the parched inland basins and plateaus support little but sage bush and cactus, a type of vegetation known as chaparral, reminiscent of the maquis of Mediterranean lands.

Moose and beaver occur along valley bottoms; elk, white-tailed deer and mule deer are characteristic of meadows in the forest zone; while rocky mountain goat, bighorn sheep, hoary marmot and pica typify the fauna of the higher alpine pastures. Carnivores include both the grizzly and black bear, as well as cougar, lynx, wolverine and the wolf.

Birds, both resident and summer visitors, are abundant, among them the mountain chickadee, white-tailed ptarmigan, mountain bluebird and northern three-toed woodpecker. Transients include birds moving to and from their northern nesting grounds, or those that nest around the northern fringes of the prairies and winter on the Pacific coast.

Three-fifths of British Columbia is covered with productive forest, and forestry is the province's principal industry. The largest timber mills in the world lie at the mouth of the Fraser River. Much of the timber is processed into newsprint.

Excessive exploitation has been the pattern of the past, but reafforestation, fire control and conservation are now accepted policy. The Pacific salmon, renowned for their quality, have also been heavily over-exploited. An international Salmon Fisheries Commission has been established to decide upon open seasons and regulate fishing methods. The Rockies are a vital source of water and conservation of water is one of their most important functions. Almost every potential dam site has been taken up for water storage and for generating hydroelectric power. But the demand is so great, and constantly rising, that water resources are stretched to the limit.

The most immediate threat comes from the millions of people visiting the Rockies each year to hike, camp or ski. Intensive tourism inevitably causes disturbance and leads to problems of litter and waste disposal at campsites and along hiking trails.

North America's finest mountain scenery is to be found in the Rockies, and the

## REDWOOD AND SEQUOIA

Arguably the most imposing forests in the world today are the sequoia forests of North America. Fossil remains of sequoias have been found in many parts of the northern hemisphere, but the two living forms – the redwood and the giant sequoia – are restricted to the Coast Ranges of northern California and southern Oregon from Monterey Bay almost to the Klamath Ranges, where they grow on seaward-facing slopes and valleys lying within the humid coastal belt, invested almost daily with warm sea fog.

Ranking among the tallest trees in the world, many redwoods exceed 90 m / 300 ft – one reaching the astonishing height of 112.1 m / 367.7 ft, the tallest tree now standing. Saplings are branched to the ground, but mature trees lose their lower branches and are limbless to a great height. Despite their bulk, redwoods stand surprisingly close together; their huge trunks, encased in fire-resistant bark more than a foot thick, reach up like gigantic columns as if to support the sky, while their outspread crowns are so dense that the sun can scarcely penetrate the canopy. At their feet a light ground cover of shrubs and ferns is interspersed with occasional saplings, for although redwoods yield a prolific crop of seed cones, the germination rate is low and the forests contain few seedlings.

Natural oils make the heartwood strongly resistant to termites and other insects, and render it practically rot-proof. The wood is light but strong, knot-free, fine-grained, and easily worked. These factors make redwood an ideal multi-purpose timber, used for almost everything, from railway sleepers to fence posts, and from house construction to furniture making.

The insatiable demand for redwood timber resulted in large-scale exploitation, particularly during California's boom years in the latter half of the 19th century. Commercial logging cut swiftly and deeply into much of the easily accessible forest. Most stands of redwood were already in private ownership before the general public realized that the big trees were in danger of being exploited out of existence.

Inspired by the Save the Redwoods League, and over a period of many years, individuals and organizations throughout the United States set about the formidable task of acquiring every available acre of redwood forest. Many of these groves are named after the person or group who saved them from axe and saw. These efforts continue. A major advance occurred in October 1968, when the already existing Jedediah Smith, Del Norte and Prairie Creek state parks were linked together by a long (60 km / 40 mls) narrow strip of coastal parkland to form Redwood National Park (424 km² / 164 sq mls). The need for the park and a sign of the speed with which new areas were being opened to logging could be seen from the US National Park Service's inability to obtain several thousand hectares originally intended for inclusion in the park, but which had already been felled by the time the park was established.

# YOSEMITE

Yosemite National Park (3,082 km²/ 1,190 sq mls) embraces one of the most spectacular sections of the Sierra Nevada. The Park centres on the Yosemite Valley, a cleft about 11 km/7 mls long and over 1.5 km/1 ml wide, carved by glacial action and forming an extension of the Merced River canyon. The valley floor is the bed of an ancient glacial lake that has been gradually transformed by depositions of silt into lush meadows and parkland. Perpendicular cliffs topping 900 m /3,000 ft, their height accentuated by the valley's relative narrowness, are capped by the granite domes of Glacier Point, Cathedral Rocks, El Capitan, and Sentinel Rock. The upper end of the valley is dominated by Half Dome, its immense form reflected in the placid waters of Mirror Lake, one of 300 lakes in the park.

High above the valley floor, tributary streams flow through 'hanging valleys' to plunge over the escarpment in a series of spectacular waterfalls (see page ii), the highest being Ribbon Fall, with a clear drop of 491 m/1,612 ft, and Upper Yosemite Fall which cascades for a total of 739 m/2,425 ft. Among other large falls are Bridalveil, Sentinel, Vernal and Nevada, all of which are at their zenith in May and June when bearing their full measure of melted snow.

Those seeking solitude can find it along the network of trails radiating from the Tuolumne Meadows towards the distant crest of the Sierra Nevada. Here in the high country are lakes, streams and rugged wilderness set among valleys and forested mountain slopes where pine and juniper provide an evergreen backdrop against which aspen, oak and dogwood can be seen in all their glory.

Yosemite derives its name from a tribe of transhumant Amerindians who seasonally occupied the valley. Although discovered in the 1830s, the valley remained hard to reach for many years. It was not until the coming of the automobile that the park could be visited by large numbers of people. This has posed many problems. Yosemite is only about 300 km/200 mls from San Francisco, and the park has become so popular that a small township containing hotels, lodges, restaurants, stores and other facilities has sprung up in the valley to cater for visitors' needs. Visitor pressure is so intense that in the summer months the guy-ropes of tents in camp sites on the valley floor virtually overlap one another.

Pressure is not confined to the valley: the backcountry areas, including trails and even the High Sierra itself, have been affected by excessive numbers of people. A major road through the Sierra has cut the park in two, while the construction of hydroelectric and water storage facilities at Hetch Hetchy and Lake Eleanor have caused substantial local disturbance. Many of these developments conflict with the purpose of the park: 'to conserve the scenery and the natural and historic objects therein and to … leave them unimpaired for the enjoyment of future generations'.

# THE SIERRA NEVADA

California's Pacific mountain system (sometimes erroneously included in the Rocky Mountains) consists of two parallel north-south ranges: the Coast Ranges extending unbroken along the Pacific shore, and the Sierra Nevada-Cascades inland. California's rich Central Valley separates the two. Extending over 640 km/400 mls, the Sierra Nevada reaches its crestline at Mount Whitney (4,414 m/ 14,482 ft), the highest point in the continental United States.

Fur trapping, logging, mining for gold and silver − starting with California's celebrated 'forty-niner' gold rush − have all left their imprint. Scars caused by indiscriminate use of high-pressure water jets to slough down hillsides and expose the gold-bearing quartz remain visible to this day. The railroads of the 1860s were laid on wooden sleepers; once they were built, the sawmills switched to logging. Sawdust dumped in streams caused siltation, destroying the spawning beds of the indigenous trout and bringing about their virtual extinction. The native animals were similarly exploited. By 1924, the grizzly bear was extinct, while coyote, cougar, wolverine and marten, once abundant in the Sierra Nevada, were reduced to a remnant. Even the bald eagle, America's national symbol, is among the endangered.

None of the Sierra's natural resources has been more consistently exploited than water. The insatiable demands of California's continually growing population necessitate ever-larger abstraction, much of it from the Sierra Nevada, as in the diversion of water from Owens Valley to augment Los Angeles' water supply. Rice-growing under irrigation in the Central Valley alone consumes huge quantities of water. Erosion arising from deforestation and misuse of the land have degraded water to the extent that one-third of it no longer meets standards of quality laid down by the state. Much of the forest is, moreover, affected by acid rain and other forms of air pollution originating from nearby industrial areas. Runoff adulterated by chemical fertilizers and pesticides drains into lakes, contaminating aquatic plants and animals.

A network of highways and roads has made the Sierra Nevada more accessible to the hordes of people who now visit the high country. Tourism has become so dominant a feature of the region that parts of the range, the lower slopes in particular, have been permanently disfigured.

Growing awareness of what is happening is fortunately taking effect. Hunting, fishing and hiking are now regulated by permit; logging has been made selective; and a moratorium has been imposed on all new building at Lake Tahoe, one of the Sierra Nevada's scenic wonders.

establishment of national parks and national forests has done much to preserve it. The importance and popularity of Yellowstone (8,991 km²/3,471 sq mls) in Wyoming, the first national park to be established, have been recognized by its designation as a World Heritage Site. Four Canadian national parks – Banff (6,641 km²/2,564 sq mls), Jasper (10,878 km²/4,200 sq mls), Kootenay (1,377 km² / 532 sq mls) and Yoho (1,313 km² / 507 sq mls) – have been jointly designated the Canadian Rockies World Heritage Site.

● Giant sequoias in the Sequoia National Park, California. (WWF / Henry Ausloos)

● The Kodiak bear of Alaska is usually considered to be the largest of the carnivores, measuring around 2.5 m / 8.25 ft from nose to tail. Like other large animals — especially predators — it needs an extensive territory in order to survive. (Matthew Hillier SWLA)

# THE HIMALAYAS

LOCATION: Running along India's northern boundary and towering high above the Ganges /Brahmaputra plain, the immense mountain barrier of the Himalayas extends in a huge arc for more than 2,500 km/1,500 mls from Nanga Parbat (8,124 m/26,660 ft) in Kashmir through Nepal, Sikkim, Bhutan and the North East Frontier Agency to Namcha Barwa (7,782 m/25,531 ft) in extreme western China. The impression of height is enhanced by mountains of 7,000 m/23,000 ft or more rising from a plain only 100–200 m/325–650 ft above sea level

AREA: The highlands included within the Himalayan region cover an area of about 3.4 million km² /1.3 million sq mls.

ALTITUDE: This very complex mountain system can most conveniently be considered as three separate longitudinal zones: the Outer Himalaya, the line of foothills averaging about 1,000–1,250 /3–4,000 ft in height; the Lesser Himalaya, averaging 3,500–4,500 m/12–15,000 ft; and the Great Himalaya, the main range, averaging 6,000 m/20,000 ft. North of the Great Himalaya (the Zaskar Range), and quite distinct from it, lies the Karakoram Range, forming part of the watershed between the river systems of Xinjiang (Sinkiang) and India. The Himalayan–Karakoram–Hindu Kush–Pamir range, with 104 peaks over 7,315 m/24,000 ft, is the greatest mountain system in the world. The peaks include Mount Godwin Austen (K.2) (8,610 m/28,250 ft), the world's second highest mountain. At its foot, the Indus flows between the flanks of the Karakoram and Zaskar ranges before rounding the foot of Nanga Parbat and altering course to the south.

The Himalayas can be further subdivided laterally into the Western, Central and Eastern zones. The Western Himalaya lies almost entirely within Kashmir. The Central Himalaya stretches from Sikkim to Himachal Pradesh, the greater part being in Nepal. This sector contains the highest Himalayan peaks: Everest (8,863 m/29,078 ft); Dhaulagiri (8,167 m/26,795 ft); Annapurna (8,091 m/26,546 ft); Nanda Devi (7,816 m/25,643 ft); and Kamet (7,756 m/25,447 ft). The terrain has been heavily affected by glacial action and is scored by deep-bedded ravines cut by swift-flowing torrents.

The Eastern Himalaya, comprising the mountains of Sikkim, Bhutan, Assam and Western Xinjiang, is cut by the deep cleft of the Tsang-po (Brahmaputra), which makes a sharp change of course around the base of Namcha Barwa. This sector contains several peaks over 7,000 m/23,000 ft, the highest being Kangchenjunga (8,593 m/28,208 ft).

The Himalayas are the product of tectonic uplift extending over many millions of years. During that time the Indian plate has under-thrust the Eurasian plate in an on-going process that continues to raise the Himalayas by 2–5 cm/1–2 in a year.

CLIMATE: So huge a mountain barrier inevitably has a marked effect on weather conditions: by obstructing the northward movement of the monsoon, it brings extremely heavy rainfall to the foothills but very little to the Tibetan Plateau above, leaving it in a rain shadow. The variation can be seen by comparing Ladakh ('Little Tibet'), where the annual rainfall does not exceed 75 mm/3 in, with the Eastern Himalaya which receives the full force of the monsoon, and where normal rainfall is in excess of 2,000 mm/80 in, though 5,000 mm/200 in is not unusual. Some places get considerably more: Cherrapunji in Assam receives 11,600 mm/457 in – the highest rainfall in the world. On the other hand, the Himalayas protect the plains of India from the bitter cold that is such a marked feature of the Tibetan Plateau.

Lying between the Greater and the Lesser Himalaya is the Valley of Katmandu, the heart of Nepal. Although only about 24 km/15 mls long, this narrow plain contains the country's three principal towns, including the capital, Katmandu, and supports more than half a million people. The valley is fertile and intensively cultivated, a notable feature of Nepalese agriculture being the extensive use of terracing to utilize every square metre of arable land.

To the south of the Zaskar Range and within the Lesser Himalaya lies the Vale of Kashmir. This valley – 130 km/80 mls long, 40 km/25 mls wide, and a little over 1,500 m/5,000 ft above sea level – has an annual rainfall of only 660 mm/26 in, much of it concentrated into a short period. As this is marginal to cultivation, most crops have to be grown under irrigation; the countryside is covered with a network of canals and waterways channelling water to the fields, while the 'floating gardens' of the Dal Lake enable produce to be grown on rafts of vegetation some distance from the shore.

Rainfall in the eastern sector

## THE HIMALAYAN PHEASANTS

The pheasants – of which there are 49 species, some with numerous subspecies, in 16 genera – are especially well represented in the Himalaya–Tibet–Sichuan region. They include the Impeyan pheasant, resplendent in plumage that glistens like burnished copper; blood pheasant, decked in green and grey with black and white markings and daubed with crimson; satyr tragopan, which is among the most handsome of pheasants – the cock's head is adorned with a pair of fleshy protuberances, and a sac under his bill can be expanded into a beautifully patterned bandana; the monotypic snow partridge; and red junglefowl, the game bird ancestor of the domestic fowl.

Among the rarer Himalayan pheasants are the cheer, with crested head and barred tail, which has become increasingly scarce in recent years; western tragopan, an inhabitant of the western Himalaya, which is both rare and localized; and Blyth's tragopan from the eastern Himalaya. The Himalayan mountain quail, an inhabitant of the western foothills, has not been recorded since 1868, but may possibly still exist.

gives rise to dense tropical rainforest, merging into deciduous forest. At the higher elevations conifers and, later, giant rhododendron and magnolia become dominant, up to the level of alpine pasture at about 4,000 m / 13,000 ft. The permanent snow line lies at about 5,000 m / 17,000 ft, descending as low as 2,000 m / 6,500 ft in winter. The luxuriant forests in this region are renowned for their wealth of orchids: Sikkim alone is reputed to have more than 1,000 species.

The Central Himalaya lies within the monsoon belt but, except at the higher elevations, rainfall is less torrential than in the east. Humidity is very pronounced in the foothills. Here the lower deciduous forest is succeeded by *Pinus roxburghii*; above 1,500 m / 5,000 ft oak and conifers become dominant, giving way to birch at the higher elevations.

Lying between the tropical Oriental Region and the temperate Palaearctic Region, the Himalayas form one of the world's most imposing natural frontiers. The fauna of the higher parts of the Himalayas includes such large mammals as the argali, the most impressive of all living sheep; the Himalayan tahr, beardless but having a conspicuous soft neckruff; the markhor, the largest of the wild goats carrying magnificent spiral horns; the bharal or blue sheep, which ranges to great heights; and the snow leopard, whose pelt ranks among the most beautiful and valuable of all wild felines'.

The Himalayan deer include the Kashmir stag or hangul, which inhabits forest in and around the Vale of Kashmir. Estimated to number 3–5,000 sixty years

## SAGARMATHA NATIONAL PARK

Located in northeastern Nepal, and covering 1,148 km² / 443 sq mls, the park covers the upper catchment of the Dudh Kosi river system, a distinct geographical entity enclosed by high mountain ranges. The park's northern boundary lies along the main divide of the Great Himalaya, coinciding with the Tibetan border.

With its lower boundary at 2,845 m / 9,334 ft and containing seven peaks over 7,000 m / 23,000 ft, the park reaches to the summit of Mount Everest (8,863 m / 29,078 ft), the world's highest mountain. The park's four main glaciers – Chhukhung, Khumbu, Gokyo and Nangpa La – feed deeply incised valleys which drain into the Dudh Kosi and its tributaries, to form part of the Ganges River system. In the upper reaches of the Gokyo Valley, the Ngozumpa Glacier – at 20 km / 12 mls, the longest glacier in the park – impounds 17 small glacial lakes, with clear, turquoise-blue waters, used by waterfowl during the spring migration.

The park is important as a high-level breeding ground for birds: of the 162 recorded species, 36 breed in the park, among them the blood pheasant, robin accentor, white-throated redstart and grandala. Other notable birds include the Tibetan snowcock, snow partridge and Himalayan monal.

Sagarmatha has relatively few resident mammals, but they include the Himalayan musk deer, serow, goral, Himalayan tahr, red panda, Asiatic black bear and snow leopard. The short-tailed mole, Tibetan water shrew, Himalayan water shrew and woolly hare are among the smaller mammals of particular interest.

Forest destruction, which increased with the arrival of Tibetan refugees in 1959–1961, accelerated still further with the growing popularity of mountaineering and trekking. The relative affluence generated by tourism – now an integral part of the local economy – has encouraged evironmental degradation by providing the means for greater investment in livestock. This has led to overgrazing of high mountain pastures, while increased demand for timber – for building material and firewood – has noticeably affected the forests, leading to soil erosion and resulting in water becoming unfit to drink. Waste disposal is a further problem.

Normal management criteria have been modified in order to reconcile the special demands made on the area by mountaineering and tourism with the requirements of conservation and with the needs of the park's 3,000 resident Sherpas – including insulating their culture and religion from adverse Western influences – at the same time safeguarding the interests of the many people in Nepal and India whose welfare is affected by the condition of the river catchment. The traditional system of controlling the cutting of timber and hunting of wildlife by local committees – *shinga nawa* – has been revived to deal with infringements of the forest regulations. Two Strict Nature Protection Areas, in which all human activity is prohibited, have been set aside in the southern part of the park, and proposals have been made for an eastward extension into the Makalu-Barun area – the Barun Valley being regarded as exceptionally important for forest birds. This would add a further 1,400 km² / 540 sq mls to the park which was listed as a World Heritage Site in 1979.

## THE *TERAI*

The *terai* – or 'moist land' – is a narrow strip of marshland, about 25–30 km / 16–20 mls wide, in northern India and southern Nepal. It lies on the Gangetic Plain, running along the base of the Himalayan foothills from the Yamuna (Jumna) River in the west to the Brahmaputra in the east, and is fed by springs along its northern edge which are the source of several rivers, the most substantial being the Ghaghara (Gogra).

The dense vegetation of the *terai* is ideal wildlife habitat. Besides such common species as wild boar, fishing cat and otter, the fauna includes a number of rare and endangered species, among them tiger, leopard and occasionally elephant, gaur, wild Asiatic buffalo, swamp deer (or barasingha) and chitral (or axis deer). The *terai* is the last stronghold of the great Indian rhinoceros, as it is of several smaller species, notably the diminutive pygmy hog and the hispid hare.

Birds, too, are numerous. The *terai* provides a haven for immense numbers of wintering waterfowl, and is the ultimate refuge for the white-winged wood-duck and the pink-headed duck, last recorded in the 1930s. Other notable species include swamp partridge, Bengal florican, crested kingfisher and Pallas's fish-eagle.

Mosquitoes abound in the *terai*, and malaria is endemic. The presence of hordes of insect pests made the *terai* uninhabitable but, starting in the 1950s, a malaria control programme using pesticides succeeded in eradicating the mosquito, opening the way to settlement. Land-hungry settlers poured in and, within a very short time, much of the *terai* had been drained and cultivated.

The influx of settlers rushing to stake claims to such exceptionally fertile land led, of course, to the construction of villages and townships – and even to minor industrial undertakings – without proper provision for such facilities as sewage or rubbish disposal. This has resulted in the rivers and streams feeding the *terai*, particularly those with a small dry-season discharge, becoming severely degraded.

An evaluation of the *terai* is urgently needed to determine what can be done to safeguard the few unspoiled areas that remain. Some sanctuaries – chiefly Dudhwa National Park (490 km² / 189 sq mls), Manas Wildlife Sanctuary (391 km² / 151 sq mls) and Kaziranga (429 km² / 165 sq mls) – have already been established, but others are needed to create an adequate system of *terai*-based sanctuaries without which this specialized habitat, together with its associated fauna, already greatly reduced, will cease to exist.

---

ago, it has since been reduced to about 150. Another distinctive race of the red deer, the shou, lives 1,300 km / 800 mls to the east in the forested Chumbi Valley and adjacent parts of Bhutan and Tibet. The shou has disappeared from the southern face of the Himalaya but may persist in the Tibetan sector of its range.

The bird fauna of the Himalayas is extraordinarily rich. It ranges from brilliantly coloured sunbirds and trogons, such as the Nepal sunbird and the red-headed trogon, to the rose-breasted rose finch which breeds far above the tree line at altitudes up to 5,500 m / 18,000 ft – higher than any other songbird.

Kashmir's numerous lakes and *jheels* are seasonally crowded with waterfowl which fly in from their distant breeding grounds in the far north. Many of these Siberian-bred ducks and geese reach their winter quarters in Kashmir by way of the mountain passes between the Hindu Kush and the Karakoram; others follow the valleys of such rivers as the Lena, Irtish and Yenisei, overflying the Tibetan Plateau to cross the Himalayas east of the Karakoram Range.

The southern slopes of the Himalayas have a profuse insect fauna, among which are butterflies of spectacular beauty. Of

## NANDA DEVI NATIONAL PARK

Lying within the Garhwal Himalaya and covering 630 km² / 243 sq mls, the park protects the huge glacial Nanda Devi Basin. Split into a series of parallel north-south ridges, the entire basin lies above 3,500 m / 11,500 ft, and is encircled by mountains on all sides except the western where the deep and virtually inaccessible lower Rishi Gorge descends to 2,100 m / 6,890 ft. Around the high mountain rim are about a dozen peaks above 6,400 m / 21,000 ft, among them Changbang (6,864 m / 22,520 ft), Dunagiri (7,066 m / 23,183 ft) and Nanda Devi East (7,434 m / 24,391 ft). Dominating the basin is Nanda Devi West (7,817 m / 25,647 ft), named after the 'Blessed Goddess', consort of Shiva. As well as being one of the most scenically spectacular parts of the Himalayas, the basin is of great glaciological interest as it possesses an unusually diverse array of glaciers at different stages of development.

The park's 'inner sanctuary' remained unexplored until 1934, when it was opened to mountaineering. By 1982, when about 4,000 mountaineers and trekkers visited the park, dumping of litter, vandalism and disturbance had reached such serious levels that the park was closed to foreign visitors. A preliminary management plan has been drawn up but has not yet been sanctioned. Among the recommendations is that the Pindari and Sundadhunga valleys at the southern edge of the Nanda Devi massif should become a designated sanctuary to protect the large populations of wild herbivores and pheasants said to occur there. Although the bird fauna is little known, only some 57 species having so far been recorded, the park's 14 species of mammals include Himalayan musk deer, serow, Himalayan thar, bharal and its principal predator the snow leopard, Himalayan black bear and brown bear. The park forms the nucleus of a much larger area extending northwards to the Dhauli Ganga, which has been proposed as a biosphere reserve.

particular interest is the relict Himalayan dragonfly, one of only two survivors of the dragonfly suborder Anisozygoptera, and among the very few animals to have remained virtually unchanged since the Triassic period, over 200 million years ago, when the dinosaurs were still at an early stage of development and the mammals and birds had yet to evolve.

The Himalayan foothills are among the most densely populated regions on earth; the ripple effect inevitably reaches into the highlands. There, the main threats arise from loss of protective forest and cultivation of steep hillsides, leading to erosion and increased sedimentation. Settlement is accompanied by tree-felling, livestock encroachment into reserved areas, overgrazing and burning. Habitat depletion in conjunction with excessive hunting is certain to cause wildlife to decline.

Since the end of the 19th century, conservation in India has been based on a system of forest reserves designed to safeguard timber, soil and water resources, with a small number of national parks added later. But large gaps remain in the system: neither the Gangetic plain, the Assam hills, nor Ladakh, for example, have representative parks or reserves. Protected areas are often poorly managed: fewer than half have management plans, many of which are in any event deficient. As management of protected areas is bedevilled by inadequate finance, this is hardly surprising. Bhutan's Wildlife Division is almost paralysed by lack of staff. Bhutan has nevertheless succeeded in maintaining a relatively pristine environment – largely because the northern part of the country has been incorporated into the Jigme Dorji Wildlife Sanctuary (7,905 km² / 3,052 sq mls). The southernmost forest belt, on the other hand, has been almost entirely cleared for human settlement and the uplands are being steadily degraded.

● A view towards the Annapurna Massif from Pokhara, Nepal. The lower slopes have largely been stripped of their trees, a process that in some parts of the country has led not only to major erosion but to massive flooding of the Ganges Delta hundreds of miles away. (WWF / Gerald Cubitt)

● The huge increase in the numbers of Western trekkers and climbers visiting the Himalayas over the last two decades has dramatically altered the way of life of the local people, many of whom now find employment as guides and porters. (WWF / Galen Rowell)

# THE TIBETAN PLATEAU

LOCATION: The Himalayas form the southern bastion of Central Asia's dominant geographical feature: the vast elevated Tibetan tableland, which rises from the Kali Gandaki, the gigantic trough immediately north of the Himalayas and parallel to them, a major geotectonic feature, about 1,125 km / 700 mls long and 50 km / 30 mls wide, extending from Ladakh to Sikkim. The plateau is the source of virtually every major river in China and Southeast Asia: the Irrawaddy, Salween, Mekong, Chang Jiang (Yangtze) and Huang He (Yellow) rivers all originate in this region. Together with the Indus, Brahmaputra, Ganges and Sutlej, which also rise in the same general area within a short distance of one another, they water some of the most densely populated places on earth. Immediately north of the Himalayas, the Tibetan Plateau descends in a series of steppes to the lowlands of Xinjiang (Sinkiang) and Nei Mongul (Inner Mongolia).

AREA: Covering 1,221,600 km² / 471,660 sq mls, a substantial part of Tibet is taken up by the vast treeless plateau of the Chang Tang, or northern plain. The northern periphery of the Chang Tang, flanked by the Kun Lun Range, forms an almost unbroken chain of mountains from the Karakoram Range in the west to the Tsing Ling Shan in Shaanxi (Shensi), where it swings north to link up with the Greater Khingan Mountains.

ALTITUDE: Lies at a height of 4,700–5,000 m / 15,500–16,500 ft, with snow-capped peaks rising to 7,000 m / 23,000 ft.

CLIMATE: The Himalayas effectively block the monsoon rains originating in the Indian Ocean, leaving most of the Chang Tang with very low rainfall. Aridity increases from southeast to northwest and from south to north, from 250 mm / 10 in in the southeast to as little as 60 mm / 2.25 in in the northwest. Rain falls mainly in the summer months. The plateau is frost-free for only three or four months of the year. Temperatures range from as low as –44 °C / –47 °F to 22 °C / 71.6 °F, with a very wide diurnal range, from sub-zero in the early morning to 21 °C / 70 °F at midday. The rapid rise in temperature has the effect of creating gale force winds which rage across the plateau.

The high Tibetan plateauland forms a succession of habitat types ranging from gravel and wind-blown soil, and desert and subdesert steppe, to comparatively rich alpine pastures in the high valleys. Low *Salix* scrub occurs along some of the rivers. Even the steppes sustain only sparse vegetation: tufts of feather grass, milk vetch, and bushy plants, with the better areas carrying couch grass, wormwood and cinquefoil.

While much of the Tibetan plateau is waterless, there are nevertheless at least 30,000 km² / 11,500 sq mls of mainly brackish lakes, fed by intermittent streams and snowmelt. Some are surrounded by bare salt flats, shingles and sand beaches. Most of the lakes were at one time larger. Their diminishing size is attributable to a lowering of rainfall resulting from the heightening of the Himalayas: the more the Himalayas rise, the more they block the moist trade winds from the Indian Ocean. As the climate of the Tibetan Plateau becomes more arid, the depth of

Lake Koko Nur (Qinghai Hu) (4,583 km² / 1,770 sq mls), for example, falls; in recent years it has dropped from 31 m / 102 ft to about 27 m / 88.5 ft. From mid-December to early April, Koko Nur is frozen to a depth of 40–60 cm / 15–24 in, but a group of warm springs at its west-

ern end maintains areas of open water throughout the winter. The *cirque* lakes of the Tibetan Plateau provide valuable staging posts for wildfowl undertaking the autumn broad-front migration over some of the most forbidding terrain imaginable. Among the large numbers of

## MISSIONARY NATURALIST

Several of Sichuan's endemic species were made known to the western world by the eminent French naturalist Père Armand David, the first western naturalist to visit Sichuan. In the course of a year, 1865, he made a remarkable collection of mammals new to science. Among his discoveries were not only the Sichuan golden snub-nosed monkey and the giant panda, but also such insectivores as the mole shrew, which lives in montane forests in Sichuan, Yunnan and northern Burma at a height of 1,200–4,500 m / 4–15,000 ft; the Sichuan burrowing shrew, which inhabits Sichuan, northwestern Burma and northern Vietnam at elevations of 1,500–3,000 m / 5–10,000 ft; and the Tibetan water shrew, which frequents fast-flowing mountain streams in the eastern Himalayas, Tibet and western China. None of these insectivores is especially uncommon, but their natural history remains virtually unknown.

## SNOW MONKEY

The mountains of the Chinese provinces of western Sichuan and southern Gansu are the home of the Sichuan golden snub-nosed monkey. Two distinct populations are known: to the north and east respectively of the Song-koi (Red River) Basin.

An inhabitant of mixed deciduous and evergreen forest between 1,500 and 3,400 m/5–11,000 ft, where it lives among dense rhododendron thickets, the golden snub-nosed monkey is mainly arboreal, rarely coming to the ground except to drink. Its habitat is under snow for half the year – indeed, one of its vernacular names is the 'snow-monkey' – yet, despite the snow, it seldom leaves the mountains, and then only when exceptionally heavy winter snowfalls force it to descend to the valleys and foothills. In recent years movement has, in fact, tended to be in the opposite direction: human pressures have compelled this monkey to retreat to the crests of the high mountains.

Because of its alleged power of preventing rheumatism, the pelt of the golden snub-nosed monkey was considered the most valuable obtainable in Sichuan. Under the Manchu emperors cloaks made from this monkey's fur were reserved exclusively for mandarin dignitaries. High price and strong demand caused the animal to be relentlessly hunted, to the extent that at the time of its discovery by the great French naturalist Père Armand David, in 1869, it was already scarce. The species is now protected under Chinese law, although skins are sometimes still offered for sale in Beijing. Current efforts to conserve the giant panda in the Wolong Nature Reserve are fortuitously benefiting the golden snub-nosed monkey also.

All four species of snub-nosed monkeys constituting the genus *Rhinopithecus* are either rare or endangered: the Sichuan golden snub-nosed monkey numbers fewer than 15,000; the Tonkin snub-nosed monkey, endemic to northern Vietnam, has been reduced to the low hundreds; the grey or Guizhou snub-nosed monkey, occurring only in the mountains of northeastern Guizhou, mainly on Mount Fanjingshan, totals no more than 500–670; and the black or Yunnan snub-nosed monkey, occupying a narrow mountainous strip between the Jinshajiang (upper Yangtze) on the east and the Lancangjiang (Mekong) on the west, and extending through northwestern Yunnan into southeastern Tibet, is estimated at 600–800.

---

migratory waterfowl frequenting the Tibetan lakes are such species as whooper swan, lesser sandplover, wood sandpiper, and Temminck's stint. Breeding species include the black-necked crane, bar-headed goose, shoveler, ruddy shelduck, great crested grebe, goosander, red-crested pochard, redshank, ferruginous duck, ibisbill and Pallas's fish eagle. For all but two months of the year, when they are frozen over, these lakes also serve to reduce the period during which a significant proportion of the wildfowl wintering in India and Bangladesh are under heavy pressure from hunting.

Some of the lakes support an abundant fish fauna, principally of the family Cyprinidia, which hibernate not only in winter but to some extent in summer as well, avoiding intensive solar radiation by day and low temperatures by night.

Except for nomadic pastoralists who visit the region to hunt, fish and graze their sheep, yaks and horses, most of the Tibetan Plateau is uninhabited. Such open country makes perfect wildlife habitat, allowing an unobstructed view over a wide area, allowing plains game to spot their principal enemies – man and wolf – at a distance. Among the animals that have become acclimatized to the rigorous environmental conditions of the Tibetan Plateau are the Tibetan gazelle or goa, Mongolian gazelle, Tibetan antelope or chiru, distinguished by long horns, large saiga-like nose, and thick protective coat, kiang or Tibetan wild ass, and wild yak, which survives only in the northeastern Chang Tang. The greatest threat to wildlife arises from the construction of roads which open up hitherto inaccessible areas to development and exploitation.

Rigorous climatic and environmental conditions and political constraints have effectively discouraged all but a few naturalists from visiting the region, with the result that its fauna remains relatively little known. The long-term prospects

## WILD YAK

Largest of all wild cattle, adult male yaks stand up to about 1.80 m/6 ft at the shoulder hump, and weigh up to 1,000 kg/2,200 lb. Females are only about one-third as large. Long shaggy coats (which reach almost to the ground) and dense underfur enable the yak to endure the bitter cold of the Tibetan Plateau.

The yak's historical range extended from the Karakoram Mountains in northeastern Ladakh through Kashmir to the Nan Shan Range in China's Gansu Province. Today, the species is limited to the more remote parts of the Tibetan Plateau, principally the immense treeless plateauland of the Chang Tang, or northern plain, where it lives at heights of up to 6,000 m/20,000 ft.

Local hunters at one time made their livelihood almost entirely from hunting yak, thereby reducing it from abundance to scarcity, and leaving only a scattering of survivors. Yak hide makes exceptionally durable leather and has long been in considerable demand for saddles, bridles, boots and many other items; the heart and blood are used for medicinal purposes, the long hair is spun into ropes, while the tail is hung up as a religious offering or used as a ceremonial fly whisk. The recent establishment of the Chang Tang Wildlife Reserve in northwestern Tibet has succeeded in securing a large part of the wild yak's remaining range.

● The takin, a curious 'ox-sheep', whose nearest relative is the musk-ox, is found along the eastern fringes of the Tibetan Plateau. Although there may be only a few hundred takins in existence, they are thought to be relatively secure, as their habitat of dense bamboo and rhododendron thickets is virtually impenetrable to humans. (WWF / G. Schaller)

● Tibetan wild ass or kiang at 4300 m / 14,000 ft in the basin of Lake Aqik in the Arjin Shan Nature Reserve. (WWF / Ron Petocz)

● The golden snub-nosed monkey, one of the many species first made known to the West by the missionary-naturalist Père Armand David, who visited Sichuan in 1865. (WWF / John MacKinnon)

for wildlife are nevertheless probably better on the Tibetan Plateau than in almost any other part of Central Asia. The establishment of the Chang Tang Wildlife Reserve (240,000 km² / 92,400 sq mls), covering a large part of the Northern Plain, and the second largest wildlife sanctuary in the world, is an earnest of the authorities' intentions. Checkposts have been established to control illegal hunting, notably of wild yak for meat and Tibetan antelope for wool. The size of the reserve is effectively increased by the Arjin Shan Reserve (45,000 km² / 17,300 sq mls) in the Xinjiang Autonomous Region, which adjoins it in the northeast.

At about 92° E the Chang Tang merges into the mountains of western China. It is here that the east–west Himalayas meet the north–south mountain ranges of Sichuan and Yunnan. Because they are geologically older than the Himalayas, China's western highlands retain among their fauna a number of unique or primitive forms, some of which have gradually colonized the Eastern Himalaya; representatives of the Himalayan fauna have similarly infiltrated into western China. And because the confluence of these two great mountain systems marks the meeting point of the Palaearctic and Oriental zoogeographical regions, there has been an intermingling of temperate and tropical species. These factors account for the remarkable faunal diversity which makes the border area between China and Tibet one of the richest zoological regions in the world.

## CHANG TANG

Some 240,000 km²/92,400 sq mls of the Chang Tang, or northern plain, forming the northwestern part of the vast elevated Tibetan tableland has been set aside as the Chang Tang Reserve. It is the largest wildlife sanctuary outside the Arctic or Antarctica. The reserve protects a number of species that are unique to the Tibetan Plateau. It covers a large part of the remaining range of the wild yak, once widespread but reduced by remorseless hunting. Other mammals of particular interest include the Tibetan antelope or chiru, Tibetan gazelle, Tibetan wild ass or kiang, Tibetan brown bear, snow leopard, Tibetan argali (or wild sheep), and bharal or blue sheep. Of the birds, 67 breeding species, mostly of Palaearctic origin, have been recorded, including six that are endemic: black-necked crane, giant babax, Kozlov's babax, Kozlov's bunting, Roborovsky's rose finch, and Taczanowski's snow finch.

The reserve still retains substantial herds of wild animals and is sufficiently large to contain the seasonal movements of migratory species such as the Tibetan antelope and wild ass. Establishment of the reserve will help to control illegal hunting, in particular the commercialized slaughter of the wild yak for its meat and the Tibetan antelope for its wool. Consideration is being given to extending the reserve westward to incorporate the Memar Lake region, containing at least 1,000 wild yak as well as being a traditional calving ground of the Tibetan antelope. In addition, the Arjin Shan Reserve (45,000 km²/17,375 sq mls), bordering the Chang Tang Reserve in the northeast, is to be extended to the west. These proposals will have the effect of consolidating an immense tract of about 390,000 km²/150,000 sq mls of land into a single protected block.

# THE ALPS

LOCATION: Western Europe's principal mountain system, the Alps form a barrier between southern Germany and northern Italy, separating the Po Valley and northern Adriatic from the Baltic and the North Sea.

AREA: Extending for 800 km/500 mls from the French Mediterranean coast of Provence across Switzerland, France, Italy, Germany and Austria to the outskirts of Vienna and the Danube, the Alps cover an area of about 165,000 km²/64,000 sq mls.

ALTITUDE: The central Alps form a complex system of mountain ranges cut by deep glacial valleys, the most clearly defined being that of the Rhône which runs between snow-capped peaks 4,000 m /13,000 ft high. Attaining their maximum height in the west, where they culminate in Mont Blanc (4,807 m/15,772 ft), the mountains fall away to 2,500 m/8,000 ft in the east.

CLIMATE: Lying between temperate northern Europe and the subtropical Mediterranean, climatic conditions vary considerably. The southern Alps come under Mediterranean influence, with high temperatures, winter precipitation, and dry summers, in contrast with the northern Alps, which have a central European climate – low temperatures, precipitation throughout the year, and no dry season.

Featuring prominently in the alpine ecology are glaciers such as the Mer de Glace below Mont Blanc, the Pasterzee below the Grossglockner, the Eiger and Jungfrau in the Bernese Oberland. The Aar and Aletsch are the longest glaciers in Switzerland. Glaciers and snowmelt help sustain some of Europe's largest rivers, including the Rhine, Danube, Rhône and Po. Spectacular mountain scenery is enhanced by magnificent lakes – Geneva, Como, Maggiore, Lucerne, Thun and many others.

The alpine forests include both deciduous and coniferous trees - oak, beech, and pine among them. As the southern slopes of the valleys receive more sun and warmth than the northern, vines are grown on the south face and deciduous fruit, woodland and grass on the north. The lower slopes form a patchwork of vineyards, orchards, forests and fields, merging above the tree line into the broad alpine meadows and pasturelands traditionally used for spring and summer grazing.

Included in the alpine flora are a number of relict species of both plants and animals that survived the last Ice Age in ice-free pockets. Plants that have disappeared from other parts of Europe in the course of intensive agricultural and industrial development also have their last refuge in the Alps.

Characteristic alpine birds include the alpine chough, snow finch, alpine accentor, black woodpecker, ptarmigan, black grouse and capercaillie.

During the last half century the status

of many species has, generally speaking, improved: chamoix, red deer and roe deer have substantially increased in number; ibex, reduced almost to extinction by excessive hunting, remains uncommon; but marmot, at one time heavily hunted for its alleged medicinal properties, has staged a comeback; while beaver and lynx have been successfully reintroduced.

Today, the danger to wildlife comes less from hunting than from dramatic alteration of the environment, not least from changes induced by acid rain originating from industrial pollution, the northern part of the Alps being more severely affected than the south.

With the alpine passes offering the only through routes, the Alps were historically of immense strategic and economic importance. Modernization of the transport system, beginning in the middle of the 19th century, coincided with industrialization based on the development of hydroelectric power. The introduction of the region's first railway revolutionized the development of the Alps. Railways offered the only means of transportation through the Alps until the mid-1960s when high-speed motorways were built to meet the demands of transit traffic and of tourism.

Industrialization is limited by the availability of the power required to activate it. Every river in the Alps without exception has accordingly been developed for generation of hydroelectric power or flooded to facilitate storage of water. Indeed, the entire Alps is in effect a huge reservoir: heavy precipitation stored in glaciers and snow throughout the winter is gradually released in summer, a factor of particular importance on the drier Mediterranean side of the Alps. Italy and France rely on alpine water both for intensive irrigation in the Po and Rhône valleys and for urban water supplies.

The exploitation of alpine water resources took another leap in the 1980s when heavy winter demand for electricity was met by using atomic power to pump water into high-altitude reservoirs. Nowadays, alpine power stations are usually housed deep inside mountains for security purposes. National security is also served by carving aircraft hangars out of the mountain side, usually alongside motorways which, in the event of war, can immediately function as operational runways.

## SWISS NATIONAL PARK

Located in the Canton of Grison in eastern Switzerland, where it adjoins Italy's Stelvio National Park (1,482 km²/572 sq mls), the Swiss National Park lies at a height of 1,500–3,174 m/5,000–10,414 ft and covers an area of 168 km²/65 sq mls. The park stands in a magnificent alpine setting amidst splendid peaks, among them Piz Quattervals (3,165 m/10,384 ft), separated by deep, narrow valleys. The Engadine Valley, on the park's northern boundary, drains into the Danube.

Broadly speaking, about one-third of the park is forest, one-third grassland, and one-third bare rock and scree. Dwarf mountain pine is the dominant tree species. At the lower elevations are Scots pine, arolla pine, larch and Norway spruce. Broadleaved trees are scattered throughout the park, principally white birch, green alder, aspen and rowan. Above the tree line – at about 2,350 m/7,710 ft – are many herbaceous plants, including edelweiss and anemone – more than 640 species of plants have been recorded.

Except for the wolf, brown bear and lynx, which have been exterminated, the park retains almost all its indigenous fauna. As well as a summer population of red deer, the fauna includes resident chamois and ibex. Apart from a small enclave at Il Fuorn owned by the Hotel Il Guron, the park is uninhabited. But it is crossed by a busy main road which is responsible for numerous road kills.

Of about 100 species of birds, 60 nest in the park, among them capercaillie, ptarmigan, black woodpecker, nutcracker and golden eagle. The smaller birds include wall creeper, ring ouzel, alpine accentor, snow finch and alpine chough. The park is a model of good management. Decisions are based on careful research; visitors are restricted to marked paths, roadsides barricaded to prevent parking of cars and every car park provided with explicit information boards, while the administrative centre – containing a lecture theatre, library and research facilities – lies outside the park, at Zernez.

## AN ENDANGERED INSECTIVORE

The semi-aquatic Pyrenean desman is a member of the mole family that has become adapted to swimming rather than burrowing. A rudder-like tail and long, webbed, hind feet propel the desman rapidly through the water; dense waterproof fur keeps it dry; and a long, tubular, prehensile snout serves as a highly mobile organ of touch and smell. Valves at the tip of its proboscis close the nostrils when swimming, or it may swim with only the tip of its snout protruding above the surface like a miniature snorkel.

The Pyrenean desman lives only in cold, swift-flowing, well-oxygenated streams in the Pyrenees and mountains in the northwest of the Iberian Peninsula. The nature of its habitat and its secretive nocturnal way of life have inhibited study of this highly specialized animal.

But enough is known to understand that its survival is endangered by the construction of dams for water storage and hydroelectric power and by river pollution. As well as being hunted for its soft pelt, the desman is killed by fishermen who allege that it eats fish. In fact, it feeds mainly on aquatic larvae, worms and small crustaceans. But the most immediate threat arises from the escape of American mink from a Spanish fur farm and their establishment in a region of prime desman habitat.

Degradation of habitat, fragmentation of its range, predation or competition from feral mink, and the apparent indifference of the Spanish scientific community combine to seriously threaten the survival of one of Europe's most fascinating montane animals.

A recent phenomenon is mass tourism – which in Europe is closely associated with the Alps. Starting in the 1950s as a by-product of post-war affluence, tourism rose rapidly to a position of enormous economic importance, the equable climate and breathtakingly beautiful scenery drawing tourists in winter and summer alike. So closely indeed are the Alps related to tourism that their character has inevitably been affected. Traditional farming systems are being abandoned, and there is a drift from the mountains to the towns, a phenomenon that is applicable not only to the Alps but to other mountain regions – the Pyrenees, for example, or southern Portugal where the uplands are littered with deserted villages.

This is partly attributable to the younger generation's disenchantment with the tedium of the traditional mode of life, and partly to farmers selling out to take advantage of the opportunities offered by tourism. The proliferation of weekend and holiday homes has resulted in hitherto unproductive land becoming valuable. Skiing alone brings 50 million people a year to the Alps; cable-cars and ski-lifts now dominate the alpine scene; chalets, hotels, restaurants and other amenities have sprung up everywhere. Modern technology has made it possible for whole regions to be modified on a gargantuan scale; four-lane motorways cut swathes through the mountains; lengthy tunnels – such as the Arlberg tunnel linking Switzerland with Austria – marvels of engineering ingenuity, burrow through mountains, while elevated causeways carry dense traffic high above valleys where only a few years ago transport was limited to the backs of pack animals.

Mass tourism, founded on low-cost transport, is now so firmly rooted that it ranks among the most lucrative industries in Europe. The Alps are of such economic importance that there can be no question of mothballing them to preserve their natural beauty. In the interests of the economy it is essential that their productivity be sustained, and this requires an adequate proportion of the wealth from tourism being returned to the mountains that generated it.

National parks have been established in each of the alpine countries as, for example, Italy's Abruzzo National Park (439 km$^2$ / 170 sq mls) and Gran Paradiso National Park (700 km$^2$ / 270 sq mls); France's Vanois National Park (528 km$^2$ / 204 sq mls); and the Swiss National Park (168 km$^2$ / 65 sq mls). An Alpine Convention, currently being negotiated by the environment ministers of Europe's seven alpine nations, is an attempt to set standards and agree guidelines relating to the management of the alpine environment in ways that are acceptable to all the countries concerned. This most important initiative could become a model for montane regions elsewhere in the world.

● Ibexes in one of their principal Alpine refuges, the Gran Paradiso National Park – formerly a hunting preserve for the kings of Italy. (WWF / G. Marcoaldi / Panda Photo)

● A male capercaillie displaying. Rarely seen, this large relative of the grouse is a denizen of dense coniferous or mixed woodland from Scandinavia to the Pyrenees, including various locations in the Alps. (WWF / Eric Dragesco)

# THE EQUATORIAL MOUNTAINS OF AFRICA

LOCATION: Straddling the Equator and flanked by the Great Rift Valley, three great mountain massifs rise from the African plain: Kilimanjaro, Mount Kenya and the Ruwenzori Range. These are the only mountains on the African continent to produce glaciers and be capped with permanent ice and snow: others, such as the Atlas Mountains and Mount Cameroon, have recurrent but transient falls of snow. Besides these great peaks are such lesser mountains as Elgon, the Aberdares, Meru (distinguished by a sheer drop of almost 1,000 m / 3,000 ft into a heath-filled crater), the Virunga Volcanoes and many more.

AREA: Mountains and highlands make up about 10% of the African continent's surface area – equivalent to approximately 3 million km² / 1.16 million sq mls – but are very unevenly distributed. The highland regions nevertheless contain three-quarters of all the continent's most productive land and are inhabited by about 60% of the population.

ALTITUDE: At 5,595 m / 19,340 ft, Kilimanjaro is the highest mountain in Africa; Mount Kenya reaches 5,199 m / 17,058 ft; and the Ruwenzori Range has its crest studded with peaks rising to Mount Ngaliema (Stanley), 5,117 m / 16,794 ft.

CLIMATE: Conditions vary according to altitude and aspect. Annual rainfall on the wetter southern slopes of Kilimanjaro at 800 m / 2,625 ft is about 800 mm / 31 in, increasing to 1,500 mm / 59 in at 2,500 m / 8,200 ft. Above that height, rainfall decreases with altitude, the summit receiving less than 400 mm / 15 in. Temperatures alternate between burning sunshine by day and piercing cold at night, varying between midday and midnight by as much as 15–21 °C / 60–70 °F – equivalent to going from summer to winter every day of the year. Having quickly dispelled the night's frost, the early morning sunshine soon gives way to mist which envelopes the summit. Hail or snow generally follows in the afternoon but usually clears again in the evening, though Ruwenzori remains hidden for weeks, even months, at a time.

● A scarlet-tufted malachite sunbird feeding on a giant lobelia on Mount Kenya. (Planet Earth Pictures / Keith Scholey)

● Lake Kitandara in the Ruwenzori Mountains. In tropical Africa, the alpine zone (above 4,000 m / 14,000 ft) is dominated by giant forms of normally insubstantial plants, such as lobelia and groundsel. (WWF / Mark Halle)

Although today the mountains of East Africa are isolated from one another, they have been separated only since the end of the last Ice Age, about 10,000 years ago. At that time alpine vegetation descended to lower elevations and spread over a much wider area, linking the mountains together. The retreat of the ice severed these links, thus segregating the mountains and enabling many of the plants to evolve into distinctive endemic forms. The retreating ice was replaced by an immense block of equatorial rainforest spreading across the continent from west to east. With the passage of time the forest contracted in its turn, leaving communities of plants and animals cut off from each other.

Ascending either Kilimanjaro, Kenya or Ruwenzori, the grass-covered foothills between 1,200–1,500 m/4,000–5,000 ft give way to high forest dominated by lofty cedar, podocarp and camphor trees, with emergents up to 50 m/160 ft tall forming a dense canopy. At about 2,400 m/8,000 ft the rainforest merges into a well-defined belt of mountain bamboo growing to a height of 12–15 m/40–50 ft in stands so dense as to exclude the sunlight. At about 3,000 m/10,000 ft the bamboo gives way to sub-alpine moorland typified by sedges and coarse tussock grasses and studded with tree heaths forming open forest 9–12 m/30–40 ft high, their branches festooned with mosses, liverworts and ferns, and trailing long ribbons of *Usnea* lichens. Higher still, and reaching to 4,000 m/14,000 ft, the moorlands merge into the alpine zone, a world of lakes, tarns and frozen waterfalls, dominated by some of the most extraordinary vegetation in the world: plants that elsewhere are no more than herbs of no particular significance are here metamorphosed into gigantic, often bizarre, arborescent growths: giant groundsel 9 m/30 ft tall, with thick, cork-like bark and a rosette of fleshy leaves at the end of each branch; giant lobelia, its spiky flower emerging from a rosette of elongated leaves, standing taller than a man, and growing amidst short alpine grasses, mosses and an abundance of everlasting flowers.

Africa's montane forests and grasslands carry a rich fauna: the high forest is inhabited by such animals as elephant, black rhino, bushpig and various forest duikers. Primates include the leaf-eating black-and-white colobus and bushbaby, one of the few nocturnal primates. Tree hyrax are more often heard than seen. Forest glades, carpeted with grass and moss and watered by ice-cold streams, are the haunt of buffalo, bushbuck and, in Kenya, giant forest hog and bongo. Leopard are abundant. Other than elephant, which feeds on seedling bamboo, and mole rat, few animals live in the bamboo. The giant heath zone marks the normal upper limit for most large animals except for duiker and steinbok; but, higher still, the open moorland abounds with small animals, mainly rodents and hyrax, which at night emerge from their rockpiles to browse on lobelia leaves. Elephant and buffalo sometimes cross the moorland to reach other feeding grounds, but seldom stay long.

Birds, too, have become adapted to the higher reaches of the mountains: the forest rings with the harsh call of the red-headed parrot, while Hartlaub's turaco reveals its presence by a sudden flash of colour. The crowned hawk eagle preys on forest monkeys, and Verreaux's eagle takes advantage of the upcurrents to soar effortlessly among the peaks intent on swooping on unwary hyrax. The scarlet-tufted malachite sunbird lives in association with giant lobelia and is believed to play a part in its pollination.

In temperate countries, mountains are generally less attractive for settlement than lowlands. The reverse holds good for Africa. Whereas pastoralists keep to the more arid open grasslands, agriculturalists gravitate to the highlands where the combination of fertile soil, equable climate and relatively high rainfall is conducive to growing crops. Only 36.5% of Kenya's total land area has more than

## THE ROLE OF THE LOWER PLANTS

The mountains of East Africa are remarkable for their floristic richness. The Afro-montane region as a whole sustains about 4,000 plant species, of which 80% are endemic. Mount Kilimanjaro's flora alone includes over 1,800 species of angiosperms (flowering plants) and 700 of lower plants (mosses, liverworts and lichens) – an unusually high level of diversity. The lower plants (bryophytes) have an important role: at the higher elevations they intercept a substantial amount of rainfall, releasing the water slowly and reducing loss of water through evaporation, particularly from the surface of the soil. As well as preserving soil moisture and keeping the soil in place, thereby preventing erosion, this has the effect of continually replenishing watercourses, and of maintaining dry season flow.

## A NATURAL POPULATION CONTROL

Until the 1960s, Kenya's South West Mau Forest Reserve (430 km²/166 sq mls) contained the heaviest concentration of bongo in East Africa. Apart from leopard, this tract of mainly bamboo forest contained few resident predators. As a consequence the bongo were subjected to little predation, allowing numbers to build up rapidly. Control was instead maintained by a plant, the *setyot* vine, which in this particular forest grows profusely in association with bamboo and forms an important part of the bongo's diet. The flowering of the *setyot* occurs only once every seven years when it signals the occasion for the Kipsigis and other tribes to hold their circumcision ceremonies. After flowering the vine dies back. The second year after die-back the *setyot* becomes highly toxic, causing scouring and bloat resulting in heavy mortality among both bongo and giant forest hog. Laboratory analysis confirms that the *setyot* is indeed toxic, an extract of 1 gm being sufficient to kill a mouse after causing it chronic diarrhoea. The population crash is followed by steady recovery which continues for a further seven years, until the next flowering of the *setyot* causes the cycle to be repeated.

## MOUNTAIN NYALA

Endemic to the eastern highlands of Ethiopia, the mountain nyala occurs in several separate groups living isolated from each other between the Chercher (Harerghe) and Balé mountains. The establishment of the Balé Mountains National Park (2,200 km²/850 sq mls) has done much to protect the mountain nyala from the excessive hunting and destruction of habitat that had reduced the overall population to an estimated 2–4,000. The effectiveness of protective measures was for a time reflected in a steady numerical increase. But, on the overthrow of the Ethiopian Government in 1991, the position deteriorated rapidly, with poaching recurring on a large scale. The current position is far from clear but, when order is again restored, the species would unquestionably benefit from spreading the risk by establishing a second protected area, possibly in the Arssi or Harerghe mountains.

## AFRICAN MOUNTAINS AT RISK

**Mount Nimba** (1,752 m/5,748 ft) at the confluence of the Guinea/Liberia/ Côte d'Ivoire borders. The limited protection currently given this biologically rich area is inadequate.

**The Cameroon Highlands** (which extend into northeastern Nigeria and the offshore island of Bioko) has 53 mountain bird species, of which 22 are endemic, as well as 55 endemic species of amphibians and 10 reptiles. Places of special importance include mounts Oku, Cameroon and Kupe in Cameroon; Mount Malabo on Bioko; and the Obudu Plateau in Nigeria. All are in urgent need of improved protection.

**Ethiopian Highlands** The largest area in Africa above 2,000 m/6,500 ft, with 20 mammal, 30 bird and 13 amphibian species endemic to the region. Balé (2,471 km²/954 sq mls) and Simien (179 km²/69 sq mls) are the only montane areas protected. Continued political unrest leaves the present position uncertain. Southwestern Ethiopia's medium altitude forest (1,200-1,800 m/4,000-6,000 ft), though relatively little known, is believed to be of great biological interest, but is threatened by an influx of Ethiopian refugees fleeing from war and famine.

**Kivu–Ruwenzori Highlands** The Ruwenzori Range and adjoining highlands of the Albertine Rift have a particularly rich fauna, including at least 37 endemic bird species (32 of them from the Itombwe Mountains) and 14 species of butterfly. Despite its exceptional importance, this area is at present unprotected except for a small part in Virunga National Park.

**Imatong–Usambara** The Imatong Hills in southern Sudan, Mount Elgon on the Kenya/Uganda border, and the Usambara Mountains in northeastern Tanzania are all under threat from logging, charcoal burning, commercial forestry plantations and tea growing. None is adequately protected.

**Uluguru–Mulanje** The Uluguru and Uzungwa mountains in eastern Tanzania and Mount Mulanje in northern Malawi all harbour interesting endemic species of birds, reptiles and amphibians. Protection is everywhere inadequate.

**Angolan Highlands** Western Angola includes a region of rainforest with marked Afro-montane affinities, part of which is thought to be a southern extension of the Central African rainforest, and to have a distinctive endemic bird fauna. This area is completely unprotected.

**Drakensberg Mountains** An important area for plants: of a total of about 1,800 plant species, 300 are endemic, while four bird species are endemic to the open montane grassland. A significant portion of the range – about 2,000 km²/775 sq mls – is protected in national parks and forest reserves.

---

500 mm/20 in rain a year, the minimum level of rainfall for cultivation.

Under the traditional slash-and-burn method the land was used for a few seasons before its fertility diminished, when it was abandoned to revert back to forest. But demographic pressure is resulting in land remaining under permanent cultivation, and in clearance spreading higher into the mountains. The only alternatives available to the peasant farmer are to cultivate marginal – that is, low rainfall – land where low yields are a certainty and crop failure a distinct possibility, or to abandon the land and drift to the cities to join the thousands of landless people living in the shanty towns that have mushroomed throughout the Third World.

Between 1948 and 1988 the number of people living on Kilimanjaro tripled. The conversion of forest to farmland has resulted in the forest now being confined to elevations between 1,820-3,050 m/6,000-10,000 ft. Forest on Mount Kenya once began at about 1,300 m/4,500 ft but now starts at 1,800 m/6,000 ft. As a general rule unspoiled forest persists only in forest reserves and other specially protected areas, and even

these are under threat. Such forest as remains is being steadily degraded by felling, burning, exposure to livestock grazing and conversion to softwood plantations. The combination of increasing isolation, changes to the forest and loss of critical habitat is adversely affecting wildlife.

Despite covering only 2% of the land surface of Tanzania and no more than 3% of Kenya, montane rainforests are nonetheless a primary source of water, and thus have an importance out of all proportion to their size. Their degradation affects places far beyond their own boundaries. The cumulative effect of cultivating steep slopes and other poor farming practices around Mount Kenya, for example, can be seen from the quan-

tity of silt choking the mouth of the Sabaki River near Malindi. It is in the country's vital interests that montane catchment forests should be effectively conserved. This could best be expressed through an integrated management and conservation programme with emphasis on sustainable development.

A number of African mountains (mainly above the treeline) have been designated national parks, among them Kilimanjaro, Mount Kenya and parts of the Ruwenzori Range. But for that, more of the African alpine zone would have gone the way of Ethiopia's ruined highlands. Many other Afro-montane areas of great hydrological and biological importance do not have even nominal protection.

# THE SIMIEN MOUNTAINS

Located in the Begemder Province of Ethiopia, about 120 km/75 mls north-east of Gondar, at an altitude of 1,900–4,430 m/6,234–14,535 ft, and covering 179 km²/69 sq mls, Simien Mountain National Park fringes the northern edge of the Amhara Plateau which forms part of the Simien Massif. Cut by gorges up to 1,500 m/5,000 ft deep, the massif includes Ethiopia's highest peak, Ras Dashan (4,620 m/15,158 ft). From north to south the plateau is bisected by the Mayshasha River, while on the south and northeast it is bounded by the Tacazze River and its tributaries. The habitat is a blend of Afro-alpine woods, heath forest, montane savanna and moorland, including such species as tree heath, giant lobelia, and everlastings.

The 21 species of mammals recorded in the Park include both the gelada baboon and hamadryas baboon. Two endangered species endemic to Ethiopia and restricted to the Simien Mountains – the Walia ibex and Ethiopian wolf – are of particular concern.

The anarchic conditions existing in Ethiopia during 14 years of Mengistu's misrule, including more than a decade of war with Eritrea and Somalia, forced the Park to close. There are no reliable reports of its present condition. Proposals for its rehabilitation include a reduction in current levels of exploitation, the reclamation of land degraded by agriculture, livestock grazing and tree felling, and a programme of reafforestation. The plan envisages establishing seven separate zones, among them a prime protection zone of 8,700 ha/21,500 acres to protect walia ibex habitat and as a precaution against its possible hybridization with free-ranging domestic goats, and a buffer zone of 40,000 ha/100,000 acres covering the lower watersheds and intended for eventual integration into a proposed biosphere reserve.

● The Simien Mountains of Ethiopia, threatened home of the walia ibex. (WWF / J. Stephenson)

# KILIMANJARO

At 5,894 m / 19,340 ft, Kilimanjaro is the highest mountain in Africa, its height enhanced by the low-lying plain, only 1,018 m / 3,340 ft above sea level, from which it rises. It is also one of the world's largest volcanos, 3,885 km² / 1,500 sq mls in extent, although it has remained inactive since the Pleistocene epoch. Kibo, the highest point on the mountain, and Mawenzi (5,148 m / 16,890 ft) to the east are linked by the Saddle, a plateau of high-altitude tundra 36 km² / 14 sq mls in extent. Kibo has a permanent ice cap from which the Penck Glacier – Kilimanjaro's longest glacier – descends to about 4,500 m / 14,765 ft.

Except for mosses and lichens, few plants can tolerate the severe conditions above 4,600 m / 15,000 ft, although various species of everlasting flowers are found among alpine grasslands and upland moor where the vegetation is mainly heath and scrub. Giant groundsel (of which there are two distinct forms) occurs on the upper parts of the mountain; in alpine bogs it grows in association with giant lobelia. Below the treeline, the mountain's wetter southern slopes are dominated by podocarp and camphor-wood, with an understorey of ferns; but are replaced on the drier northern slopes by cedar and olives.

While a surprising number of Kilimanjaro's animals, including elephant and buffalo, have been recorded above the treeline, most prefer the forest. The fauna includes three primates: black-and-white colobus, blue monkey, and galago, as well as such species as mountain reedbuck and eastern tree hyrax. Both Abbott's duiker and Abbott's starling are endemic to Kilimanjaro and neighbouring mountains. Of the few birds, the most conspicuous are the raven and the scarlet-tufted malachite sunbird, the latter confined to alpine moorlands and the upper forest edge. The butterfly known as the Kilimanjaro swallowtail is restricted to Kilimanjaro, Mount Meru and Ngorongoro.

Kilimanjaro National Park (755 km² / 291 sq mls) is confined to the upper

● Sunrise on Kilimanjaro. Africa's highest mountain (5,595 m / 19,340 ft). Despite lying virtually on the equator, Kilimanjaro is capped with permanent ice and snow. (WWF / Fritz Polking)

parts of the mountain above the treeline, together with six corridors extending downwards through the belt of forest encircling the mountain between the 1,520 and 2,740 m / 5,000 and 9,000 ft levels. With more than 60,000 visitors a year, the park's popularity has led to problems of litter and waste disposal. Despite the Presidential Decree of 1984, illegal hunting, tree felling, and incursions by domestic livestock still take place, while honey-gathering and grass burning lead to frequent fires. Protection at present stops short of the forest. The park should be extended to incorporate all the montane forest, thereby recognizing Kilimanjaro's importance as a catchment and at the same time affording maximum protection to its indigenous flora and fauna.

# ANTARCTICA

*What freezings have I felt, what dark days seen!*
*What old December's bareness everywhere!*
Shakespeare, Sonnet XCVII

LOCATION: Lying at the hub of the southern hemisphere.

AREA: Twice the size of Australia, Antarctica covers 14 million km²/5.4 million sq mls, equivalent to about one-tenth of the world's land surface. The Antarctic continent is overlain by an ice cap, averaging some 2,500 m/8,000 ft thick, and containing about 90% of the world's ice, estimated at 30 million km³/7 million cu mls. The ice cap is only 10–12 million years old. Before that, Antarctica bore a resemblance to present-day southern Chile, New Zealand's South Island, and southwestern Tasmania, with southern beech *Nothofagus* the predominant tree. Earlier still – some 250 million years ago – Antarctica formed part of the great southern continent of Gondwanaland that included Australia, South America, Africa and India.

ALTITUDE: Antarctica consists of two distinct geological regions: Greater (or East) Antarctica, fringed by the Transantarctic Mountains, rising to about 4,500 m/15,000 ft and extending over a distance of 4,800 km/3,000 mls, and the smaller (and geologically younger) Lesser (or West) Antarctica. The latter, an elevated inland region, contains the Ellsworth Mountains – including the Vinson Massif (5,140 m/16,864 ft) Antarctica's highest peak – which continue northward into the mountainous Antarctic Peninsula. These ranges, protruding towards Cape Horn, are an extension of the Andes, to which they are linked by a submarine ridge, the peaks emerging from the sea as a series of islands, among them the South Orkneys, South Georgia and South Sandwich Islands.

CLIMATE: Temperatures in the Antarctic, even in summer, seldom rise above freezing. Mean temperatures on the elevated central plateau average –50° to –60°C/–58° to –76°F. The coastal regions are much warmer, –10° to –20°C/14° to –4°F. The lowest temperature ever recorded was –89.2°C/–129°F at the Russian South Pole Station in 1983. Gale force winds are a feature of the region – gusts of up to 200 kph/125 mph have been recorded. These polar winds exert an influence that is felt far beyond Antarctica.

The Antarctic is in many respects almost the exact reverse of the Arctic. The basic difference between them is that the Arctic is an oceanic basin lying beneath a thin layer of ice, whereas the Antarctic is a land continent buried under a permanent ice cap and encircled by sea. A phenomenon of particular importance is the flow of ice from Antarctica to the ocean. In some places – the west side of the Antarctic Peninsula, for instance – glaciers on reaching the sea terminate in ice cliffs which break away in the conventional manner to form small icebergs. But over about one-third of Antarctica's continental coastline numerous glacial outfalls come together to form an almost continuous ice shelf some 2–300 m/650–1,000 ft thick – a floating ice sheet extending for hundreds of kilometres that rises and falls with the tide, and covers an extensive area of ocean. The seaward edge of this ice sheet is continually breaking off as it sheds icebergs, some very large, which are carried on the ocean currents. It has been estimated that Antarctica spawns more than 1,450 km /348 cu mls of icebergs each year. Sea ice, formed by the freezing of the sea's surface and spreading outwards from Antarctica, may cover 3 million km²/1 million sq mls of the Southern Ocean in summer and as much as 20 million km²/ 7.7 million sq mls in winter – an area greater than the continent itself.

Around the fringes of the Arctic are substantial areas of land that are ice-free for part of the year and on which plants can grow during the brief summer. But the Antarctic continent is permanently frozen – for all practical purposes no vegetation can grow there. Only 2% of the surface is ice-free, and in that small area lichens are the most abundant plants.

Although terrestrial plants are almost non-existent, plants nevertheless form the broad base of the Antarctic food pyramid, for the seas around Antarctica contain a wealth of plant life that makes the Southern Ocean one of the most biologically productive regions on earth. The contrast between the barren land and the high productivity of the sea is very marked and fundamental to the ecology of the region. This is attributable to the Antarctic Convergence, a belt of water about 40 km/25 mls wide, where cold currents radiating from Antarctica meet subtropical waters. The upwelling of nutrient-rich water and the mixing of the different water masses creates changes not only in surface temperature but also in the ocean's chemical make-up. This is reflected in the marine and bird life which differ from one side of this natural boundary to the other. Many biologists regard the Antarctic Convergence as Antarctica's true boundary.

The Antarctic Convergence generates conditions that are ideal for the production of phytoplankton – minute aquatic plants (mainly unicellular algae) that form the broad base of the food chain and sustain Antarctica's web of life. Zooplankton, the most important of which is krill, feed on the phytoplankton. Fish, crabs and other marine creatures feed on the plankton, and are themselves preyed upon by larger fish and squids. The fish and squids are eaten by seals, toothed whales, penguins and many other birds. They in their turn are devoured by the larger predators – leopard seal and killer whale – at the top of the food chain.

Krill swarms on or just beneath the surface, and is extraordinarily abundant. It forms the staple food of many polar animals – seals, penguins and the great whales among them – which have evolved in ways that enable them to tap this prolific source of food. Seals, penguins and many kinds of oceanic birds spend the greater part of their lives at sea, coming ashore only during the austral spring and summer to moult and bear young. For a few months the shores and coastal waters of Antarctica are thronged with thousands of temporary residents, most of whom withdraw on the approach of winter.

Despite its image as pristine wilderness, Antarctica has in fact been subjected to more than two centuries of exploitation. Its original attraction lay in the lucrative trade in seal skins that were there in huge numbers for the taking. The fur seals were systematically hunted, to the extent that within a few years they had been reduced almost to extinction. The sealers then set about exploiting the elephant seal for oil, until it too became scarce. It was then the turn of the whales, the whalers initially concentrating on the biggest and most valuable species –

blue, fin and sei.

The demise of the whales is being followed by intensive exploitation of Antarctica's remaining marine food resources – fish, krill and squid – which it is claimed could exceed the world's total fish catch. Commercial krill fishing, currently accounting for about 350,000 tonnes of krill a year, is still at a comparatively low level. But should this figure increase significantly the impact on Antarctica's marine ecosystem could be far-reaching. Antarctica's marine ecosystem is very fragile, and excessive exploitation of krill, the key link in the food chain, could precipitate an ecological

disaster. The sustainable yield of krill is undoubtedly high, but little is known about it; any attempt at exploitation must therefore be carefully regulated or the entire Antarctic food chain could be affected.

Apart from marine resources, fears have been expressed about the possibility of exploiting minerals lying buried under Antarctica's ice cap. Iron and coal exist in moderate quantities and there are small deposits of tin, gold, silver, zinc, manganese and cobalt – but none is regarded as commercially viable. No mineral exploitation has yet taken place and it is to be hoped that it never will.

## FEAST OR FAMINE

Krill forms the staple food of numerous polar animals. But because plankton production is activated by solar energy, the processes giving rise to it cannot take place during the polar winter, a time of almost perpetual darkness, when it largely disappears. Animals dependent almost exclusively on krill for food thus have to adapt to a feast and famine regime.

The seasonal availability of krill governs the migrations of the great whales. They spend the summer gorging themselves on krill – a full-grown whale consuming up to 4 tonnes a day – in the process acquiring a thick layer of body fat, which insulates the animal against cold as well as serving as a reserve of food on which to draw in times of scarcity.

In autumn, as the stocks of krill begin to diminish, the pregnant female whales migrate to their breeding grounds in tropical waters, a journey of several thousand kilometres which takes them away from their feeding grounds for at least four months. Throughout that time, and despite feeding her calf large quantities of milk, the female herself does not eat.

Only four species of seal remain in Antarctica through the winter. To do so, they have had to adapt to living under some of the most rigorous conditions on earth. While the Ross seal, leopard seal and crabeater range far out among the pack ice, the Weddell seal remains close to the coast where it lives under the sea ice, hauling out to bask on the ice when

the weather is suitable. As the sea freezes, the seals cut breathing holes through the ice. Much of their time is spent gnawing away at the ice to keep these airholes open. Despite having no protective blubber, the Weddell pups are born at the coldest time of year. The females remain with them until weaning, at about seven weeks, during which time the females have nothing to eat. While the pups grow plump on their mothers' rich milk, to the extent that they can scarcely move, the females become as haggard as scarecrows.

The only penguin to hatch its eggs at the height of the Antarctic winter is the emperor penguin. On the approach of winter, the adults leave the sea and congregate in large rookeries, usually on sea ice in the shelter of an ice cliff, where they go through an elaborate courtship. A few weeks after pairing, the female lays a single large egg, which her mate incubates by cradling it on his feet and tucking it into a fold of abdominal skin under downy feathers. She then returns to the sea to resume feeding. For 63 days the male incubates the egg in temperatures well below zero. Such is his dedication that he continues to do so even when almost buried under snow, his only concession to comfort being to huddle together with other males for mutual warmth and support. The female returns when the egg is on the point of hatching. Only then is the male relieved of his responsibilities, after being two months without food.

A comparatively new phenomenon is the rise in Antarctic tourism. Cruise liners and special-purpose cruise ships bring 2–3,000 tourists to Antarctica each year. In addition, there were by 1990 forty permanent scientific stations in Antarctica and others on subantarctic islands. The servicing of this human presence necessitates the construction of buildings and airstrips, importation of equipment and supplies, increased sea and air traffic, noise disturbance by helicopters and vehicles, and waste disposal. So far tourism has been at a modest level

● Midnight sun over an iceberg, Antarctica.
(WWF / Jack Stein Grove)

● Adult sperm whales with a calf. In spring, adult males migrate from tropical waters to the Arctic and Antarctic, while the females and young remain in temperate waters.
(International Fund for Animal Welfare)

and conducted in a responsible manner, but any unregulated increase in tourist activity could be damaging. The scientists themselves are not always above criticism for neglecting their impact on environmentally sensitive areas, or for the inadequacy of their waste disposal systems.

Antarctica is the only undeveloped continent and the last unspoiled wilderness of any great size remaining on earth. Its scientific importance lies in its extensive ice cap which reflects incoming solar radiation, exerts a big influence on the world's weather systems and activates the principal ocean currents. These factors make Antarctica an immense natural laboratory, providing unparalleled opportunities for the study of such matters as ice conditions, atmospheric and

● Weddell Seals near Rothera Station, Antarctica. Although the continent itself supports virtually no life, the seas around it teem with fish and crustaceans, providing abundant food for seals, whales, penguins and sea birds. (WWF / Cassandra Phillips)

oceanic circulation, and marine biology. The global nature of these issues is exemplified by the ozone 'hole' caused by the dramatic thinning of the layer of ozone shielding the earth from the damaging effects of ultraviolet radiation.

Recognition of Antarctica's global importance and of the need for greater understanding of its role reached a turning point in 1957–1958, International Geophysical Year, when 12 nations established a network of permanent scientific stations. This was followed by a number of positive measures, notably the signing in 1961 of the Antarctic Treaty, designed to ensure that Antarctica, including the surrounding oceans to latitude 60° S, remains devoted to peaceful and disinterested scientific purposes. As a result, the continent is effectively demilitarized, and the testing of nuclear devices and dumping of nuclear wastes prohibited. A further international treaty, signed in 1987, limits the manufacture of chlorofluorocarbons (CFCs), the substance primarily responsible for activating the processes leading to depletion of the ozone layer, while the Environmental Protocol to the Antarctic Treaty, signed in October 1991, commits the 40 signatories to a ban on all mining and oil drilling 'for at least the next 50 years'.

The importance of Antarctica's role in monitoring and giving early warning of global change is now universally recognized. If this purpose is to be achieved, it is essential that Antarctica is retained intact by according protection that is truly comprehensive. The various agreements and conventions relating to Antarctica have gone far in that direction. The time has come for them to be incorporated into a single legal instrument binding on all nations, with mechanisms for ensuring compliance, as already proposed by IUCN.

## KRILL

Largest of the planktonic crustaceans, krill are shrimp-like creatures about 5 cm/2 in long, and are circumpolar in distribution, occurring in waters south of the Antarctic Convergence, which forms a natural barrier to more northerly movement. They spend the Antarctic summer close to the surface of the ocean, appearing in huge concentrations, sometimes covering hundreds of square kilometres. By day shoals of krill tint the surface of the ocean red; at night they become luminescent in the darkness.

The standing biomass of krill has been calculated at 600 million tonnes, representing about half the entire Antarctic plankton production, and forming the most important constituent of the Antarctic food chain. Krill are the principal source of food for several species of seals, besides penguins and other birds, squid and fish; they even satisfy the giant appetite of the whalebone (baleen) whales. But only in summer: in winter they descend to greater depth or vanish beneath the pack ice.

Declining whale stocks have resulted in some whaling vessels being converted to catch krill. The commercial krill fishery currently takes about 350,000 tonnes a year. The effect on the marine ecosystem is uncertain, but the likelihood is that any significant increase could adversely affect the food chain. Depletion of the ozone layer could, moreover, result in increased levels of ultraviolet radiation reaching the sea and penetrating deeper than previously beneath the surface. As photosynthesis is intolerant of ultra-violet radiation, it is feared that this could bring about a reduction in photoplankton (microscopic plant life), thus endangering the entire Antarctic food chain.

# RARE AND ENDANGERED SPECIES

*Nature does require
Her times of preservation which, perforce,
I her frail son, amongst my brethren mortal,
Must give my tendence to.*
Shakespeare, *Henry VIII*

Extinction is an integral part of the process of natural selection. Species have a finite life span and, since life first appeared on this planet, plants and animals have evolved into different forms. While some have vanished without issue or have not adapted quickly enough to changing environmental or climatic conditions, others have died out for such reasons as over-specialization, inability to compete with other more successful species for the necessities of life, or because of some natural cataclysm such as the advance or retreat of glaciation.

The fossil record shows that since the Cambrian period, which ended some 510 million years ago, there have been a number of mass extinctions, the best known being the demise of the dinosaurs 65 million years ago. Many theories have been put forward to account for these mass extinctions, but opinion remains divided both as to the reasons and the period of time over which they occurred. Our inability to measure periods of geological time shorter than about 1 million years means that it is impossible to know for sure whether these extinctions were a gradual process spread over thousands of years or whether they were more or less instantaneous.

Perhaps the greatest assemblage of wildlife the world has ever seen was during the Pleistocene epoch. The late Pleistocene, which ended about 10–15,000 years ago, saw the extinction of a high proportion of the large mammals of the Americas, Eurasia, Australia and, to a lesser extent, Africa. Among those that died out were animals such as the mammoth, woolly rhinoceros, sabre-toothed cat, beavers the size of bears, sheep as big as oxen, elephant-sized bison, and a gigantic ground sloth. It is sometimes asserted that man played a significant part in their decline; but responsibility is more likely to lie elsewhere. The world at that time was so thinly populated and weapons so primitive that it is difficult to believe people capable of the level of mass destruction necessary to bring about the extinction of continental faunas. Indeed, it would have been contrary to early man's own inclinations and interests, for he was essentially a subsistence hunter, killing only sufficient to satisfy his own immediate needs. The North American Indians, to take a recent example, had an almost symbiotic relationship with the bison; their entire culture and way of life revolved around it. While killing all they wanted, they left the basic stocks unharmed. Not until the arrival of Europeans with their sophisticated weaponry were the once-vast herds almost obliterated. Present-day subsistence hunters, such as the Australian Aborigines, recognize that their own well-being, indeed their very survival, is dependent on the wildlife with which they share their habitat. Overkill would not be to their advantage.

## Human Influence

But although early man may have been little more than a bystander in the great Pleistocene extinctions – a theme that can only be speculative – there is no disputing his prime responsibility for the extinctions that have occurred within historical times. Island faunas have been particularly severely affected. This is because island species are often restricted to single islands, and their numbers are correspondingly low. Having, moreover, evolved in isolation, they are ill-equipped to withstand competition and are highly vulnerable to predation, either by man himself or by such animals as rats, cats and pigs that have been either inadvertently or intentionally introduced. The demise of the dodo and solitaire are, of course, the classic examples.

The Polynesian colonization of the Hawaiian archipelago, in the 1st century AD, is believed to have been responsible for the extermination of about 50 endemic species of land birds – half the bird fauna of the islands. The same can be said of New Zealand. Between the arrival of Polynesian colonists in about AD 750 and Cook's first voyage in 1769, the entire order of moas (Dinornithiformes), comprising about 20 species of huge, flightless, ostrich-like birds standing up to 3 m / 10 ft high, was wiped out. Another order of large flightless birds, the Aepyornithiformes, the so-called elephant birds, was exterminated in Madagascar – as were at least 14 species of lemurs, most of them large ground-dwelling animals – since the island was first colonized by immigrants of Indonesian origin around 2,000 years ago.

The Age of Discovery sparked off another round of extinctions. Island faunas were again the first to be affected. More than 90% of the bird species that have become extinct since 1600 have been island forms. Nearer our own day, the continental faunas have become increasingly severely depleted, exemplified by the passenger pigeon of North America, numbered in millions until well into the 19th century, but exterminated by the second decade of the 20th. The species known or believed to have become extinct since 1600 can be summarized as follows:

| Molluscs | 191 | Amphibians | 2 |
| Crustaceans | 4 | Reptiles | 23 |
| Insects | 59 | Birds | 122 |
| Fish | 29 | Mammals | 60 |

The African faunas in particular, which stood comparison with the great Pleistocene assemblages, have within our own lifetime been reduced from abundance to relative scarcity, a trend that continues at an ever-quickening pace and is a direct consequence of the growth in human numbers crowding out every species but our own. From about 800 million in 1750, the world population has risen to 5.3 billion today, and is expected to increase by another billion (one thousand million) by the end of the present century. Demographic pressure has impelled people to spread over the face of the earth and to occupy land that was previously vacant, causing widespread environmental change and having far-reaching repercussions on the distribution and status of plants and animals.

Loss or degradation of habitat, a prime cause of extinction, has affected the oceans as badly as the land. The seas have become the world's rubbish tip and have even been used as a dumping ground for nuclear wastes. Overfishing has reduced once-abundant marine species to scarcity. Pollution reaching into the atmosphere is depleting the thin layer of ozone surrounding the planet that protects all life on earth from the harmful effects of solar radiation.

## Conserving Life

In general people are more interested in conserving wild animals than in conserving soil, water and vegetation. But the interrelationship between an animal and its habitat is of such fundamental importance that it is impossible to think in terms of conserving one without the other. The injunction to 'look after the habitat and the animals will look after themselves' may be trite but is nonetheless an elemental truth.

The best way to safeguard rare species of animals and plants is to ensure the conservation of the biotic community of which they are a part. Rarity is not in itself a cause for concern. Some species are inherently rare, often because they occupy a highly specialized ecological niche. But few species can survive outside their natural habitat, and if the habitat is threatened – or illegal hunting, introduced predators, or disease cannot be adequately controlled – there may in the final resort be no alternative but to capture the last survivors and bring them into captivity. While this last-ditch solution is tantamount to an admission of failure, it provides the opportunity to build up stocks until conditions are right for the species to be reintroduced back into its natural habitat, as happened with the Arabian oryx and Père David's deer.

Some species, certain primates in particular, are for psychological or other reasons incapable of breeding under captive conditions. Others – such as the blue whale – would obviously be impossible to keep in captivity. And even with those species that can adapt successfully to captivity a dilemma arises if the end result – the release back into the wild – proves impossible because suitable habitat no longer exists. What happens, for example, to the golden lion tamarin and the woolly spider monkey when the few hundred hectares of Brazil's Atlantic rainforest – all that remains of their once-extensive habitat – runs out? Or what can be done to safeguard the world's rhinoceroses if they cannot be adequately protected in the wild?

Scattered remnants of a species can sometimes be brought together and concentrated in one part of the natural range to create a viable breeding nucleus and provide more efficient protection, as was done with the bontebok and the mountain zebra in South Africa. An alternative is to translocate a species from its original habitat and establish it in an entirely new area, as happened in 1963 when a herd of 30 hirola or Hunter's antelope was captured on the Walu Plains on the north bank of the Tana River and airlifted to the Tsavo National Park.

Clearly, the aim should be to forestall such emergencies by setting aside suitable areas and according them effective protection. One of the purposes of national parks is, of course, to provide sanctuary for rare and endangered species; but the existing network of national parks is not sufficiently comprehensive to cover every eventuality. Responsibility for ensuring that the national parks system is adequate for the purpose rests squarely with the government of the country concerned. International pressure or influence may sometimes be necessary to encourage a government to meet its obligations. If a country cannot afford the cost, international funding may be necessary. There is growing awareness that schemes for the conservation of individual species are unlikely to succeed unless they can be integrated into broader land-use strategies involving complete ecosystems. It is also vitally important that conservation projects should enlist the support of the local people, without whose active collaboration – and, wherever possible, participation – little of permanence will be achieved.

While loss or degradation of habitat is a prime cause of declining wildlife, inordinate hunting is a major contributary factor. Hides and skins have long been a valuable article of commerce, and the insatiable demand for 'bush meat' provides the impetus for commercialized poaching. Trade in wild animals poses a threat to many species – chiefly birds but also other animals and even plants. Trade takes several forms, ranging from furs, feathers and other animal products to live animals for the pet trade and for medical research.

## Gaps in our Knowledge

A surprising number of species have become endangered by default – through inefficient administration, ignorance of their ecology, or simply because nobody seems to have either the interest, authority or responsibility for taking constructive action to safeguard them.

While it is true that our knowledge of the factors responsible for the decline of species has improved in recent years, this applies mainly to birds and mammals – and even for them information is often woefully inadequate. The lower vertebrates and invertebrates – which have a vital role in maintaining the ecosystem – remain poorly known. Concern has, for example, been expressed at the dramatic decline in amphibian populations currently taking place in different parts of the world simultaneously. The remarkable stomach-brooding frog from Queensland, Australia, has not been seen since 1979. The golden toad of Costa Rica has not bred in the Monteverde Cloud Forest Reserve since 1982, while many species of frogs have all but disappeared from parts of California, and are also becoming increasingly uncommon in the United Kingdom. Amphibians are important constituents of many ecological communities and, as their life cycle is partly terrestrial and partly aquatic, they are particularly sensitive to water quality, pollution and atmospheric change. Some scientists believe their decline to be indicative of significant man-made changes to the environment in general and the biosphere in particular.

The first essential in taking remedial action to halt the decline of species is to acquire accurate data from which reliable conclusions can be drawn. In countries where a great deal of information is obtained by armies of amateur naturalists it is relatively easy to make reliable assessments, but where that advantage is lacking it is not always possible to be so sure.

Data on rare and endangered species is collected through a global network of volunteer groups with specialist knowledge of the species for which each is responsible, and collated by the IUCN Species Survival Commission (SSC), with the International Committee for Bird Pres-

## IUCN RED DATA BOOK CATEGORIES

The number of species and subspecies listed in the *1990 IUCN Red List of Threatened Animals* stands at 698 mammals, 1,047 birds, 191 reptiles, 63 amphibians, 762 fishes, and 2,250 invertebrates, while an estimated 60,000 plant species are under threat of extinction or genetic degeneration. These are, of course, only the ones that are known. Many others, particularly invertebrates, are disappearing before they become known to science. A representative sample of mammals, birds, reptiles and invertebrates is included in this chapter; others are mentioned throughout the book.

Species on the *Red List* are categorized according to the degree of threat, thus allowing firm conclusions to be drawn about species for which there are reliable data, but to be less positive about those for which information is inadequate. Placing a species in either the *Vulnerable* or *Insufficiently Known* categories, for example, indicates the need for further information, and stimulates attempts to obtain it.

**Extinct (Ex):** species for which there have been no definite reports for 50 years.

**Endangered (E):** species in danger of extinction and whose survival is considered unlikely as long as the underlying causes continue.

**Vulnerable (V):** species that are thought likely to become endangered unless the adverse factors that threaten them can be removed.

**Rare (R):** species at risk through occurring in low numbers, or which have a very restricted range, or are sparsely distributed, but are not at present endangered.

**Indeterminate (I):** species known to be Endangered, Vulnerable or Rare, but where information is insufficient to know which of these categories is appropriate.

**Insufficiently Known (K):** species suspected of being eligible for one or other of the preceding categories, but where lack of information makes confirmation impossible.

**Threatened (T):** species comprising several subspecies that are threatened in different ways and are thus placed in different categories.

**Commercially Threatened (CT):** species whose survival is threatened by commercial exploitation.

## INTERNATIONAL CONVENTIONS

Of the various conventions that have been adopted in recent years, the following are of particular relevance to this chapter:

CITES. Convention on International Trade in Endangered Species of Wild Fauna and Flora. Promulgated in 1973 to regulate international trade in species of plants and animals listed in three Appendices:

Appendix 1 lists species whose survival is threatened by trade. Special permits are required from both exporting and importing countries, and will be issued only in exceptional circumstances.

Appendix 2 lists species that are likely to become endangered unless trade

is regulated. Species that are difficult to distinguish from those threatened are also included on the list in order to close a potential loophole.

Appendix 3 lists those species which any signatory to the convention considers are in need of regulation in areas under its jurisdiction, that is to say at the national as distinct from the international level.

ICRW. International Convention for the Regulation of Whaling, 1946. Signatories agree to introduce international regulations for whaling to ensure that all cetaceans are protected from over-fishing, that whaling operations are restricted to species able to sustain exploitation, and that provision is made for the recov-

ery of depleted species, thereby ensuring the orderly development of the whaling industry. This convention led to the establishment of the International Whaling Commission.

RAMSAR. Convention on Wetlands of International Importance Especially as Waterfowl Habitat, 1971. The principal intergovernmental forum for promoting international cooperation for the conservation of wetlands.

ACCN. African Convention on the Conservation of Nature and Natural Resources, 1968.

CMS. The Bonn Convention on the Conservation of Migratory Species of Wild Animals, 1979.

ervation (ICBP) responsible for the birds. Until 1979, the results were summarized in the *Red Data Book*, the official international register of the world's rare and endangered species. The first two volumes – *Mammalia* and *Aves* – appeared in 1966, followed by companion volumes on amphibians and reptiles, fish, and invertebrates, as well as one on plants compiled by the Royal Botanic Gardens, Kew, England.

The value of the *Red Data Book* was underscored when the USA, the former USSR, and a number of other countries

published their own national *Red Books*. In 1979, responsibility for the compilation of the animal volumes of the RDB was taken over by the World Conservation Monitoring Centre (WCMC) at Cambridge, England. WCMC also collates information on trade in wildlife and on protected areas, making it available to those responsible for national and regional conservation programmes, as well as to development banks and technical assistance agencies engaged in funding and supporting conservation schemes all over the world.

The main thrust is now directed at compiling a series of Action Plans as part of a World Conservation Strategy. Emphasis has shifted from single species to a much broader regional approach, and Action Plans have been drawn up for such diverse groups as the Australian Marsupials and Monotremes, Lemurs, Primates, Antelopes, Otters, Parrots, Island Plants, Orchids, Fungi and many more. Other specialist groups cover such themes as Captive Breeding, Reintroductions, and Sustainable Use of Wild Species.

# MAMMALS

## Thylacine or Marsupial Wolf

*Thylacinus cynocephalus*
**Order:** Marsupialia
**Family:** Thylacinidae (of which the thylacine is the only representative)
**Size:** Stands 47 cm/18.5 in at the shoulder, and measures more than 1.5 m/5 ft from nose to tail.
**Range:** Southwestern Tasmania.
**Habitat:** Originally inhabited savanna woodlands or open sclerophyll (eucalypt-dominant) forest, where kangaroos and wallabies, its chief prey, were most abundant. Later confined to dense bush.
**Red Data Book category:** Extinct?
**Threat:** To clear land for pasturage, the early European settlers burnt off the bush and scrub, in the process not only radically altering the character of the countryside, but also destroying much of the wildlife associated with it. Deprived of its natural prey, the thylacine turned sheep-killer. The settlers retaliated by systematically hunting it down, using packs of hounds for the purpose. Regarded as vermin, and with the encouragement of a government bounty, the thylacine was slaughtered at every opportunity and by any means. Yet, despite ceaseless harassment, numbers remained at a reasonable level until 1910. In that year the animal went into sudden decline, attributable to disease, possibly distemper. By 1933, when the last known

thylacine was caught in a trapper's snare, the species was assumed to be extinct.
**Conservation:** The possibility nevertheless remains that a relict population managed to survive in some of the more remote and inaccessible parts of Tasmania. From time to time footprints and even sightings are reported. Some prove false, others appear genuine. In 1961, a young male thylacine was accidentally killed at Sandy Bay on the west coast of Tasmania. An expedition sent into the area found two lairs that had evidently been used by thylacines. The Tasmanian Wilderness World Heritage Area (10,813 km²/4,175 sq mls), established to consolidate a number of national parks and other protected areas, covers a large part of the country in which the thylacine is thought most likely to have survived. The Tasmanian Parks Wildlife and Heritage Authority investigates and analyses all reports on the species. In 1990, there were 13 sightings, three of them classified as 'very good'. Perhaps one day we will know for sure whether this fascinating animal has escaped the fate of the dodo and so many other creatures.

## Indri

*Indri indri*
**Order:** Primates
**Family:** Indriidae (leaf-eating lemurs)
**Size:** Head and body (male) up to 70 cm /28 in; tail 3 cm/1.25 in; weight about 6

kg/13 lbs.
**Range:** Confined to the northern and central sectors of Madagascar's eastern rainforest from the latitude of Sambava (a little to the north of the Bay of Antongil) southwards to the Mangoro River.
**Habitat:** Eastern rainforest from sea level to 1,500 m/5,000 ft.
**Numbers:** Although occupying an extensive range, distribution is uneven and the population density low.
**Red Data Book category:** Endangered
**Threat:** Mainly from the indiscriminate destruction of forest, partly for timber and firewood and partly to make way for agricultural development. Over the last few decades destruction of the indigenous forest has been widespread, with corresponding reduction of habitat suitable for the indri which, moreover, has a low reproductive rate. Females do not become sexually mature until 7–9 years of age, and bear only a single young every third year.
**Conservation:** Represented in several reserves, including: Zahamena (731 km² /282 sq mls), the largest sanctuary in the eastern rainforest; Betampona (22 km² /8.5 sq mls); and Anjanaharibe-Sud (321 km²/124 sq mls). The Analamazaotra (Périnet) Reserve (810 ha/2,000 acres) was set aside specifically for the protection of the indri, but has the disadvantage of being not only small but isolated from

other areas of forest, and is therefore only of limited conservation value. Proposals have been made for the establishment of a new national park a little to the north of Périnet, in the vicinity of Mantady, with the management of the two areas being combined. In some parts of Madagascar the hunting of indri is taboo, but this traditional belief is becoming increasingly outmoded; the extent to which it still holds good is difficult to assess. What is certain, however, is that the indri's survival is dependent on the few reserves in which it still occurs. Unfortunately, inadequate staff and lack of supervision mean that, more often than not, protection is only nominal, with the result that forest destruction and hunting of lemurs takes place even in the reserves set aside for their protection. It is vital that a concerted effort be made to ensure that these protected areas are competently administered, failing which the outlook for the species is bleak.

## Orang-utan

*Pongo pygmaeus*
**Order:** Primates
**Family:** Pongidae (great apes)
**Size:** Males stand about 137 cm/4.5 ft and weigh 70 kg/154 lbs or more. Females much smaller, about 1.15 m/3.75 ft high and half the weight of the male. Arms exceptionally long and powerful.
**Range:** Borneo and the northern part of Sumatra. The species is most abundant in Kalimantan (Indonesian Borneo), particularly along the east coast, where extensive areas of forest still remain. In 1980, Kalimantan's forests were estimated to cover 353,950 km²/136,660 sq mls – some 65% of the total land area.
**Habitat:** Primary rainforest, with a preference for lowland areas. In Sumatra the species has been forced to move into upland forest.
**Numbers:** There may be as many as 180,000 orang-utans in Sumatra and Kalimantan, with another 4,000 or so in Sabah and Sarawak. But although these figures are higher than earlier estimates, numbers are believed to be declining at the rate of several thousand a year. The only orang-utans that can be regarded as reasonably secure are the 20,000 or so estimated to occur in established reserves; but these reserves are in need of more rigorous protection and higher standards of management.

**Red Data Book category:** Endangered
**Threat:** Since the end of the Second World War, the number of orang-utans has substantially declined. This is chiefly due to widespread loss of habitat, orang-utans being particularly vulnerable to habitat destruction and hunting pressure. Habitat destruction has arisen from the need to provide cultivated land for the constantly expanding human population, and from extensive commercialized exploitation of the primary forest in which the orang-utan lives. Fire is a further hazard: in 1983, a huge fire destroyed about 30,000 km²/11,500 sq mls of forest, including 8,000 km²/3,000 sq mls of primary forest. Most of the orang-utans in Sarawak and Sabah occur in 'forest reserves', a designation implying protection, but the term is misleading. Forest reserves are expressly earmarked for licensed timber extraction by contractors from whom the government draws a royalty. Clear felling of the forest has had the effect of splitting the orang-utans into small, often isolated, groups, making their survival difficult. Heavy losses have also been incurred in capturing orang-utans both for medical research purposes and for the pet trade. There is a lucrative market for baby orang-utans in many parts of the Far East, and the high prices paid prove an irresistible temptation to smugglers. The capture of baby orang-utans generally involves slaughtering their mothers; few of the young survive the separation. Young orang-utans are delicate animals and are susceptible to the same diseases as human beings. Captured animals are generally kept under unhygienic conditions and forced to exist on an unnatural diet, with the result that most of them die from malnutrition or disease. For every young orang-utan that survives in captivity, ten die.

**Conservation:** Trade in young orang-utans is now illegal, with the governments of Singapore and Hong Kong prohibiting their import and export. The International Union of Directors of Zoological Gardens has also introduced stringent regulations governing the acquisition of orang-utans. Biological field stations have been established in both Borneo and Sumatra with the purpose of rehabilitating confiscated pets and smuggled orang-utans for reintroduction to the wild. Certain medical research organizations in the US have undertaken to establish their own private breeding cen-

## THE THYLACINE

The largest contemporary carnivorous marsupial is a remarkable creature, superficially resembling the wolf but in no way related to it. The thylacine, whose scientific name means 'pouched dog with wolf head', is an interesting example of evolutionary convergence; that is to say, the occurrence in widely separated parts of the world of animals that although entirely unrelated have adapted to similar ecological niches. In the process they have come to bear a superficial resemblance to each other not only in habits and lifestyle but in physical appearance also.

Slightly smaller than a wolf, the thylacine has a massive wolf-like head, and jaws with an exceptionally wide gape, almost to the ears. The tail, on the other hand, resembles a kangaroo's, being very broad at the base and so rigid that it cannot be wagged. Indeed, if seized by the tail the animal is incapable of turning on its captor. In pursuit of its prey, the thylacine runs rather like a dog, following its quarry at a steady trot, culminating in a final burst of speed. Early accounts said that when really hard pressed the thylacine would rise up on its hind legs and bound over the ground in the manner of a kangaroo.

Before Tasmania became separated from Australia the thylacine occurred on the Australian mainland, where it is believed to have been exterminated by the dingo, the wild dog of Australia, brought there by the Aborigines. By the time the first Europeans arrived, the thylacine was confined to Tasmania.

tres in an attempt to minimize the need for wild-caught animals. These measures have succeeded in almost completely halting the illegal trade in orang-utans. But there remains the need to establish a series of large and well-managed orang-utan sanctuaries, in addition to Sumatra's existing Gunung Leuser National Park (9,464 km²/3,654 sq mls).

The only area in Sabah to have a legally protected population of orang-utans is the Tabin Wildlife Reserve (1,205 km²/465 sq mls). The species' prospects have been greatly improved by the establishment of the Danum Valley Conserva-

# MAN OF THE WOODS

The orang-utan – whose name, appropriately, means 'man of the woods' – is the only great ape living outside Central Africa. It is also the most arboreal of the great apes, and well adapted for life in the trees. It moves through the forest swinging by its arms from branch to branch, a style of movement that has led it to develop arms that are half as long again as its legs; when hanging loosely they reach almost to its ankles. The orang-utan's long, narrow hands and feet are, similarly, adaptations for grasping branches. On the ground it moves awkwardly on its hind legs with arms held over its head. When moving more quickly it uses its long arms like crutches, pivoting on clenched fists and swinging its body between them. Usually it lives either singly or in twos, occasionally in small groups of up to four individual animals. Old males live apart except briefly when mating. The orang-utan feeds mainly on fruit – notably the evil-smelling but pleasant-tasting fruit of the durian tree – supplemented by leaves, bark, birds' eggs, freshwater crustaceans, and insects.

Females and young sleep in the trees 10 m / 30 ft or more above the ground. Each female makes a nest in the form of a simple platform in the fork of a tree, put together in a matter of minutes. Nests are seldom used more than once.

The orang-utan has a low reproductive potential; females do not become sexually mature until about ten years of age, and reproduce only every fourth year. The single young is suckled for at least 12 months and does not become completely independent for four years. There is moreover a high (40%) infant mortality rate. Thus a female orang-utan may succeed in raising no more than two or three young during her lifetime.

● The orang-utan – the 'man of the woods' – tends to be of a solitary disposition. Every night it makes a fresh nest to sleep in. (WWF / Michel Terrettaz)

● A rare glimpse of a blue whale's fluke as it dives into the Indian Ocean. (WWF / Elizabeth Kemf)

tion Area (427 km²/165 sq mls) in eastern Sabah, an area unsuited to agriculture which is believed to contain a substantial number of orang-utans, as well as a group of Sumatran rhinoceroses.

The government of Sarawak is considering extending the Lanjak-Entimau Wildlife Sanctuary (1,687 km²/651 sq mls), created primarily for the protection of the orang-utan in 1983, and is also proposing to establish the Batang Ai National Park adjoining the Lanjak Entimau Sanctuary's southern boundary. The Brunei government has agreed in principle to set aside a substantial area as 'conservation forest'. These protected areas (both existing and proposed) cover more than 8,000 km²/3,000 sq mls of largely unspoiled habitat, which, if adequately protected, should go a long way towards ensuring the species' survival in Sarawak.

## Blue Whale
*Balaenoptera musculus*
**Order:** Cetacea
**Family:** Balaenopteridae (baleen whales, rorquals)
**Size:** The largest animal that has ever lived, dwarfing even the dinosaurs. *Brachiosaurus*, a giant among dinosaurs, was less than half the weight of the blue whale – about 50 tons compared with the blue whale's average of 106 tons. A female caught in 1947 weighed 190 tonnes/187 tons. Length averages about

26 m / 85 ft, with a record length of 33.58 m / 110 ft 2.5 in.

**Range:** The oceans of the world, with the principal populations in the southern hemisphere. The blue whale migrates between its summer feeding grounds among the Arctic and Antarctic pack ice to its winter breeding grounds in subtropical waters. It is essential for births to take place in warm waters as the newborn calf has only a thin layer of blubber.

**Habitat:** Polar waters and tropical seas.

**Numbers:** The original southern stock of blue whales is believed to have numbered about 250,000. By the mid-1930s this figure had dropped to about 40,000; by the 1950s to 10,000 or less; and by 1963 to no more than 1,000. Thus, since the start of whaling in Antarctic waters in 1904, the number of blue whales has been reduced by 99%.

**Red Data Book category:** Endangered

**Threat:** Until little more than a century ago whaling was conducted from open boats, which limited the catch to the smaller whales. It was not until the invention of the harpoon gun in 1864 that man had a weapon capable of killing the large rorquals. Besides marking the beginnings of the modern whaling industry, this invention also carried the seeds of the industry's own destruction. Rapid technological advance, including the development of ocean-going catchers as vehicles on which to mount the harpoon gun, so improved the efficiency of hunting that stocks of whales in the northern hemisphere were quickly exhausted. Around the start of the 20th century, the whaling industry began to exploit the hitherto untouched seas of the southern hemisphere. By 1910 there were six shore stations and 14 factory ships operating from the Falkland Islands Dependencies; other shore stations were established in Chile, South Africa and elsewhere. The development of the factory ship enabled the radius of operations to be greatly extended. From the early 1920s, the accepted method of conducting whaling operations centred on ocean-going factory ships, each with its covey of catchers. Such a task force was capable of remaining at sea for long periods, transferring the accumulated oil to attendant tankers for shipment to the home port. Modern refinements include helicopters and sonar equipment for spotting purposes. As the blue whale was the largest and most valuable commercial species – yielding about twice as much oil as the fin whale – it became the principal quarry of the whaling fleets. Between 1927 and 1936 – the peak years of Antarctic whaling – blue whales formed 80% of the catch.

**Conservation:** In 1946 the International Whaling Commission was formed and the International Whaling Convention adopted, with the object of regulating the whale catch and of ensuring the effective conservation of the world's whales. Although now fully protected by international agreement, the ban on hunting the blue whale was not imposed until after the species had been reduced to the level at which it was no longer of any commercial significance. Only when it had been exploited to the point of extinction did the whaling industry agree to its protection. The prohibition on hunting is, moreover, applicable only to those governments represented on the International Whaling Commission. Some countries remain outside the agreement. Coming at a time when environmental awareness was growing, the IWC's failure to fulfil its mandate aroused worldwide condemnation. Japan (a member of the IWC) served her own interests by surreptitiously arranging for several non-member countries, and ships sailing under flags of convenience, to hunt whales for the Japanese market. Such blatant poaching threatened not only the whales but the IWC as well. While some countries – and, since the break-up of the Soviet Union, Russia in particular – have recently become more amenable to conserving the whales, Japan remains the principal whaling country and the main market for whale meat.

In 1982, the Indian Ocean north of 55° S became a designated whale sanctuary in which all commercial whaling is prohibited, a decision that was ratified in 1992. That same year, France proposed giving comparable status to the entire Southern Ocean south of 40° S. This suggestion comes at a time when protection of Antarctica and the seas around it is in the forefront of discussion. A positive decision is expected at the Commission's forthcoming meeting in 1994. Between them, these two huge sanctuaries should protect the whalebone whales throughout their annual cycle: the Indian Ocean sanctuary covering their breeding ground and the Southern Ocean sanctuary their feeding zone. There could be no more suitable place for the rehabilitation of the great whales.

## LEVIATHAN OF THE DEEP

The blue whale's migrations may cover thousands of kilometres. Remarkably, one marked blue whale is known to have travelled over 3,000 km / 1,900 mls in 47 days. And because the breeding migration takes the animal away from its feeding grounds, it eats virtually nothing for at least four months, during which time it lives on reserves accumulated in its own body.

The newborn blue whale calf is gargantuan – about 7 m / 23 ft long and weighing 2 tonnes or more. It consumes more than half a tonne of milk a day, doubling its birth weight within the first week of life. The female suckles her calf for about 5–7 months; by the time it is weaned it will have doubled its birth length to about 15 m / 50 ft.

The blue whale feeds almost exclusively on krill, a small shrimp-like crustacean occurring in immense concentrations in polar waters, sometimes in such densities that the surface of the sea has the appearance of reddish-brown porridge. Opening its mouth to the maximum extent, the whale swims through the krill. Having taken a large mouthful, it closes its mouth and raises its tongue, forcing the water through the horny baleen plates that hang suspended from either side of the roof of its mouth, thereby expelling the water and leaving the krill trapped inside.

## African Wild Dog

*Lycaon pictus*

**Order:** Carnivora

**Family:** Canidae (dogs)

**Size:** Shoulder height up to 78 cm / 2 ft 6.75 in; head and body length up to 112 cm / 3 ft 8 in; weight about 25 kg / 55 lbs.

**Range:** Originally widely distributed over the open savanna lands of Africa south of the Sahara (with relict populations in the

Saharan massifs). In recent years the species has been eliminated from much of its former range, to the extent that it is either extinct or close to extinction in 19 of the 32 countries in which it formerly occurred. Only six countries are believed still to hold reasonably stable populations: Botswana, Ethiopia, Kenya, Tanzania, Zambia and Zimbabwe. Even in these countries the number of wild dogs declined by about one-third during the 1980s. The possibility remains that a viable population still survives in the Sudan, but that country has been wracked by war and famine for so many years that the present position is uncertain.

**Habitat:** Found in a wide variety of habitats ranging from open grasslands to bush country and even forest fringes.

**Numbers:** The overall total probably does not exceed 2,000.

**Red Data Book category:** Endangered Regarded as the most endangered large carnivore in Africa.

**Threat:** Like all carnivores, the African wild dog lives by preying on other animals; but the widely held notion that it is a ruthless and indiscriminate killer of both wild and domestic animals is without foundation. Like the wolf, its efficiency as a predator has led it to be falsely accused of being a menace to wildlife. Even game wardens have been known to shoot wild dogs in what they misguidedly perceived to be the best interests of the wild fauna. Pastoralists, too, detest the wild dog, which they regard as vermin to be killed at every opportunity, often by setting poison baits. In point of fact, the wild dog fulfils an important role in maintaining ecological equilibrium. Disease, notably canine distemper and rabies to which it is particularly susceptible, has contributed to this carnivore's decline.

**Conservation:** Legally protected throughout much of its range, but law enforcement is uniformly inadequate. The wild dog is a particularly difficult animal to protect. Although represented in a number of national parks, reliance on protected areas is unlikely to be a sufficient safeguard: its strongly nomadic habit makes the wild dog so wide-ranging that none but the very largest sanctuaries can contain it. The survival of this species calls for an educational campaign aimed at discouraging its wanton killing by pastoralists and others.

## AN ABERRANT CARNIVORE

The giant panda is a carnivore that has adapted to a vegetarian diet. It eats meat whenever it can, but being a poor hunter, must generally rely on carrion. Yet the panda retains a carnivore's stomach: taken in combination with the low nutritional value of bamboo, this obliges the animal to spend about two-thirds of every day consuming the immense quantities of bamboo – up to 45 kg/100 lbs a day – that are necessary to sustain it. Even so, it lives on a nutritional knife-edge. This heavy intake of food necessitates eating continuously for eight hours, followed by a four-hour sleep, a routine that goes on day and night, interrupted only by the need to drink: on such a diet water is essential, and the panda drinks four or five times a day. Most of its time is spent squatting on the ground chewing bamboo. In this it is helped by a special claw acting as a sixth digit which enables the panda to grasp bamboo almost as effectively as a finger. Despite being so firmly grounded, it remains a competent climber. Reacting to danger, it takes refuge in the nearest tree where, squatting in a fork, it remains still and silent until the threat – whether from brown bear, leopard, wild dog or human – has gone.

## Giant Panda

*Ailuropoda melanoleuca*
**Order:** Carnivora
**Family:** Ursidae (bears)
**Size:** Body length up to 1.5 m/5 ft; tail 12.5 cm/5 in; weight about 120 kg/265 lbs. Female slightly smaller.
**Range:** Central China, where it is confined to a few isolated mountain ranges – notably the Qionglai, Daxiangling, Xiaoxiongling and Liang mountains in Sichuan Province, the Min Mountains in Sichuan and Gansu provinces, and the Qinling Range in Shaanxi Province.
**Habitat:** Bamboo thickets at elevations of 1,500–3,000 m/5,000–11,500 ft.
**Numbers:** A three-year survey, conducted between 1985 and 1988, concluded that the overall population stood at 1,120 plus or minus 240. The most recent estimate places the total at about 750.
**Red Data Book category:** Endangered
**Threat:** The giant panda has long been hunted for its pelt, which the Chinese believe has the power to protect them against evil spirits. In recent years, mounting human pressure on the land has compelled the panda to retreat deeper into the mountains. A specialist bamboo feeder, the panda's distribution is governed by the availability of this plant. A characteristic of bamboo is the infrequency with which it flowers – perhaps only every 50 years or so, according to variety. After blossoming and seeding it withers and dies. Die-back marks the end of the panda's food supply, and persists until the seed has germinated and new shoots have sprouted. In the past pandas were free to migrate to unaffected areas, but this option is no longer open to them. The most recent die-back occurred in 1983–84 when at least 200 pandas are known to have died of starvation.
**Conservation:** The Chinese people regard the panda with a mixture of pride and affection, and support any measures necessary for its protection. Officially designated a 'national treasure', the animal is rigorously protected, hunting being totally prohibited and carrying severe penalties. Twelve special sanctuaries have been established, with the capacity for 500–600 pandas, the largest – the Wolong Nature Reserve (2,000 km²/770 sq mls) – having a population of 100 pandas. Chinese and American scientists are currently working in the reserve studying the animal's natural history and initiating a comprehensive conservation programme. One of the practical results of this collaboration is the Chinese Government's decision to increase the number of panda reserves to 25. This will mean that of the total 11,000 km²/4,247 sq mls of existing panda habitat, almost half – 5,000 km²/1,930 sq mls – will be protected. The plan envisages establishing 'bamboo corridors' to link otherwise isolated patches of panda habitat, while outside the reserves forestry practices will where necessary be modified in the panda's favour. The recent discovery,

● The African wild dog or hunting dog, falsely accused of being a menace to wildlife, in fact has an important role in helping to maintain ecological balance. (Matthew Hillier SWLA)

## A MUCH-MALIGNED PREDATOR

The African wild dog has an unusually complex social organization. It lives communally in packs varying in size from half a dozen to as many as five dozen, taking up residence in an abandoned burrow, usually an aardvark's. Mating is the prerogative of the dominant male and female, and she alone will rear pups. She and her litter – generally numbering 7–10 but sometimes as many as 19 – remain in the den, where they are fed on food brought by other members of the pack and regurgitated.

The pack is highly nomadic and roams over a hunting territory that may cover as much as 1,500 km²/580 sq mls. The communal way of life extends to hunting, medium-sized antelopes such as gazelle being the usual prey. Wild dogs are essentially long-distance runners. Although over a short distance their quarry can outrun them, they possess great staying power, which they use to wear down their prey. Singling out an animal from a herd, a pair of dogs sets off in pursuit, covering the ground at a steady lope and driving the prey towards a fresh pair of dogs, who take over and continue the chase. Eventually the quarry tires, whereupon the dogs close in for the kill.

announced in 1990, of a method of making bamboo flower to order could clearly have an enormous bearing on conserving the panda. Zoos are collaborating in developing a captive breeding project aimed at restocking suitable areas of natural habitat with zoo-bred animals.

## Asiatic Cheetah
*Acinonyx jubatus venaticus*
**Order:** Carnivora
**Family:** Felidae (cats)
**Size:** Length (excluding tail) 137 cm / 4.5 ft; height at shoulder 76 cm / 2.5 ft; weight up to 60 kg / 130 lbs. The cheetah is the only member of the cat family to lack sheaths into which the claws can be retracted. It hunts by running down its prey, its long legs being specialized adaptations for sprinting; indeed, it is faster than either the racehorse or the greyhound, achieving speeds (over short distances) of up to 120 kph / 75 mph. The general consensus is that only two races are recognized: *A. j. jubatus* in Africa and *A. j. venaticus* in Asia. The differences between the Asiatic and African cheetahs are slight. The Asiatic form is reputedly smaller (although such evidence as exists on this point is inconclusive), its fur is thinner, and it is said to lack a mane. The characteristics on which the two races are separated are so minor and based on such inadequate comparative material from the Asiatic sector of the range that it is questionable whether separation is justified.

**Range:** Originally reaching from Syria and Arabia eastwards to India and into the southern part of Russian Turkestan, extending northwards at least as far as the Syr Darya and Amu Darya. It has disappeared from most of this huge region, and is now known for certain only in Iran, where it is believed to occur in the eastern part of the country, in Sistan and along the frontier with Afghanistan and Pakistan. In these sparsely inhabited areas gazelles abound and, in Iran at least, the cheetah's distribution appears to coincide with the gazelle's. A few individuals may still linger in some of the more remote parts of Turkmenistan, Afghanistan and Pakistan seldom visited by man. In India, the cheetah was thought to have been exterminated in 1951, when three were shot in one night. But, in 1990, a sighting was reported from India's Simlipal Wildlife Sanctuary (1,355 km² / 523 sq mls) in Orissa State.

**Habitat:** A creature of arid and semi-arid open grasslands and shrublands. In Turkmenistan it was said to inhabit the dense *tugai* vegetation along the banks of the Syr Darya and Amu Darya.

**Numbers:** In 1977, the population in Iran was estimated at about 200. Since then, numbers are believed to have declined, but the current position is not known.

**Red Data Book category:** Endangered
**Threat:** The presumption that the cheetah preyed on domestic livestock led pastoralists to equate it with the leopard and to hunt it at every opportunity. It was also heavily hunted for its attractive skin; when motorized hunting became the vogue, the cheetah stood no chance. Remorseless hunting and development of pastoral lands were responsible for

● A herd of African elephants grazing in the Ngorongoro Crater, Tanzania. (WWF / Rick Weyerhaeuser)

the decline of its natural prey, the gazelles. The cheetah is of a nervous, timid disposition, unable to tolerate disturbance, and quickly disappears from areas actively exploited by people.

**Conservation:** Listed in Appendix 1 of CITES, and legally protected in Iran and the former USSR. Protected areas have been established in both countries, in particular the Badkhyz Reserve (876 km² /338 sq mls) in the extreme south of Turkmenistan, close to the border with Iran and Afghanistan.

## African Elephant

*Loxodonta africana*
**Order:** Proboscidea
**Family:** Elephantidae (elephants)
**Size:** Males may stand up to 3-4 m/10-13 ft at the shoulder and weigh 5-6 tons. Both males and females carry tusks. The heaviest pair of tusks ever recorded weighed 102 and 109 kg/225 and 240 lbs respectively.
**Range:** Tropical Africa.
**Habitat:** Open savannas to rainforest.
**Numbers:** The overall African elephant population has declined from over 2 million in 1973 to 625,000 today.
**Red Data Book category:** Vulnerable
**Threat:** Ivory is a valuable commodity and the quest for it has gone on for centuries. But only in the last two decades has the African elephant been hunted so intensively as to threaten its survival. During the last 20 years more than 1 million elephants have been slaughtered.

Between 1979 and 1987, the number of elephants in Tanzania, for example, fell from 316,000 to 85,000. In almost the same period (1977-1986) the elephant population in the Selous Reserve in southern Tanzania halved from 110,000 to 55,000; three years later it had halved again to 30,000.

This huge reduction in the numbers of elephants is entirely attributable to man's greed for ivory. During the 1970s the value of ivory on the world market soared, sparking off a massive upsurge of poaching. This involved elephant massacres, with the ivory being flown out by the planeload, often with the collusion of highly placed officials. Sometimes even the ministers in charge of the government departments responsible for protecting wildlife were themselves implicated in the corruption.

## ELEPHANT SOCIETY

Elephant society is based on small, strongly matriarchal family units. By killing the mature animals (which carry the largest ivory and are thus the prime target of the poachers) the social structure is inevitably destroyed. Without an experienced leader the survivors are weak and vulnerable. No calf can live without its mother; and, if the matriarch is killed, the herd's chances of survival are greatly reduced. A report written in 1988 states that more than 70% of elephant family groups in Tanzania's Mikumi National Park consisted almost entirely of immature animals: the adult females had practically all been killed. Moreover, the slaughter of adult bulls (which do not become active breeders until about 35 years of age) leaves a shortage of sexually mature males.

## THE IVORY TRADE

Measures already taken to protect the dwindling number of African elephants include placing the species on Appendix 1 of CITES. In 1990, most of the signatory nations voted for a ban on all international trade in ivory. Some countries – among them Kenya and Tanzania, which lost more than two-thirds of their elephants to poachers – support a total ban on ivory trading, believing that as long as the trade exists, illegal ivory will find its way into the system. Southern African countries – South Africa, Zimbabwe, Zambia, Mozambique, Botswana and Malawi – oppose the ban, claiming that the ban on ivory trading will force up the black market price and thus stimulate a revival of poaching. They favour the sustainable management of their elephants, arguing that they should not be penalized for other countries' misfortunes, and take the view that because of the need to maintain numbers at optimum levels, they should be allowed to continue selling ivory culled in the course of managing their herds, with the proceeds used to help defray the costs of conservation. This argument would be more plausible if money from the sale of ivory and rhino horn did in fact benefit conservation; but of course it goes into general revenue, and no matter how much cash ivory may fetch, the National Parks' annual subvention remains unchanged.

Sustainable use is the sum and substance of conservation, and a ban conflicts with that principle. Whether or not a ban is imposed on the sale of ivory, elephant numbers will in the normal course of events steadily rise, and management will necessitate culling a proportion of the herds to keep them within the carrying capacity of the land. If animals have in any event to be shot, it makes no sense to deny local people the benefits. Indeed, elephant management provides a practical opportunity for the people most directly concerned to see for themselves the tangible advantages deriving from conservation. Rigorous international control over the ivory trade is absolutely vital, but the need for control should not be confused with the principle of sustainable use that forms the keystone of conservation. But the attitude of the local people is unlikely to alter as long as southern Africa's military forces, police and national park personnel continue to be deeply implicated in illegal ivory dealings. The Zimbabwe National Army and South African Defence Forces, with the complicity of well-placed government officials, were heavily embroiled in killing large numbers of elephants and rhinos in Mozambique, while military involvement in poaching and smuggling ivory and rhino horn was responsible for the slaughter of tens of thousands of elephants to help finance the UNITA forces fighting in Angola. Indeed, during the Angolan war, the upper echelons of the Cuban military establishment became so deeply implicated in ivory smuggling (along with drugs and precious stones) that Castro was forced to act. The Cuban military commander, General Ochoa, and three other high-ranking military officers were executed and six officials of the Cuban Interior Ministry 'resigned'.

**Conservation:** It is self-evident that the first step in conserving the African elephant lies in the elimination of poaching, for unless the wholesale killing is prevented, or at least greatly reduced, it is only a matter of time before the elephant ceases to exist. The most intractable aspect of the poaching problem is corruption in high places, with those responsible elevated above the law. With such an example before them, it is hardly surprising that corruption filters down through every level of the hierarchy to the game rangers in the field.

A further fundamental requirement is the establishment of an adequate network of national parks and game reserves in which the interests of the elephant are paramount. But the world's largest living land animal needs plenty of space, and it is not always possible to provide the herds with reserves that are sufficiently large for year-round living. This is especially true in the dry season when elephants have to trek long distances in search of food and water. Attempts to confine them or restrict movement will cause them either to ruin their habitat or to break out and encroach on tribal land.

In 1989, Kenya took an encouraging step when President Moi appointed Richard Leakey as Director of Wildlife Conservation with responsibility for safeguarding the country's fauna, and thus the tourist industry (worth over US$400 million a year). He at once set about weeding out corrupt officials – half his department had to go – and pursuing poachers: 60 poachers (mainly Somalis), equipped with automatic weapons, were killed in the Tsavo Park alone during the first six months of anti-poaching operations. If this initiative can be maintained there is at last hope for Kenya's elephants. But it will be necessary for other countries to follow suit before the African elephant can be said to be secure. And if we cannot ensure the survival of the elephant, what hope can there be for lesser creatures?

## Great Indian Rhinoceros

*Rhinoceros unicornis*
**Order:** Perissodactyla
**Family:** Rhinocerotidae (rhinoceroses)
**Size:** Stands up to 1.8 m/6 ft at the shoulder; length 4.2 m/13 ft; weight 2.5 tons.

---

## THE GREAT INDIAN RHINOCEROS

The great Indian rhino's most distinctive feature is its loosely folded hide, giving the impression that its body is clad in armour-plate, an illusion enhanced by the rivet-like tubercles with which the sides and upper parts of its legs are studded. Like the black rhinoceros, it has a prehensile upper lip, with which to grasp the grasses, shoots and water-plants on which it feeds. Its lower jaw carries a pair of incisors that have developed into tushes, which it uses for slashing with devastating effect on the rare occasions when it is aggressive. Despite its size and intimidating appearance, the great Indian rhino is a timid, inoffensive animal, which seldom behaves belligerently unless wounded or defending its young.

Water is an essential factor in the life of the great Indian rhino. It is never far from water in which to bathe or mud in which to wallow – a practice that helps alleviate attacks by hordes of swamp-dwelling insects.

---

**Range:** At one time the great Indian rhino was distributed across the entire northern part of the Indian subcontinent from Peshawar and Kashmir in the west, along the foothills of the Himalayas eastwards as far as India's frontier with Burma, and southwards through the Ganges river system to the Sundarbans. It may also have occurred in Burma, southern China and parts of South-East Asia, but such evidence as exists may refer to one of the other two Asiatic rhinos. Today, the species is confined to the floodplain of the Brahmaputra River in northeastern India and the marshlands of the Terai in south central Nepal.

**Habitat:** Seasonally flooded grasslands and wetlands.

**Numbers:** About 1,700 all told, of which 1,300 are in India and the remainder in Nepal.

**Red Data Book category:** Endangered

**Threat:** Until little more than a century ago the species remained abundant. So common was it that, in 1876, the government of Bengal offered a bounty of 20 rupees to anyone shooting a rhino in the interests of crop protection. But within a few years, excessive hunting, together with increasingly widespread expropriation of habitat to provide land for the expanding human population, had reduced the animal almost to extinction. By 1908, the survivors were confined to Kaziranga in northeastern India and Chitwan in Nepal.

**Conservation:** The rehabilitation of the great Indian rhinoceros is one of conservation's successes. The initial step towards safeguarding the species was taken in 1908 when the Kaziranga National Park (430 km²/166 sq mls) was first set aside. For 30 years it remained a closed area, with the Assam Rifles occasionally being called in to combat well-armed gangs intent on obtaining rhino horn. Despite many setbacks, India's rhino population gradually increased from a few dozen in 1908 to 1,250 by 1980.

Nepal was equally successful. Chitwan National Park (932 km²/360 sq mls), Nepal's premier rhino sanctuary, lies in the Rapti River Valley. Up to the early 1950s the valley was sparsely populated, and used as a hunting ground by the powerful Rana family. Among the many wild animals inhabiting the area were an estimated 800 Indian rhinos. Following the overthrow of the hereditary Rana prime ministers in 1952 and the period of political instability that ensued, large numbers of land-hungry people moved into the valley. Poachers from India took advantage of the prevailing confusion to kill several hundred rhinos. Strong representations to the government of Nepal led to the setting up of a Land Settlement Commission empowered to evict illegal squatters and resettle them. By 1965, some 22,000 people had been removed from the Rapti Valley, including 4,000 from the Chitwan rhinoceros sanctuary itself. Having cleared the area of squatters and taken firm measures to control poaching, the number of rhinos steadily increased from the 60 estimated to have remained in Chitwan in 1960 to 360–380 in 1987.

Since then, the massive increase in the value of rhino horn – the retail price of Indian rhino horn in Taiwan, for example, is currently US$54,000 a kilo – has given rise to heavy losses of rhinos by poachers armed with sophisticated weapons infiltrating into Assam from nearby Nagaland. Between 1982 and 1985, more than 230 rhinos were killed in India by poachers. Strong anti-poaching measures were necessary before order could be restored. In Nepal, on the other hand, there has been little poaching, mainly because rhinos are the personal property of the King and are guarded by several hundred soldiers of the Nepalese army, stationed in 35 guard posts in and around the Chitwan National Park.

The IUCN/SSC Asian Rhino Specialist Group (ARSG) has drawn up a plan for conserving the species, which has as its object the maintenance of a total of at least 2,000 great Indian rhinos in the wild state, spread over six key sanctuaries: Kaziranga, Manas, Orang and Dudhwa in India, and Chitwan and Bardia in Nepal. The number of both sanctuaries and rhinos will be increased where and when suitable opportunities occur, provided that each sanctuary has a minimum carrying capacity of 100 rhinos, and can be adequately protected. Translocation of rhinos from Kaziranga to Dudhwa National Park has already started.

India's rhino sanctuaries nevertheless remain under unremitting pressure as much from land-hungry people as from poachers, pressures that are certain to increase. Ceaseless vigilance will therefore be necessary if the great Indian rhinoceros is to survive.

## Scimitar-horned Oryx

*Oryx dammah*
**Order:** Artiodactyla
**Family:** Bovidae (cattle, antelopes, gazelles, sheep, etc.)
**Size:** Shoulder height 1.2 m / 4 ft; horns up to 1.27 m / 4.16 ft; weight 200 kg / 450 lbs.
**Range:** The Sahel, along the southern fringes of the Sahara from the Atlantic coast to the Nile.
**Habitat:** Arid grasslands.
**Numbers:** The size of the surviving population is not known, but at best is believed to be in the low hundreds.
**Red Data Book category:** Endangered
**Threat:** Once abundant over much of the Sahara, the oryx was heavily hunted for its meat and hide, for which there was great demand. Large numbers were slaughtered by desert nomads hunting in the tradi-

---

## THE RHINO CRISIS

No family of large mammals gives cause for greater concern than the rhinoceroses. Each of the five living species, three in Asia and two in Africa, has undergone a massive decline and been brought perilously close to extinction – and for similar reasons. With rhino horn fetching more than its weight in gold on the world market, dealing in horn has become as lucrative as pushing drugs; the temptation to poach is irresistible.

As well as the great Indian rhinoceros, the Asian species include the two-horned Sumatran rhino, the smallest member of the family. Scattered in small groups through Peninsular Malaysia and Sumatra, with a few still surviving in Sabah, Sarawak and possibly Kalimantan, its overall numbers have been reduced to an estimated 500–900.

The Javan rhino, once widespread in Southeast Asia, is now confined to Udjung Kulon National Park (761 km² / 294 sq mls) at the extreme western tip of Java, where it numbers about 50. Early in 1989, a hitherto unknown group of 10–15 Javan rhinos was found in Vietnam. The Vietnam Government has responded positively by increasing the size of the Nam Cat Tien Reserve (365 km² / 141 sq mls) and stationing armed guards in the area, at the same time establishing a special Rhinoceros Conservation Group to oversee a rehabilitation programme.

In Africa, the square-lipped or white rhino has been all but wiped out in the northern part of its range, between the west bank of the Nile and Lake Chad. The last stronghold of the northern race was in Zaïre where, until 1963, about 1,000 were concentrated in Garamba National Park (4,920 km² / 1,900 sq mls). In that year, rebel soldiers overran the park, killing all but a dozen rhinos. The southern race, 3,000 km / 2,000 mls to the south in the Umfolozi/Hluhluwe Game Reserve in Zululand, was once in a similarly precarious position, but its rehabilitation has been so successful that the park is in the almost embarrassing position of having a surplus.

Thirty years ago, Africa's black rhino was by far the most numerous of the five species. But from a total of 100,000 in the early 1960s, overall numbers have fallen to below 4,000, leaving the animal in deep crisis. In a matter of only two decades, Kenya lost more than 95% of its rhinos, numbers falling from 20,000 to barely 400. If the black rhino is to have any prospect of survival it is essential that the international ban on trade in rhino products, to which practically every country in Asia has subscribed, should be rigorously enforced. Four countries are conspicuously failing to enforce the ban: South Korea, Taiwan, China and Thailand. Unless they can be persuaded to conform, the world's rhinos are unlikely to long outlive the 20th century.

---

tional manner. The leading hunters were the Haddad, a tribe inhabiting northern Chad, whose entire way of life revolved around the oryx. The Haddad perfected a method of hunting that involved mounted men driving the animals into nets. But while traditional hunting has taken its toll, it was the introduction of the four-wheel-drive vehicle and modern firearms in the hands of prospectors and the military that finally tipped the scales and reduced this species from abundance to scarcity. Within the last 30 years the scimitar-horned oryx has been eliminated throughout its vast range except in central Chad. By the mid-1970s, 95% of the entire world population (then amounting to several thousand) was to be found in Chad's Ouadi Rime–Ouadi Achim Faunal Reserve (80,000 km² / 30,000 sq mls). The outbreak of war between government forces and Libyan-backed rebels in 1978 made continued protection of the reserve

impossible. The number of oryx fell from several thousand to a few hundred. Chad is nevertheless the only country still possessing a viable free-living population. The survival of the scimitar-horned oryx in the wild is largely dependent on the Ouadi Rime–Ouadi Achim Reserve. The cessation of hostilities in Chad should allow the protection of this important reserve to be resumed.

**Conservation:** Re-establishment of the scimitar-horned oryx in suitable parts of its range can be accomplished through captive breeding programmes. In 1966, when the species was still relatively abundant, 41 were caught in western Chad to form the nucleus of a captive breeding herd. The descendants of these animals now number more than 2,000. A start on reintroducing them has been made in Tunisia where, in 1985, ten oryx were released into the Djebel Bou-Hedma National Park (164 km² / 63 sq mls).

# RARE BIRDS

## Grey-necked Picathartes

*Picathartes oreas*
**Order:** Passeriformes
**Family:** Muscicapidae
**Subfamily:** Picathartinae
This strangely beautiful bird – with distinctive grey underparts, a green back, a large eye and partially-bald-partially-red head – is remarkably prehistoric-looking. Together with the only other member of its genus, the white-necked picathartes *P. gymnocephalus*, it forms a super-species. The taxonomic position of the two species has long been disputed. Originally placed in the crow family, they were later thought to be more closely related to starlings, and are now considered aberrant babblers.
**Range:** Cameroon, Equatorial Guinea and Gabon in Central West Africa.
**Habitat:** Restricted to relatively undisturbed tropical rainforest where there are caves or cliff faces close by. The grey-necked picathartes has the unusual distinction of breeding in small colonies in caves and on rock faces under rainforest canopy. It builds a peculiar nest of mud and plant-fibre, rather like a swallow's. It feeds on insects, sometimes alone, but occasionally in small groups, usually on the forest floor close to its breeding cave.
**Numbers:** The precise status, distribution and numbers of grey-necked picathartes remain a mystery. Where its specialist habitat requirement is met, the bird is not uncommon. Indeed, its secretive habits and its practice of nesting in inaccessible locations suggest that it may be less rare than generally supposed.
**Red Data Book category:** Vulnerable
**Threat:** The greatest threat to picathartes' survival is forest clearance: there are no records of this bird from areas where forest has been destroyed. It is highly sensitive to disturbance. The species is known to be hunted for food in some parts of Cameroon and almost certainly in Gabon and Equatorial Guinea as well; but the extent to which local hunting represents a threat is not clear. Collecting for zoos and for captive breeding purposes is also a matter for concern.

**Conservation:** Little is at present being done to safeguard either the grey or white-necked picathartes. The clear priority must be the establishment of protected areas in rainforest with suitable areas of caves and cliffs, and where the canopy remains undisturbed. Aided by the World Wide Fund for Nature (WWF), Cameroon has established the Korup National Park (1,260 km²/486 sq mls), and is currently gazetting Dja National Park (5,260 km²/2,031 sq mls), both parks housing viable populations of the grey-necked picathartes. The species also occurs on Mount Kupe, where the International Council for Bird Preservation (ICBP) is currently working with the Cameroonian authorities on introducing measures for protecting the forest. Although captive breeding has been attempted for both species in a number of zoos around the world, it has regrettably proved unsuccessful.

## Seychelles Warbler

*Acrocephalus seychellensis*
**Order:** Passeriformes
**Family:** Muscicapidae
**Range:** Three islands in the Republic of Seychelles. This small, brown, very weak-flying species was confined for a long time to the tiny (27 ha/67 acre) Seychelles island of Cousin in the Indian Ocean. It is the only species of warbler found in the inner Seychelles, and one of 17 endemic species in the archipelago.
**Habitat:** Scrub to mature *Pisonia* woodland. The Seychelles warbler is unique among birds in that its territory is the smallest recorded for any perching bird. Once carrying capacity was reached on Cousin, an unusual social system developed. Young birds which, through lack of suitable habitat, were unable to establish their own territories remained in the parental territory and assisted with caring for subsequent broods. This system of breeding resulted in greater numbers of young being produced from territo-

ries with helpers compared with territories occupied by single pairs; it also provided the young birds with 'training' in family life which equipped them for eventual parenthood in their parents', or an adjacent, territory.
**Numbers:** Precise numbers were not known until 1959, when a survey of Cousin revealed just 26 birds. By that time, only tiny pockets of native woodland remained on Cousin, although the island was still free of alien predators.
**Red Data Book category:** Rare
**Threat:** First discovered on Marianne Island in the Seychelles, and described in 1873, the Seychelles warbler was later reported from several other islands including Cousine and Cousin. But within a few years, it had disappeared from every island except Cousin as a result of the wholesale clearance of native forest for planting with coconuts. Destruction of its natural habitat, combined with the introduction of alien predators, notably cats and rats, brought the Seychelles warbler to the brink of extinction.
**Conservation:** The danger of its imminent extinction led to Cousin Island being bought for ICBP to manage as a nature reserve, with the aim of securing the future of this single-island endemic bird and other native Seychelles flora and fauna. Management concentrated on eradication of the coconut palms and regeneration of the island's natural vegetation. Hunting and other forms of human intrusion were prohibited and rigorous measures taken to exclude exotic species. Gradually the Seychelles warbler recovered. By 1975, an estimated 274 warblers occupied 120 territories on Cousin; and, by the mid-1980s, the island appeared to have reached its carrying capacity of between 300 and 350 birds.

Throughout the 1980s detailed ecological studies were carried out on Cousin by a number of ICBP resident scientists and managers who were able to build up a detailed picture of the behaviour and ecology of this delightful bird. They determined the types of vegetation required by the species and compiled detailed information on diet, territory size and breeding behaviour. Despite its successful recovery, the Seychelles warbler remained at risk as long as the entire population was concentrated on a single island. Investigations of other islands within the Seychelles archipelago that might support Seychelles warblers began in the late 1980s. Two potentially suitable islands were identified – Aride Island, run by the UK-based Royal Society for Nature Conservation, and Cousine Island, which is privately owned – both containing substantial tracts of native vegetation and free from the most destructive introduced predators. In September 1989, following a detailed appraisal of Aride's insects and vegetation, ICBP translocated 29 Seychelles warblers from Cousin to Aride. Within two days, two of these birds had paired and begun nest building. All the birds survived translocation and were breeding within a few weeks of the transfer. The population doubled within the first six months and, by December 1991, just over 200 were counted on Aride. A further translocation to Cousine Island, made a year later, shows every sign of being successful. Thirty years ago the Seychelles warbler was near to extinction; its subsequent recovery means that its beautiful, melodic, robin-like territorial song can again be heard throughout the native woodland of Cousin Island.

The conservation of the endemic Seychelles warbler is a remarkable achievement. From being a single-island species in the late 1950s, totalling no more than 26 individuals, the bird is now established on three islands, two with populations of over 200. It can reasonably be considered out of danger and can thus be withdrawn from the *Red Data Book* – an extremely rare occurrence for any globally threatened species.

## Madagascar Little Grebe

*Tachybaptus pelzelnii*
**Order:** Podicipediformes
**Family:** Podicipedidae
**Range:** Closely resembling the little grebe inhabiting much of Europe and Africa, this small grebe is endemic to Madagascar, where it occurs from sea level to 1,800 m/5,900 ft.
**Habitat:** Lakes, pools and slow stretches

of river. For reasons that are not well understood, the Madagascar little grebe appears to prefer waters that are rich in aquatic plants, where it feeds on fish, aquatic invertebrates and even young birds.
**Red Data Book category:** Insufficiently known
**Threat:** The decline of this little-known species highlights a range of unusual threats to some globally endangered birds. Until recently, it was common and widespread throughout Madagascar, but in the last 30 years has dramatically declined. In addition to water pollution, which is obviously a problem for any aquatic animal, five further threats have been identified, including an unusual type of habitat destruction: the introduction of *Tilapia*, a vegetarian fish, into many of Madagascar's lakes and ponds has resulted in a massive reduction in aquatic vegetation. Another exotic fish, the black bass, not only competes with the Madagascar little grebe for food but also preys on its chicks. The widespread clearance of vegetation that followed the introduction of *Tilapia* has resulted in the waterways of Madagascar becoming more suitable for another species of little grebe, *T. ruficollis*, which, since 1945, has spread rapidly, greatly increasing in number and largely displacing the indigenous species. Moreover, as it becomes more dominant, *T. ruficollis* appears to have started interbreeding with the Madagascar little grebe. Such hybridization may seriously jeopardize its prospects of survival, as well as almost certainly leading to the extinction of yet another endemic Malagasy grebe, the Alaotra grebe *T. rufolavatus* – and little can be done to prevent it. The position is not helped by the lowering of Madagascar's water table, which has substantially reduced the amount of suitable grebe habitat.
**Conservation:** No active conservation measures have yet been taken. What is needed immediately is an evaluation of this species' precise status and an appraisal of the threats outlined above to determine which are the most serious. Yet, enough is already known to indicate the need for establishing a network of vegetation-rich lakes and pools from which non-native fish are excluded. Should hybridization be found a serious problem, setting up a captive propagation programme with genetically pure birds would have to be considered.

## Northern Bald Ibis

*Geronticus eremita*
**Order:** Ciconiiformes
**Family:** Threskiornithidae
**Range:** At one time the species was distributed throughout central Europe and probably across most of North Africa into the Middle East. Evidence that it was once widespread derives from Egyptian hieroglyphics, clearly illustrating the reverence with which this ibis was held in Egypt several thousand years ago.
**Habitat:** A gregarious species, breeding in colonies that traditionally use inland cliffs near watercourses. It feeds on invertebrates in soft damp ground, most frequently on river beds, sandbanks and sandy coastal strips where vegetation is sparse. Marshes, flooded fields and grasslands are all favoured for foraging.
**Numbers:** Now on the brink of global extinction, the world population was, by 1990, reduced to two small widely separated populations, one a resident breeding group in Morocco and Algeria, the other occupying a breeding site in southern Turkey and migrating to the Red Sea coast in winter. The latter group became extinct in 1990. The population surviving in northwestern Africa consists of one small colony in Algeria and about 12 colonies in Morocco, having declined from at least 38 known earlier this century. The thousand pairs of northern bald ibis known to exist in Morocco in the 1930s are now down to less than 16. Something of a mystery surrounds the continued sightings of individuals around the Red Sea in Saudi Arabia and the Republic of Yemen, despite the bird's disappearance from Turkey. It is just conceivable that an as yet undiscovered population survives somewhere in the Middle East.
**Red Data Book category:** Endangered
**Threat:** The demise of the eastern population breeding at Biricek in Turkey occurred despite the efforts of the Turkish authorities, a number of Turkish conservation bodies and the World Wide Fund for Nature (WWF). The blame is attributable to agricultural pesticides, which started to be used intensively in southern Turkey in the late 1950s and are still used today. Disturbance of the birds at their breeding site may have been an important contributory factor. The Moroccan and Algerian populations have so far not suffered from pesticides. The reduction in their range is due to the progressive

drying out of the semi-arid plains of North Africa and the Middle East, with hunting, poaching and disturbance during the breeding season probably the most critical subsidiary factors.

**Conservation:** Numerous attempts have been made to halt this bird's decline. As early as 1923 it received legal protection in Morocco, although the law was seldom enforced. Since the mid-1970s, WWF and the Moroccan authorities have undertaken a detailed study of the species as well as starting to protect nesting sites and critical feeding areas. An ambitious conservation plan that includes creating a national park for the species has also been drawn up. Unlike many birds, the northern bald ibis breeds well in captivity. In 1975 there were 215 individuals in twenty-nine zoos; by 1981, this figure had increased to 401 in thirty-three zoos. Attempts at re-establishing captive-bred birds in the wild in Israel were unsuccessful. Propagation is one of the contributions made by zoos to bird conservation in recent years. It is not inconceivable that within a few years the species will survive only in zoological collections.

## Jackass Penguin

*Spheniscus demersus*
**Order:** Sphenisciformes
**Family:** Spheniscidae
**Range:** The jackass penguin has colonized a number of islands off the coast of Namibia and South Africa, and in one or two places on the mainland.
**Habitat:** Like all penguins, the jackass lives mainly in the sea, but spends part of its life-cycle ashore. But, unlike many other penguins, it remains for most of the year close to its breeding grounds, feeding in the nearby sea.
**Numbers:** Despite being the subject of a number of aerial surveys and intensive studies, reports on the precise numbers of jackass penguin are conflicting. Estimates from the mid-1950s suggested a total of between 200,000 and 300,000 birds. In 1978, using similar methods, the figure was estimated at just over 97,000. It seems likely, therefore, that the population more than halved in just 22 years.
**Red Data Book category:** Threatened – of special concern
**Threat:** There is little doubt that the initial decline of the jackass penguin was caused by prolonged over-exploitation

of its eggs. Early reports suggest that egg-cropping took place on a huge scale. In 1930, for example, almost 14 million eggs were taken from the island of Dassen alone. Guano – the birds' accumulated droppings – was also collected for fertilizer. Removal of the guano not only causes disturbance but makes burrowing more difficult; and birds forced to nest in the open produce fewer young. Moreover, guano extraction encourages seals to take over land formerly occupied by penguins. Seabirds, at the top of the marine food chain, are extremely sensitive monitors of oceanic conditions. Studies show that in the late 1950s three species of fish – pilchard, anchovy and mackerel – were the most important components of the jackass penguin's diet. During the 1960s, stocks of South African and Namibian pilchard and horse mackerel plummeted as the result of overfishing. Deprived of its preferred food, the jackass penguin now relies on anchovies, plus a few less favoured items (notably squid) as its principal source of food. There are indications that this change and restriction of diet have led to lowered reproduction (through chick starvation). Shortage of food is currently the main threat to this species. Oil pollution has added to the plight of the jackass penguin. Between 1970 and 1980, more than 7,000 oiled penguins were counted, and the undocumented death toll at sea is certain to have been far higher.

**Conservation:** The jackass penguin is now fully protected by law in both South Africa and Namibia. Since 1969, a ban has been imposed on collecting both eggs and guano on many islands; several nesting sites are located within a network of marine reserves; and South Africa is taking more stringent measures to deal with oil pollution. Experiments aimed at deterring penguins from entering oil-polluted areas, by playing tape-recordings of killer whale vocalizations, have met with some success.

Priorities for the future must include a complete ban on all oceanic fishing within 25 km/15 mls of penguin breeding colonies. This should improve breeding success and post-fledging survival. More colonies should also be given nature reserve status, while illegal egg-collecting and guano scraping should be treated more seriously. Given these measures, the future of the jackass penguin should be assured.

## Freira

*Pterodroma madeira*
**Order:** Procellariiformes
**Family:** Procellariidae
**Range:** The island of Madeira, in the eastern Atlantic. Arguably Europe's most threatened species of bird, the Madeira freira highlights just how little is known about many of the world's seabirds. In the spring of each year, this small black, grey and white seabird arrives on the Portuguese island of Madeira – its only breeding place. It nests in burrows on narrow ledges on the upper slopes of the almost vertical cliffs of Madeira's inhospitable central mountains, laying a single egg. After breeding, it leaves Madeira and spends the winter somewhere in the South Atlantic; no one knows where.
**Habitat:** Upper mountain slopes and the sea. The freira's ancestors are believed to have colonized Madeira during the Pleistocene epoch when the climate was cooler and wetter. Its present restriction to the cool, moist, upper slopes of the island may therefore be a natural consequence of long-term climatic change.
**Numbers:** The freira is never likely to have been common. The population now exceeds 30 pairs, and at least three juveniles were fledged from the breeding ledges in 1990.
**Red Data Book category:** Endangered
**Threat:** Within six years of its discovery in 1903, the species had almost disappeared through the activities of collectors, who gathered not only the eggs but the chicks and adults as well. Over the next 60 years only two freiras were seen in Madeira, both young birds that had come to grief on the island's coast. The combination of human and animal predation – mainly by feral cats, shepherd dogs and the black rat – brought this species to the verge of extinction. After its rediscovery in the 1960s, Madeiran ornithologists suspected rats of inhibiting breeding in the only known colony. Visits to the ledges confirmed that rats were taking eggs and young, and were responsible for reducing the freira's breeding rate almost to zero.
**Conservation:** The rediscovery of the freira, in 1968, came about through a remarkable piece of detective work. Having recorded a closely related species on the nearby island of Bugio, ornithologists played the tape to local shepherds, who sometimes spend the night in the mountains. One shepherd

immediately recognized the cry. Thus encouraged, it was not long before several fresh burrows were located. Tragically, no sooner was this information made public than collectors removed three adult freiras and six eggs, and smuggled them out of Madeira – highlighting the fragility of many of the world's threatened birds. In 1987, ICBP's representative in Madeira, in close collaboration with the Parque Natural da Madeira and Museu Municipal do Funchal, organized a Freira Conservation Project. This part-

● The northern bald ibis is no beauty, but its iridescent black plumage, pink legs and bald, pinkish-red head give it a very striking appearance. (WWF / Udo Hirsch)

● The jackass penguin, like many other sea birds, has been a victim of oil pollution. But some success has been achieved in keeping it out of waters polluted with oil by playing tape-recordings of killer whales. (WWF / Y. J. Rey-Millet)

nership was strengthened by the financial support and technical assistance of ICI Public Health and the UK Government's Agricultural Development and Advisory Service. This remarkable team of local and international conservationists drew up and implemented a rat control programme. Two years of intensive rat control were necessary before the freira succeeded in breeding again. It is hoped that EEC funding will soon be made available to purchase the entire breeding area. Besides its importance to the freira, it also contains other endemic birds and plants. Management would then be the responsibility of the Parque Natural da Madeira, and protection thus greatly enhanced. Elimination of the rat is essential if the freira is to survive. But there is no easy solution; it is almost impossible to eradicate rats from an island the size of Madeira. A control programme must be a rigorous and continuous process. Although much remains to be done, results so far are encouraging. A determined effort has succeeded in bringing back

from the brink of extinction a species that had disappeared for 60 years.

## Mauritius Kestrel
*Falco punctatus*
**Order:** Falconiformes
**Family:** Falconidae
**Range:** This beautiful little falcon has always been confined to the island of Mauritius in the Indian Ocean.
**Habitat:** An inhabitant of mature, native, evergreen forest, where it feeds on a variety of small mammals, birds, lizards and insects. At the time of its discovery, the falcon was abundant throughout the native forest that entirely covered the island. It is now confined to the few relict patches of upland forest that remain.
**Numbers:** Even as recently as the 19th century the falcon was still relatively plentiful. Since then, its decline has been rapid. By the early 1970s, it was so rare that extinction appeared imminent. Surveys made at the time revealed that the population had been reduced to as few

as eight or nine individual birds, including the only four known nesting pairs surviving in the wild.

**Red Data Book category:** Endangered
**Threat:** The main threat to this kestrel, and indeed to all forest-dependent birds in Mauritius, is the loss of natural habitat, the native forest. Forest clearance has now more or less ceased – but only because little remains to fell. The few relict patches that have managed to survive continue to be degraded by the invasion of vigorous exotic plants, browsing by introduced deer and rooting by feral pigs. To this must be added the various problems caused by monkeys, rats and the Indian mynah, such as competition for food and nesting sites, and predation on eggs and nestlings.

**Conservation:** In 1973, the World Wide Fund for Nature (WWF), in collaboration with ICBP and the New York Zoological Society, took action to save this species. The programme they devised still continues under the management of the Government of Mauritius and the Jersey Wildlife Preservation Trust, with financial support from several other organizations. The programme focuses on breeding kestrels in captivity and reintroducing them into suitable localities – one of the few instances of captive-bred birds being successfully re-established in the wild. Between 1984 and 1988, 17 pairs of kestrels held at a breeding centre on Mauritius raised 75 young, 60% of which have been released. These birds established their first three nesting territories during the 1988–89 breeding season. Since the kestrel's natural food remains in short supply, ornithologists are exploring the possibility of changing its hunting habits by conditioning young kestrels to eat small birds – which are still abundant on the island – rather than their normal diet of lizards.

An imaginative conservation programme has succeeded in giving this bird a reprieve. But, if it is to survive in the long term, efforts must concentrate on the restoration and conservation of the habitats upon which this and other endemic Mauritian species depend. The World Bank has been asked to help finance the establishment of a national park to protect as much as possible of the forest that remains. This is essential for, without adequate habitat, free-living kestrels will have no possibility of increasing their numbers.

# Bannerman's Turaco

*Tauraco bannermani*
**Order:** Cuculiformes
**Family:** Musophagidae
**Range:** The mountains and highlands of Cameroon. By the time the first specimen of Bannerman's turaco was collected in 1922, most of its original forest habitat had already been destroyed. It is now confined to the Bamenda Banso highlands in northwestern Cameroon, and is one of 20 bird species that occur only in the montane forests of Cameroon and eastern Nigeria. These montane forests constitute the most important centre of endemism for birds in West and West-Central Africa, the largest remnant in Cameroon being on Mount Oku, known locally as Kilum.

**Habitat:** A member of a most unusual family of birds, inhabiting small pockets of forest throughout tropical Africa, Bannerman's turaco is restricted to montane forest, especially favouring ravines and crater rims, where it feeds entirely on fruit and berries of native trees. This strikingly colourful bird is predominantly green with a spectacular red crest, grey face, red-and-yellow bill, blue tail and prominent red primary feathers which are seen only when it flies – and are traditionally worn in tribal headgear as an indication of social status.

**Red Data Book category:** Endangered
**Threat:** Concern for the future of Bannerman's turaco arose from surveys of Cameroon's montane forests and birds conducted by the International Council for Bird Preservation. ICBP concluded that unless the forests of Kilum were effectively protected, Bannerman's turaco and another endemic species, the banded wattle-eye, *Platysteira laticincta*, were likely to become extinct.

**Conservation:** In 1987, ICBP launched the Kilum Mountain Forest Project, funded by WWF and the British Government's Overseas Development Administration and aimed at introducing a programme for the sustainable use of the Kilum Forests. Conservation of these forests would benefit the local people who depend heavily upon them for firewood, building materials, honey, medicine and a plethora of other forest products, and would have the incidental effect of providing a sanctuary for Bannerman's turaco. The Kilum Mountain Forest Project has had to contend with a number of problems: grazing by goats and sheep

has prevented forest regeneration on the upper parts of the mountain; lower down, trees whose bark is of medicinal value have been over-exploited; clearance for agricultural purposes has resulted in widespread loss and degradation of forest; and fires started by honey-hunters cause frequent damage. These difficulties have been met with an equally diverse array of conservation measures, including the creation, in agreement with the local farming community, of a forest reserve with clearly demarcated boundaries, a soil conservation and tree planting programme, and assistance to anyone wishing to utilize forest products in a sustainable manner. Much remains to be done, but there is no doubt that without the Kilum Mountain Forest Project neither the forest nor Bannerman's turaco would survive for long.

# Great Bustard

*Otis tarda*
**Order:** Gruiformes
**Family:** Otididae
The great bustard is one of the most impressive birds to be found striding across the open grasslands of Europe. Large, robust and stately, it is also strikingly coloured: its back, wings and tail are orange-and-black, neck orange-brown and head blue-grey, with a long white moustachial spray falling backwards from the bill across its neck. The male, standing just over 1 m/3 ft 3 in high and weighing up to 16 kg/35 lbs, is twice the size of the female. This is a gregarious bird, gathering for much of the year in flocks of up to 200 individuals. During the breeding season, the males perform elaborate courtship displays as they bid to attract females for mating.

**Range:** The great bustard inhabits grasslands reaching from as far west as southern Portugal to the northeastern steppes of China and the former Soviet Union. Most globally threatened birds are confined to small areas and it is hard to imagine that a species occupying so vast a range could be at risk, but in common with other members of the bustard family, the great bustard is now seriously threatened throughout its range.

**Habitat:** Open grassland, steppes and underdeveloped farmland form its natural habitat. Clearance of Europe's forests and their replacement with extensive arable land proved to the bustard's advantage, allowing it to spread as far north as southern Scandinavia and Scotland. But, starting with the Industrial Revolution, the steady mechanization of farming caused the species to dwindle until, today, this impressive bird has disappeared from the greater part of its range.

**Numbers:** No more than a few hundred bustards remain in eastern Europe, including Romania, Yugoslavia and Czechoslovakia; the Moroccan population is estimated at less than 100; while the seemingly endless steppes of the former Soviet Union support only about 3,000 birds. The largest surviving population is in the Iberian Peninsula where the Spanish authorities estimate numbers at 12,000–14,000.

**Red Data Book category:** Rare

**Threat:** In recent years both range and numbers have been progressively reduced as the result of constant harassment, disturbance by domestic livestock, and, most importantly, intensified agricultural development. This is particularly damaging when crops are sprayed with herbicides and pesticides, thereby reducing the insects on which the bustard feeds. The planting of hedges and shelter belts and the erection of fences effectively reduces the habitat available to the bustard, for whom an unrestricted view on at least three sides appears to be essential. This species' last remaining stronghold is the Iberian peninsula. Land management in this part of the world is traditionally labour-intensive, with crops harvested largely by hand, using extremely small applications of agrochemicals. Under such conditions the great bustard flourished. But the scene is rapidly changing. Spain's entry into the European Community and the consequent investment in capital-intensive farming is transforming the landscape: much suitable bustard habitat is likely to disappear.

**Conservation:** The traditional method of safeguarding a species by protecting particular sites is inappropriate for a bird requiring such wide expanses of grassland. Low-intensity farming is ideal, but does not lend itself to a system of protected areas. The best hope for the great bustard lies in collaboration between conservationists and farmers.

## Slender-billed Curlew
*Numenius tenuirostris*
**Order:** Charadriiformes
**Family:** Scolopacidae
**Range:** Morocco to eastern Russia. Closely resembling the Eurasian curlew and whimbrel, the slender-billed curlew is a long-distance migrant, moving from its nesting grounds on the Siberian forest steppe to its winter quarters in North Africa and parts of the Middle East.
**Habitat:** Steppes and freshwater wetlands.
**Numbers:** Once extremely common, surprisingly little is known about current numbers, distribution, or the reasons for its decline. At the turn of the century it was said to be the commonest species of curlew throughout North and West Africa, Malta and Sicily. In Hungary it occurred in flocks so dense that several birds could be killed with a single shot. But overall numbers may now be down to 100 individuals.

**Red Data Book category:** Insufficiently known

**Threat:** During the course of its migrations the slender-billed curlew regularly crosses the steppelands of Hungary and Yugoslavia and the wetlands of Romania, Greece and Italy, all of which have changed dramatically, agricultural development – both intensive and extensive – having greatly reduced the amount of suitable habitat available. As the curlew evidently makes temporary use of a variety of habitats on migration, this factor is unlikely to be the sole cause of its decline. Large waders are commonly hunted throughout Europe and North Africa, and the slender-billed curlew is extremely vulnerable to hunting, partly because of its innate tameness and partly because of the narrowness of its overland migration route. Regrettably, hunting continues, although ICBP is lobbying the authorities in a number of European countries in an attempt to persuade them to tighten up and enforce their hunting regulations.

**Conservation:** In 1988, concern for this bird's plight led ICBP to start identifying the factors responsible for its decline and the measures required to conserve it. Initially, ICBP concentrated on establishing its current range and preferred habitats, and on locating sites suitable for conservation. Eighteen key sites were found, reaching from the Sea of Azov in the former Soviet Union to the Mehran Delta in Iran and the Merja Zerga wetland in Morocco, totalling more than 8,000 km²/3,000 sq mls.

Some mysteries remain unresolved: although the species' only regular wintering ground is in Morocco, a significant number of unconfirmed sightings continue to be made in Iraq and Iran; both countries could be important over-wintering areas. Further surveys are clearly required. Yet more mysterious is Siberian scientists' failure to discover evidence of breeding despite the huge areas of Siberian forest steppe that appear to be suitable. Meanwhile, the future of the slender-billed curlew hangs by a thread. It remains one of the rarest and most poorly known birds in the Western Palaearctic, and stands as an example of the ease with which a species can slip almost unnoticed into oblivion.

## Harpy Eagle
*Harpia harpyja*
**Order:** Falconiformes
**Family:** Accipitridae
**Range:** From Mexico to Brazil and northernmost Argentina.
**Habitat:** This splendid raptor frequents undisturbed lowland tropical rainforest where it preys on animals of the forest canopy – sloths and monkeys in particular – by plucking them from the topmost branches.
**Red Data Book category:** Rare
**Threat:** The harpy eagle is totally dependent on pristine rainforest and is intolerant of any activity that adversely affects it. Individuals are believed to be unusually long-lived, but the reproductive rate is low.
**Conservation:** The conservation of large raptors poses a dilemma, as not much can be done to protect them. A bird as wide-ranging as the harpy eagle requires an immense area of undisturbed forest, and few reserves are large enough to contain it. Studies in French Guyana suggest that a pair of harpy eagles occupies a territory of 100–200 km²/40–80 sq mls. This means that at least 37,500 km² /14,500 sq mls (an area slightly larger than the Netherlands) of intact forest would be needed to support a harpy eagle population of 250 pairs. Its best hope lies in growing awareness of the importance of ensuring that large tracts

of unspoiled rainforest are preserved – not only for the benefit of the native fauna and the indigenous forest peoples but also for well-known climatic reasons.

## Kakapo or Owl Parrot
*Strigops habroptilus*
**Order:** Psittaciformes
**Family:** Psittacidae
This extraordinary bird, which is classified in a subfamily of its own (Strigopinae), is substantially larger than any other parrot. Nocturnal in habit and unable to fly, it spends its life on the ground, where it feeds on leaves, shoots and berries. The kakapo lacks the bright colours generally associated with parrots, its mottled green and yellow feathers helping to make it inconspicuous. Indeed, camouflage is this inoffensive bird's main means of defence. If threatened it freezes in the hope that the danger will pass. However satisfactory this form of defence may have been during the long period of time when there were no mammals (other than bats) and few birds of prey in New Zealand, it proved inadequate against alien predators when they were introduced into the country.

Apart from meeting briefly for mating – and females mate only about every fourth or fifth year – males live quite separately from females. As part of their courtship ritual, the males assemble on special display grounds, where for nights on end they concentrate on attracting females. Part of this 'lek display' involves making a characteristic booming sound, which reverberates through the forest, drawing in females from a wide area. After mating, each female returns to her own territory where she makes her nest on the ground among the roots of trees. Being on her own, she has periodically to leave first her eggs and then her nestlings untended – when they are of course extremely vulnerable to predation – while she forages for food.
**Range:** Originally found throughout New Zealand, but now confined to a few small islands off the coast of South Island.
**Habitat:** Lives in association with *Nothofagus* (southern beech) forest.

● A Mauritius kestrel hovers above the Black River Gorges, among the last remnants of upland forest on the island. (Planet Earth / Nick Garbutt)

● The harpy eagle is claimed to be the most powerful of all the birds of prey. Its position at the top of the forest food chain makes it an ideal indicator of environmental quality. (WWF / Fulvio Eccardi)

● The ground-dwelling kakapo or owl parrot, one of New Zealand's rarest birds, survives only in the wildest and most inaccessible parts of its range. (Auscape International / Geoff Moon)

**Numbers:** The total number of known kakapos stands at 43, of which only 14 are females: 11 on Little Barrier, 29 on Codfish, and three (all males) on Maud.
**Red Data Book category:** Endangered
**Threat:** Its decline seems to have started with the arrival of the Maoris in New Zealand in the 8th century. The Maoris hunted the kakapo not only for food but also for feathers, which they made into cloaks. They brought with them the dog and the rat, both of which preyed on the kakapo. The arrival of European settlers in the mid-19th century was followed by extensive clearance of the kakapo's habitat – by the early 1920s almost three-quarters of New Zealand's native forests had been felled. Competition for food from the introduced red deer, together with heavy predation by introduced carnivores – dog, cat, ferret and in particular the stoat, as well as two more species of rat – contributed to the decline.
**Conservation:** Several attempts have been made to establish the kakapo on islands with suitable stands of *Nothofagus* – the most ambitious being the release of 370 kakapos on Resolution Island, a large island in Fiordland, around the beginning of the 20th century; but all were defeated by stoat predation. In the mid-1970s, a concerted effort was made to scour Fiordland for kakapos, using specially trained dogs for the purpose. Over a three-year period 17 (all males) were caught; three were released on Maud Island, a predator-free island in the Marlborough Sounds, at the northern end of South Island. In 1977, a hitherto unknown colony of about 200 birds was discovered on Stewart Island, where the main focus of conservation activity was for a time concentrated. But hopes of having found a safe haven for the kakapo were dashed when it became evident that substantial numbers of birds were

being killed by feral cats. Attempts to eliminate the cats were only partly successful, and as the kakapo would be at risk as long as any cats remained, New Zealand's Department of Conservation concluded that the only way of ensuring its survival lay in establishing the kakapo on a cat-free island. In 1982, 21 birds were taken to Little Barrier Island, off New Zealand's east coast to the north of Auckland. Five years later, in 1987, it was decided to spread the risk by moving the

last remaining birds on Stewart Island to nearby Codfish Island, while measures continued to round up the few birds still on Stewart Island.

In 1989, the Department of Conservation started a five-year recovery programme for the kakapo, costing the equivalent of US$1.8 million. An eight-man Kakapo Recovery Group has been set up charged with ensuring the survival of the species.

# REPTILES

## Green Turtle

*Chelonia mydas*
**Order:** Testudines
**Family:** Cheloniidae (marine turtles)
**Size:** Up to 1 m/3.25 ft long, and weighing up to 270 kg/600 lbs.
**Range:** Circumtropical, in waters with a temperature not less than 20 °C/68 °F.
**Habitat:** Distribution conforms closely with the availability of sea grass pastures on which the adult green turtle mainly feeds.
**Numbers:** Of the 150 known nesting sites, only 10-15 harbour sizeable populations – i.e. 2,000 or more nesting females. The largest rookery is on Raine Island (northern Barrier Reef) where up to 80,000 females come ashore each year to nest. Apart from Raine, the only other nesting populations not in decline are on Heron Island (southern Barrier Reef), Tortuguero (Costa Rica), Ras al Had (Oman), three islets near Sandakan (Sabah, North Borneo), and the islands of Europa, Tromelin and Glorieuse in the Mozambique Channel (all of which are dependencies of Réunion).
**Red Data Book category:** Endangered
**Threat:** Heavily hunted for its meat, oil (used in the manufacture of cosmetics), and calipee (the cartilaginous part of the lower shell). Commercial exploitation is not confined to local people: Japanese turtle boats reaching deep into the Pacific and Indian Oceans were responsible for catching over 16,000 tonnes of turtle meat in 1968. The almost insatiable demand for turtle eggs – reputed to possess aphrodisiac qualities – has been a major factor in the green turtle's decline. Whereas, in the 1930s, 2 million eggs were collected annually in the turtle islands of Sarawak, for instance, the total had fallen to fewer than 300,000 by the 1970s. Many adult green turtles are taken during the breeding season when, for a period of three months, they congregate only a short distance offshore. The females are particularly vulnerable when they come ashore to nest. While some females lay only a single clutch, others lay from three to seven clutches at intervals of 10-15 days, each occasion

involving great risk. Heavy losses are incurred by the hatchlings as they break out of the nest and make a dash for the sea. The distance from nest to sea is usually short, but every step of the way is hazardous in the extreme, for both on and above the beach are hordes of predators of every description: rats, mongooses, pigs, dogs, monitor lizards and other land animals lie in wait. Armies of crabs surge across the sand to seize the hatchlings in their claws, while flocks of gulls and other birds strike from above. Even when they have reached the surf the danger is not over: sharks, groupers and all manner of predatory fish lurk in the shallows. The running of this nightmarish gauntlet is accomplished only at prodigious cost. It seems remarkable that any hatchlings manage to survive: few, of course, do. Young turtles are believed to spend the first year of their lives in mats of *Sargassum* seaweed, and mounting evidence points to the probability of heavy mortality being incurred by accumulations of oil particles in the *Sargassum*. Yearling turtles have been found with their respiratory tracts blocked with blobs of tar. Many turtles are also caught accidentally in nets and trawls set for fish and shrimps.
**Conservation:** Legally protected over much of its range, and listed in Appendix 1 of CITES. Because of its vast range and migratory habits – one population migrates from its feeding grounds off the coast of Brazil to its nesting site on Ascension Island, 2,250 km/1,400 mls out in the Atlantic – measures to protect the green turtle require international sanction. In some places (e.g. the Galápagos and Tortuguero) national parks have been established over the nesting beaches. Malaysia has created a Marine Turtle National Park (17 km²/6.5 sq mls) covering the three turtle islands near Sandakan, and has imposed a ban on trawlers operating in their immediate vicinity. Thailand has established the Tarutao Marine National Park (1,490 km²/575 sq mls) embracing 51 islands, with ships of the Royal Thai Navy patrolling offshore during the breeding season. Some countries

license egg collectors and impose limits on the number of eggs they may take; others stipulate that a percentage of the eggs must either be left in the nest or taken to fishery stations for hatching and subsequent release. The establishment of turtle farms might be expected to relieve pressure on wild stocks; but while some authorities believe that this represents the only way of conserving marine turtles, others hold the view that it is likely to have the opposite effect by increasing the demand for wild-caught turtles.

The economic importance of the green turtle rests upon its availability as a source of easily procurable natural food found in great quantity in parts of the world where meat is otherwise scarce. With a little foresight and a modicum of management, this prolific source of high-grade protein could continue indefinitely to meet the needs of large numbers of people. But if depletion is permitted to continue and the green turtle is exploited to extinction the people living at subsistence level around the shores of tropical seas will be deprived of an incomparable and irreplaceable source of food.

## Galápagos Giant Tortoise

*Geochelone elephantopus*
**Order:** Testudines
**Family:** Testudinidae (land tortoises)
**Size:** Mature giant tortoises weigh 140-180 kg/300-400 lbs, and have a carapace length of up to 1.20 m/4 ft.
**Range:** Confined to the Galápagos Archipelago.
**Habitat:** Ranges from arid lowlands, where *Opuntia* cactus is a major source of water, to humid uplands. Where possible, giant tortoises spend long periods submerged in pools.
**Subspecies:** The different ecological conditions prevailing on the various islands of the Galápagos have led forms to evolve in ways that differ from one island to another, resulting in the evolution of 15 distinctive races, five of which are on Isabela, where they have long been iso-

lated from one another by impassable lava flows. Four of the races are extinct, while the Pinta race is represented by only a single male.

**Numbers:** At its peak the Galápagos giant tortoise was numbered in the hundreds of thousands. Fewer than 15,000 remain.

**Red Data Book category:** Endangered

**Threat:** Exploitation of the tortoises for food occurred on an increasing scale from the 16th and 17th centuries onwards. They were a convenient source of fresh meat, being plentiful, easily caught and good to eat. A further advantage was their ability to survive several months without food or water; they could therefore be stored aboard ships (usually on their backs) until required. In the 19th century tortoises were slaughtered not only for meat but also for their oil. Yet, destructive as this uncontrolled exploitation unquestionably was, it was not as devastating as the damage caused by alien animals introduced on to the islands colonized by man. As the number of tortoises diminished, their place as principal herbivore was taken by goats released on several islands. Immense damage was also caused by pigs, dogs and rats digging up the eggs and devouring hatchlings and young tortoises.

**Conservation:** Strictly protected in the Galápagos, and listed in Appendix 1 of CITES. The Charles Darwin Research Station has started a programme for raising young tortoises. This involves collecting eggs from islands where they are threatened, incubating them, and rearing the hatchlings in captivity until large enough to fend for themselves, when they are returned to their island of origin. Without this help, the Pinzón race, for example, would almost certainly be extinct. The Española race was similarly depleted until only two males and 13 females remained. In 1965, the survivors were taken to the Charles Darwin Research Station for captive breeding, and 79 young tortoises were subsequently released back on Española, with a further 68 following later.

The success of these captive-breeding programmes highlights the fact that the survival of the giant tortoise in the wild is dependent on the eradication of exotic predators, to which they are particularly vulnerable and against which they can offer no defence.

---

# THE LONG-LIVED GIANT TORTOISE

Giant tortoises may live for a century or more: the one presented to Captain James Cook by the King of Tonga in the late 1770s lived until 1927, so must have been at least 160 years old. They do not become sexually mature until 30–40 years of age. Mature females lay from 1–4 clutches of eggs each season, averaging in the order of 2–16 eggs a clutch.

Giant tortoises obtain most of their water from dew deposited on vegetation. Although capable of going for long periods without drinking, they are nevertheless extremely fond of water, and whenever possible like to submerge themselves in shallow pools. This provides some degree of protection against clouds of mosquitoes which bite between the horny plates of the carapace. In the highlands, submersion is also warmer than exposure to the cool night air.

Tortoises rarely have disputes, but when they do – usually over food – the contestants rise up on their legs, open-mouthed and necks outstretched. But this seldom leads to biting: the loser – generally the smaller of the two – concedes defeat by moving away, leaving the field to the victor.

A special relationship has evolved between finches on the one hand and tortoises (and marine iguanas) on the other, whereby the birds remove ticks and other parasites from the reptiles' skin. The finch signifies its readiness by hopping in front of the tortoise. The tortoise responds by raising itself high on its legs, neck outstretched, thus making it possible for the bird to reach the parts of the skin where the ticks are located.

---

# Aldabra Giant Tortoise

*Geochelone gigantea*
**Order:** Testudines
**Family:** Testudinidae (land tortoises)
**Size:** Carapace length of up to 105 cm/ 41.25 in.
**Range:** Aldabra Atoll 9°24′ S, 46°20′ E, approximately 420 km/260 mls NNE of Madagascar, and 640 km/400 mls from the African mainland. The atoll is about 29 km/18 mls long, and consists of four principal islands – the largest being South Island (Grande Terre) – surrounding a large mangrove-fringed lagoon.
**Habitat:** Open grassy areas with a scattering of trees and bushes and scrubland on coral limestone; also frequent mangrove swamps. Most of Aldabra's tortoises are concentrated at the east end of South Island. In the dry season, they spend the early morning wallowing in freshwater pools, retiring during the heat of the day to the shade of *Pandanus* palms. They feed mainly on 'tortoise turf', a stunted vegetation of grasses, sedges and herbs. An interesting symbiotic relationship has evolved between the giant tortoise and the green gecko, *Phelsuma abbotti*, which feeds on mosquitos attacking the tortoise's exposed soft neck skin.
**Numbers:** Total about 150,000.
**Red Data Book category:** Rare

**Threat:** Giant land tortoises were at one time to be found on every continent except Australia. Two centuries ago they inhabited numerous islands in the Indian Ocean - Madagascar, the Comoro Islands, the Mascarene Islands (Mauritius, Réunion and Rodriguez), the Seychelles, Aldabra, Assumption, Astove, Cosmoledo, Providence, Farquhar, Amirantes, and the Chagos Archipelago. But, by the middle of the 19th century, they had been exterminated everywhere except Aldabra. A related species, *G. elephantopus*, lives in the Galápagos. These two widely separated island groups – one in the Indian Ocean and one in the Pacific – are the only places where free-living giant land tortoises have survived.

The disappearance of the giant tortoise from other islands in the Indian Ocean was attributable partly to human predation - it was a readily available source of food for visiting seamen - and partly to competition for forage with goats and other introduced livestock. The only reason the tortoise survived on Aldabra was because the atoll was so inhospitable that mariners usually avoided it. Adult tortoises have few enemies other than man; but the scarcity of young tortoises is indicative of massive mortality during the first few years of life. Bird predation is severe, but the heaviest

● Two male angulated tortoises fighting for dominance at the Ampijoroa Captive Breeding Station on Madagascar. (WWF / Meg Cawler)

● A green turtle off Hawaii. An incomparable source of human food, this species is in danger of being hunted to extinction. (Planet Earth Pictures / Doug Perrine)

losses among young tortoises are caused by the robber crab.

**Conservation:** Although numbers are currently at a level at which the species appears secure – and on South Island are probably at their optimum – concentration on a single atoll nevertheless leaves this tortoise vulnerable to either natural or man-made disasters. To spread the risk, small breeding groups have been introduced onto several islands in the western Indian Ocean, among them the Seychelles, Mauritius, Réunion, Prison Island (off Zanzibar), and Nossi Bé (north-eastern Madagascar). Man-made disaster came close to reality in the mid-1960s when the British government announced its intention of constructing a military airfield and transmitting station on Aldabra, a proposal that was the more reprehensible when it became known that the base was required to meet only a temporary need and would probably be abandoned within a decade. Apprehension at the government's intentions culminated in the President of the Royal Society personally delivering a memorandum to the Secretary of State for Defence expressing the Society's concern.

This unprecedented action was followed by the despatch of a Royal Society Expedition to Aldabra to study the position at first hand. This intervention led to the airfield proposal being abandoned and to the establishment of a permanent research station on the atoll – now maintained by the Seychelles Islands Foundation on an international basis.

## Angulated Tortoise (Angonoka)
*Geochelone yniphora*
**Order:** Testudines
**Family:** Testudinidae (land tortoises)
**Size:** Carapace about 45 cm/17 in (70 cm/27.5 in when measured over the dome).

**Range:** Endemic to Madagascar, where it is restricted to a very small area around Baly Bay, on the northwestern seaboard.
**Habitat:** The vegetation around Baly Bay consists of three distinct types: mangrove, dry deciduous forest and palm savanna. The angonoka favours the zone between forest and savanna, and especially places where secondary forest and bamboo intergrade into more open grassland. This mixed habitat combines open herbaceous areas for foraging with dense *Terminalia* and bamboo thickets for concealment, protection and shelter from the heat of the day.
**Numbers:** Unlikely that more than 100–400 remain. This species has been described as the rarest tortoise in the world.
**Red Data Book category:** Endangered
**Threat:** Annual burning of the vegetation in order to improve grazing for livestock results in the land being swept by fire, killing those tortoises unable to escape the blaze and further degrading the very limited habitat that remains. The destruction of eggs and hatchlings by rooting pigs represents a major threat to the angonoka. This follows the introduction of the bushpig from Africa. As it is seldom hunted, this animal has become common in the vicinity of Baly Bay.
**Conservation:** Listed in Appendix 1 of CITES, and protected under Malagasy law. The local Sakalav people observe a taboo (*fady*) on the eating of land tortoises. Angonokas are nevertheless kept as pets in the belief that they have the power to ward off poultry disease. IUCN has recommended the establishment of a special angonoka reserve at Cape Sada on the northeastern tip of Baly Bay, an area that is currently uninhabited. A captive breeding programme has been started to enhance the reptile's prospects of survival, and an educational programme is being prepared to encourage greater awareness in the local people of conservation issues, not least in developing less destructive methods of agriculture.

## Chinese Alligator

*Alligator sinensis*
**Order:** Crocodylia
**Family:** Alligatoridae (alligators and caimans)
**Size:** A small species, attaining a maximum length of 2 m/6.5 ft.
**Range:** Restricted to the lower Chang Jiang (Yangtze) valley in Anhui, Zhejiang and Jiangsu provinces. This compares with its former range which extended along the lower and middle reaches of the Chang Jiang basin, west to the Yunmeng Swamp in Hubei Province and the Dongting Hu in Hunan Province, and northwards to the Huang He (Yellow River). More specifically, it is believed still to occur in a number of freshwater lakes draining into the Chang Jiang, among them the Shengjin Hu, a complex of about 40 interconnecting lakes, both large and small, forming an extensive area of marsh and swampland covering about 1,500 km²/579 sq mls, including the Shijiu Hu (whose surface area fluctuates in accordance with the level of the Chang Jiang from 214–340 km²/82–131 sq mls); the Gucheng Hu (243 km²/94 sq mls); and the Nanyi Hu (171 km²/66 sq mls).
**Habitat:** Of semi-aquatic and secretive habit, the Chinese alligator spends much of its time in burrows which it excavates in river banks.
**Numbers:** More than 500 were counted in 1956, but 20 years later only about 100 could be found. Today, the free-ranging population in Anhui Province is estimated at about 300.
**Red Data Book category:** Endangered
**Threats:** Heavy pollution and siltation in conjunction with loss of wetland habitat, disturbance and wanton killing brought this alligator close to extinction. It is killed at every opportunity, both for its valuable skin and for its meat, which is highly esteemed. The extent to which every man's hand is against it makes its survival seem almost miraculous.
**Conservation:** Listed in China as a 'First Class Endangered Species', and given full legal protection. Sporadic poaching nevertheless continues. In 1982, the Chinese Alligator Nature Reserve (430 km²/166 sq mls) was established in southern Anhui, 60 km/37 mls southeast of Wuhu, 30°50′–31°00′ N, 118°20′–119°25′ E. A successful captive breeding programme is being undertaken at the Xiadu Alligator Farm in Xuancheng County, Anhui Province, where about 1,000 alligators are being raised each year, among them some of the second generation to be born in captivity.

## Nile Crocodile

*Crocodylus niloticus*
**Order:** Crocodylia
**Family:** Crocodylidae (crocodiles, caimans and alligators)
**Size:** A typical specimen measures about 3.5 m/11.5 ft; exceptionally, individuals attain a length of about 5 m/16 ft. Larger specimens have been recorded in the past: one of 6.4 m/21 ft from Lake Victoria, and another of 7.9 m/26 ft from Lake Kioga (Uganda).
**Range:** Occurs throughout Africa south of the Sahara, extending northwards up the Nile as far as the Aswan Dam (Lake Nasser), but absent from the arid northeast (mainly Ethiopia and Somalia) as well as from much of southwestern Africa, notably Namibia (excepting the Cunene River on the border with Angola). It also occurs in Madagascar.
**Habitat:** Inhabits a variety of freshwater habitats, in particular large rivers, lakes and wetlands, but has also been recorded from estuaries and mangrove swamps.
**Red Data Book category:** Vulnerable
**Threats:** The decline from abundance to scarcity which has come about since the end of the Second World War is attributable primarily to excessive hunting. During the 1950s some 5–10 million Nile crocodiles a year were killed for their skins. Even in areas where numbers have been so heavily reduced that commercialized hunting is no longer economically viable, crocodiles continue to be taken by opportunist hunters. Human pressure on the land has brought widespread degeneration of habitat, stemming from such factors as clearance of riverine forest, drainage of wetlands, changes in water regimes through irrigation schemes (including increased levels of salinity), burning of reedbeds, trampling of nesting sites by cattle. The increasing use of gill nets (which are replacing the traditional nets and fish traps in many parts of Africa) is responsible for heavy mortality, the crocodiles becoming entangled in the mesh and drowning. This has been particularly evident on Lake Turkana in Kenya.
**Conservation:** Listed in Appendix 1 of CITES. Legally protected in many parts of its range, but such protection is largely nominal. The crocodile's best hope of survival lies in the various national parks and other designated sanctuaries in which it occurs. Several projects have been started for breeding crocodiles in captivity and subsequently restocking suitable rivers and lakes. South Africa pioneered the breeding of crocodiles in captivity,

and has established the Lake St Lucia Estuary Crocodile Centre for this purpose. Commercial crocodile farming is also being tried in a number of countries. In Zimbabwe, important populations on the Zambezi River and Lake Kariba are protected, with a yearly quota of 2,500 crocodile eggs being allotted to each of four licensed rearing stations. Some 5% of the crocodiles raised from these eggs are returned to the Department of National Parks and Wildlife Management for restocking purposes. The remainder are slaughtered and marketed commercially. It is important that the hide trade should continue to be regulated and monitored, failing which surviving stocks will become increasingly depleted.

## Tuatara

*Sphenodon punctatus*
**Order:** Rhynchocephalia
**Family:** Sphenodontidae (tuatara)
**Size:** This lizard-like reptile is about 70 cm / 27.5 in long. Adult males weigh about 1 kg / 2.25 lbs. Females are smaller, rarely exceeding 500 gm / 1 lb 2 oz.
**Range:** Confined to New Zealand, where it is found on about 30 coastal islands, of which three are in Cook Strait (between North and South Islands) and the remainder off the northeast coast of North Island between North Cape and East Cape. Nearly all the islands are wholly or partly fringed by cliffs and subject to sudden, and sometimes violent, changes of sea and weather conditions. Landings are thus difficult or hazardous, so few islands have been thoroughly investigated by zoologists. Subfossil remains from early Maori middens and peat bogs indicate that the tuatara's original range extended from North Cape at the northernmost tip of North Island to Bluff in the extreme south of South Island.
**Habitat:** Tuatara islands are characterized by low forest and scrub, with plants that are tolerant of salt spray. The ground is generally riddled with burrows made by one or other species of petrel. Many of the islands have also been colonized by other marine birds, often occurring in dense concentrations. High levels of manuring give rise to large numbers of small soil organisms and terrestrial insects on which the tuatara feeds.
**Numbers:** Probably in excess of 100,000, of which one-third or more are found on

---

### MONSTER OR BENEFACTOR?

Crocodiles are the surviving remnants of the Archosaurs, the 'Ruling Reptiles', that dominated the earth during the Mesozoic Era. Having lived for more than 100 million years, they have been so severely reduced during the last 40–50 years that, of the 23 living species of crocodilians, 11 are endangered.

Our attitude towards crocodiles has changed diametrically: from being revered by the ancient Egyptians, they are nowadays regarded almost everywhere with revulsion. The game regulations of some countries lend force to this prejudice by drawing a distinction between game animals that may be legally hunted only by approved methods, and crocodiles that can be hunted without restriction, like vermin. Crocodile hunting usually takes place at night, using spotlights. Needless to say, this is a highly destructive practice against which the crocodile has no defence and little chance of escape.

The work of Dr Hugh Cott and others has shown that neither in the ecological nor in the commercial sense are many of the popular prejudices against crocodiles justified. Despite their ungainly appearance and reputation for ferocity and cunning, crocodiles are neither voracious feeders nor do they consume immense quantities of fish; on the contrary, they eat little fish except in middle age, when they generally prefer the coarser, predatory species. The commercially important types of fish are seldom included in their diet.

Apart from its ecological importance, the crocodile is an animals of substantial commercial value: prime crocodile skins fetch the equivalent of many months' conventional income for a peasant farmer – which serves as much to indicate the reptile's economic importance as to explain the principal cause of its decline.

By drawing attention to the wilful dissipation of a valuable natural resource, Hugh Cott's findings helped stimulate the establishment of the IUCN/SSC Crocodile Specialist Group. This body of international specialists currently regards seven species of crocodilians as critically depleted: the Chinese alligator; the Orinoco, Philippines, and Cuban crocodiles; the tomistoma and the gharial. Four other species are endangered: the broad-snouted and black caimans; the American crocodile; and Morelet's crocodile. The different species vary greatly not only in size and appearance but also in their habitat requirements, food preferences, and reproductive behaviour.

Demand for skins remains as high as ever, and regulation of the trade – valued at more than $200 million a year – is vital. By concentrating on meeting the requirements of the trade from the product of crocodile farms, IUCN hopes to reduce the pressure on wild stocks, while at the same time doing everything possible to reduce the illegal trade in skins, which is particularly pronounced in Latin America and Southeast Asia.

---

Stephens Island (150 ha / 370 acres), which supports at least 30,000, and possibly as many as 50,000.
**Red Data Book category:** Rare
**Threat:** Until the arrival of the Maoris, the tuatara was widely distributed throughout New Zealand. By the time of European settlement, it had already disappeared from the two main islands, having been exterminated either by exotic predators introduced by the Maoris or as the result of climatic or ecological change. Although Maoris sometimes ate tuataras, there is no evidence that this was a factor in their disappearance. Ro-

dents pose a permanent threat: the tuatara cannot survive in the presence of rats. Despite every precaution, the possibility of rats gaining access to tuatara islands cannot be ruled out.
**Conservation:** Stringently protected under the law by the New Zealand Wildlife Service. All tuatara islands have been designated Wildlife Sanctuaries or Flora and Fauna Reserves, and special permission is required to visit them.

● A Nile crocodile in the Côte d'Ivoire. Once
widespread and abundant, this reptile has
undergone a dramatic decline in recent years.
(WWF / Albrecht G. Schaefer)

● The Chinese alligator has been hunted to
the verge of extinction both for its skin and
for its meat. (WWF / Urs Woy)

# ANCIENT SURVIVOR

Despite its superficial resemblance to a lizard, the tuatara cannot be classified with any of the four main groups of reptiles – turtles, crocodiles, lizards and snakes. In fact it is not a lizard but the last surviving representative of the rhynchocephalians, a very ancient order of reptiles (once numbering 23 species) that flourished before the Age of Dinosaurs, and has remained almost unaltered since it arose in the Triassic more than 200 million years ago. The tuatara is thus of exceptional zoological interest. It is the only living animal to have retained virtually unchanged the basic anatomical features of an archaic group of reptiles otherwise known only from fossils. These include a pineal (or third) eye. The precise function of this vestigial third eye remains uncertain; but in lizards it appears to act rather like a thermostat in regulating exposure to sunlight – a useful function in a cold-blooded animal.

All the islands on which the tuatara occurs have breeding colonies of petrels and shearwaters. An unusual adaptation to island living is the tuatara's practice of taking up residence in the burrows that these birds are in the habit of excavating, not only while they are at sea but often while they are incubating eggs or rearing chicks. Bird and reptile tolerate each other. The tuatara is spared the trouble of having to dig its own burrow as well as benefiting from food left over from the mutton bird's meals; occasionally it even helps itself to the eggs and chicks of its host, although beetles and other invertebrates, along with small lizards, form its more usual diet. The reasons for this bird-reptile association are obscure, as the tuatara is perfectly capable of exca-

● The tuatara of New Zealand is the only living representative of an order of reptiles that flourished some 220–150 million years ago. (WWF / Urs Woy)

vating burrows for itself. Tuataras may use several burrows, and sometimes several tuataras share the same burrow.

Growth rate is extremely slow. The tuatara does not reach maturity for more than 20 years – but may then live for more than a century. It is also a slow breeder: fertilization occurs as much as ten months before the eggs are laid – in clutches varying from 8 to 15 – in specially constructed undergound chambers or blind tunnels, which are then filled with soil and abandoned. The eggs may take a further 12–15 months to hatch.

# INVERTEBRATES

## Triton's Trumpet (Giant Triton)
*Charonia tritonis*
**Phylum:** Mollusca
**Class:** Gastropoda
**Order:** Mesogastropoda
**Family:** Cymatiidae
**Size:** Shell reaches a length of 40 cm / 16 in or more. May take up to six years to attain maximum size. Has long been used throughout the Pacific region as a trumpet.
**Range:** From the Red Sea and Indian Ocean – Mozambique and the Seychelles – to the Indo-Pacific, notably Australia's Great Barrier Reef, Philippines, Indonesia, Fiji, Guam, the Marshall Islands and Hawaii.
**Habitat:** Normally found on coral reefs. On the Great Barrier Reef it frequently occurs on seaward slopes or on coral pinnacles in lagoons, but in Indonesia it has been recorded on sand near coral. Nocturnal in habit, it feeds principally on the crown-of-thorns and other starfish. In some places, infestations of crown-of-thorns starfish have been found to correlate with low numbers of giant tritons, but the extent to which the giant triton exercises control over these starfish is not known, and other factors are certainly involved.
**Red Data Book category:** Rare
**Threat:** Its great size and attractive appearance make this mollusc very popular with the ornamental shell trade. Over-collecting, to which it is extremely vulnerable, has been responsible for its decline in the Philippines, Guam, southern Japanese waters and other parts of its range. In a three-month study of the Sudanese Red Sea, only two specimens were found, and local collectors say that they are now found only occasionally. In Indonesia, a survey of 133 sites on 92 reefs also recorded only two.
**Conservation:** Protected throughout Australian waters. In 1970, Taiwanese fishermen were fined for collecting giant tritons near the southern end of the Great Barrier Reef. Protected in Seychelles and export from Fiji banned.

## Giant Clam
*Tridacna gigas*
**Phylum:** Mollusca
**Class:** Bivalvia
**Order:** Veneroida
**Family:** Tridacnidae
**Size:** The largest living shelled mollusc, measuring up to 1.37 m / 4.5 ft and weighing over 200 kg / 440 lbs, of which 55–65 kg / 120–140 lbs is living tissue. Growth is rapid during the early stages of life but slows down with age. Giant clams may live for 100 years or more.
**Range:** Extends from the Philippines to Micronesia. More specifically, it has been recorded from the Ryukyu Islands, Philippines, Indonesia, Queensland and Western Australia, Papua New Guinea, Solomon Islands, Vanuatu (New Hebrides), Caroline Islands, Marshall Islands and Gilbert Islands.
**Habitat:** Giant clams are a dominant feature of most Indo-Pacific coral reefs, where they are confined to shallow waters at depths of between about 1-20 m / 3-65 ft.
**Red Data Book category:** Vulnerable
**Threat:** Giant clam meat is a popular food throughout much of Asia, and clams of all kinds are commercially fished. The Taiwanese in particular take large quantities. They are also collected for their ornamental shells. Giant clams are extremely vulnerable to over-exploitation, and uncontrolled collecting has led to local extinctions as, for example, throughout much of western Indonesia. In Ponape, the presence of empty shells is evidence that the giant clam once appeared there. The taking of giant clams for subsistence purposes is a further factor in the decline. In Tonga, the demand for clam meat is greater than the supply. Loan schemes for the purchase of boats and outboard motors have increased pressure on stocks, and fishermen are having to travel ever greater distances to obtain catches.
**Conservation:** Fully protected in all Australian waters including the Great Barrier Reef, where aircraft are used for surveillance purposes. Together with the introduction of the 320 km / 200 mls Australian Fishing Zone, this has succeeded in virtually eliminating clam poaching by Taiwanese on the Great Barrier Reef. In Papua New Guinea, Taiwanese fishermen caught poaching have had both their catches and their boats confiscated and their captains fined. Surviving stocks in most parts of the Pacific are in urgent need of protection. Suitable legislation needs to be adopted and enforced, and special clam sanctuaries established. Proposals have been made to include the giant clam in Appendix 2 of CITES, thus exercising control over the export trade.

## Giant Gippsland Earthworm (Karmai)
*Megascolides australis*
**Phylum:** Annelida
**Class:** Oligochaeta
**Order:** Haplotaxida
**Family:** Megascolecidae
**Size:** Among the world's largest earthworms, averaging about 1.5 m / 5 ft long with a diameter of 2 cm / 0.75 in, but attaining a length of up to 4 m / 13 ft and a diameter of 4 cm / 1.5 in.
**Range:** Confined to South Gippsland, Victoria, Australia, where it is limited to an area of about 1,000 km² / 385 sq mls, centred on Loch Korumburra, principally the hilly parts of the western Strzelecki Ranges and the undulating country around Warragul.
**Habitat:** Sloping land in open forest and pasturelands used for dairy farming. Giant worms favour damp soils, and are accordingly found close to the surface in winter when the ground is wet, but descend to greater depths during the summer. Unlike most earthworms, the giant earthworm may live colonially, constructing a permanent system of burrows, in which each worm has its own individual hole. It feeds on organic matter in the soil. Giant earthworms are believed to emit a pungent odour that repels birds, although the kookaburra is known to eat them.

**Red Data Book category:** Vulnerable
**Threat:** Giant worms and their cocoons might seem vulnerable to ploughing, but there is no evidence that they have been adversely affected by cultivation.
**Conservation:** Mount Worth State Park (164 ha/405 acres) in the western Strzelecki Ranges is the only protected area within the distributional area. Field studies are required to ascertain the status of the giant worm and to learn more of its natural history and ecological requirements, with a view to deciding what measures should be taken for its long-term protection.

## Red-kneed Tarantula

*Brachypelma smithi*
**Phylum:** Arthropoda
**Class:** Arachnida
**Order:** Araneae
**Family:** Theraphosidae
**Size:** Mature red-kneed spiders attain a diagonal span of about 14 cm/5.5 in; exceptionally, they may reach 17 cm/6.75 in.
**Range:** Endemic to Mexico west of the Sierra Madre Occidental.
**Habitat:** This spider is essentially an inhabitant of semi-desert country, and builds its burrows in sand or soil banks. It feeds mainly on insects, and occasionally small vertebrates and carrion.
**Red Data Book category:** Insufficiently known
**Threat:** While cultivation of marginal agricultural land may be reducing the range of this spider, the principal cause of decline is the demand for unusual and exotic pets which has grown enormously in recent years. Despite its ability to inflict a painful bite with its poisonous jaws, the red-kneed tarantula is highly prized on account of its colourful appearance – black with attractive red markings – in an otherwise rather drab family. The trade is channelled through American dealers, with specimens fetching £20 or more on the London market. Adults survive well in captivity: one was kept for more than 13 years. Indiscriminate collecting has resulted in the complete destruction of many colonies.
**Conservation:** Mexican legislation does not specifically protect the red-kneed spider, and there appears to be no restriction on its importation into the USA, Europe or elsewhere. The collection of this tarantula should be monitored and

studies made of its natural history, population dynamics and ecological requirements. In conjunction with trade statistics, this should enable an appraisal to be made of its status. If collecting is found to be causing a permanent decline, consideration should be given to listing the species under CITES with a view to controlling international trade. Pressure might be taken off wild stocks by instigating a captive breeding programme for the pet trade.

## Coconut or Robber Crab

*Birgus latro*
**Phylum:** Arthropoda
**Class:** Crustacea
**Order:** Decapoda
**Family:** Coenobitidae
**Size:** The largest terrestrial arthropod in the world, a distinction made possible by abandoning its gastropod shell (though juveniles retain their shells up to a certain age). Specimens measuring 1 m/3.25 ft from leg tip to leg tip and weighing 3 kg/7 lbs are not unusual. Much higher weights have been recorded.
**Range:** Found throughout the eastern Indian Ocean and western Pacific, where it is confined to oceanic islands or small islets adjacent to large continental islands. It has been exterminated over much of its erstwhile range as, for example, on Mauritius and on almost all the islands in the Seychelles group, though it remains abundant on Aldabra.
**Habitat:** Almost entirely terrestrial in habit, except that the female is obliged to return to the sea to release her eggs. The crab normally inhabits rock crevices, but can also burrow in sand, among coral rock in dense undergrowth, or among roots of *Pandanus* palms or fallen coconut fronds. The burrows serve to protect them from the sun as well as from enemies, and are defended against rival crabs. At night they emerge to forage for food such as coconuts and *Pandanus* fruit. This is taken to the burrow where they may remain feeding for several days. On Aldabra the species is most numerous in the sandy coconut grove of Anse Mais and in damp *Pandanus* thickets on South Island, although it also occurs in more barren places.
**Red Data Book category:** Rare
**Threat:** Regarded both as a delicacy and an aphrodisiac by islanders, who hunt it

persistently. Introduced animals, chiefly rats, pigs, and monkeys, prey on the younger crabs. These factors are responsible for the extermination of the coconut crab over much of its range.
**Conservation:** Fully protected on Aldabra. On some islands (e.g. Saipan) a prohibition has been imposed on the collection of crabs with a carapace width of less than 7.5 cm/3 in, as well as during the peak period of reproductive activity. Specific proposals have been made for conserving the coconut crab on the Mariana Islands, which could serve as a model for other places. There should be a moratorium on the taking of crabs on most of the islands on which it still survives to provide the opportunity for its status to be assessed, and detailed conservation proposals made.

## Relict Himalayan Dragonfly

*Epiophlebia laidlawi*
**Phylum:** Arthropoda
**Class:** Insecta
**Order:** Odonata
**Family:** Epiophlebiidae
**Features:** Discovered in the adult form only as recently as 1963, this is one of two relict species of the suborder Anisozygoptera which goes back to the Mesozoic Era. It is one of the very few groups of animals to have survived virtually unchanged since the time before mammals and birds had evolved and the dinosaurs were beginning to make their appearance. In this respect, it ranks in phylogenetic rarity and importance with the tuatara *Sphenodon punctatus*, and the coelacanth *Latimeria chalumnae*. It possesses both dragonfly (Anisoptera) and damselfly (Zygoptera) characteristics, the body resembling that of a dragonfly while the adult wings are stalked like a damselfly's. The only related living species is the Japanese dragonfly *E. superstes*.
**Range:** Confined to the eastern Himalaya of India and Nepal, between Darjeeling and the Kathmandu Valley.
**Habitat:** Appears to be restricted to high altitude: larvae have been found at heights of 2,000–2,700 m/6,500–8,850 ft, and adults at 2,400–2,732 m/7,875–8,964 ft. Larvae are believed to inhabit clear, cold, fast-flowing streams, and may take as much as seven years to reach the adult stage.
**Red Data Book category:** Vulnerable

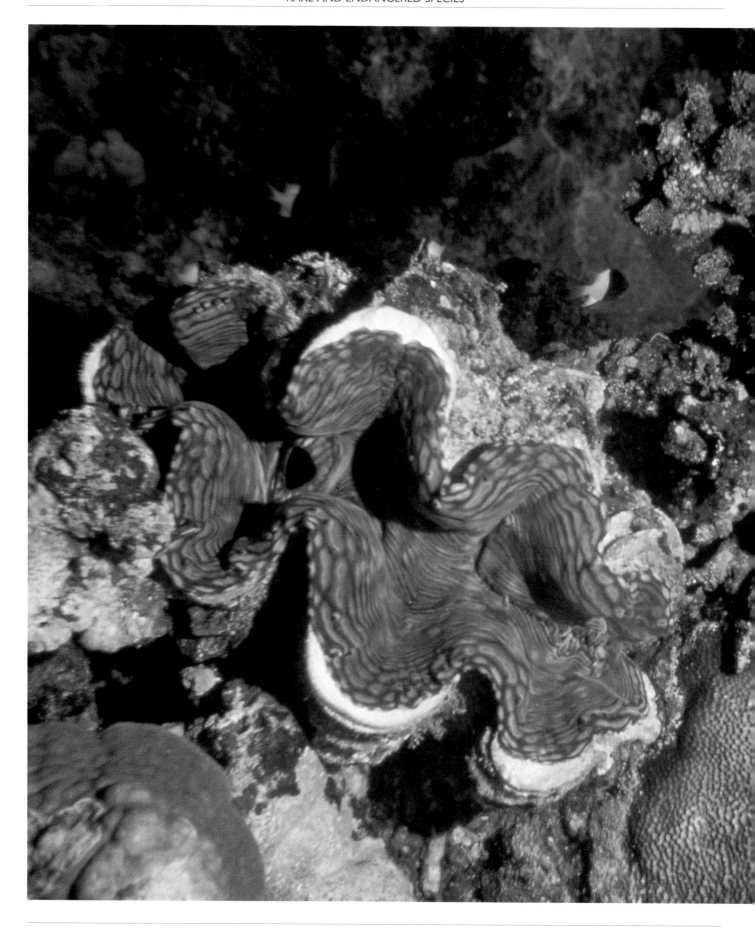

**Threat:** Changes in land-use, brought about by such factors as deforestation, high-level cultivation, soil erosion, chronic pollution of watercourses, tourism and, in particular, livestock grazing right up to the summits, inevitably jeopardize the dragonfly throughout its very constricted range.

**Conservation:** The IUCN/SSC Odonata Specialist Group has undertaken a survey of the status of the relict dragonfly and its habitat. The Group emphasizes the need to protect habitats known to harbour the species, and recommends the establishment of one or more reserves, especially around the headwaters of the Bagmati River in Nepal.

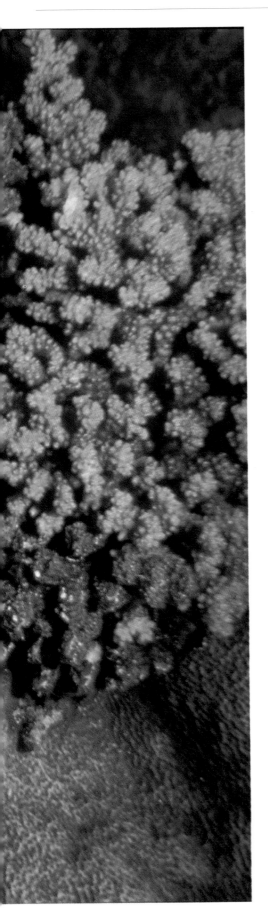

● A giant clam off New Caledonia. Giant clams are unusual among bivalves (clams, mussels, oysters, etc.) in that they obtain their nutrition by a symbiotic association with photosynthetic algae. (WWF / R. Seitre / BIOS)

● A robber crab on Aldabra – one of the few islands in the Indian Ocean where it is still found. (WWF / J. Mortimer)

## Stephens Island Weta
*Deinacrida rugosa*
**Phylum:** Arthropoda
**Class:** Insecta
**Order:** Orthoptera
**Family:** Stenopelmatidae
**Size:** One of a primitive group of four species of giant wetas that are among the largest of living insects: an adult female wetapunga, *D. heteracantha* (the largest of the four) may measure 85 × 32 mm / 3.3 × 1.25 in and, laden with eggs, weigh up to 71 gm / 2.5 oz.
**Range:** Confined to Stephens Island (41°42′ S, 174°00′ E) and Mana Island (41°03′ S, 174°45′ E) in Cook Strait, New Zealand. It may possibly still occur on Middle Trio Island, though none has been seen there for more than 20 years.
**Habitat:** The Stephens Island weta is primarily a ground-dwelling insect occurring in entanglements of *Muehlenbeckia*, on tussocks and in grassland. Of nocturnal habit, wetas hide by day beneath rocks, logs, crevices, treeholes, or among low vegetation, emerging at night to feed on shrubs, grasses,

herbs and possibly ripe fruit, and occasionally on invertebrate carrion.

**Red Data Book category:** Vulnerable
**Threat:** The absence of mammalian predators from New Zealand's indigenous fauna provided the opportunity for large ground-dwelling birds to evolve. Terrestrial invertebrates evolved in a similar manner, to the extent that wetas occupied the ecological niche that elsewhere was filled by rodents; hence the appellation 'invertebrate mice' by which wetas are sometimes known. The introductions of exotic animals that accompanied human settlement left them the prey of cats, rats and other vertebrates against which they were defenceless. The population on Middle Trio Island is thought to have been exterminated by the illegally introduced wood hen, which has subsequently been removed.

**Conservation:** All four species are legally protected. Both Stephens Island and Middle Trio Island have been designated flora and fauna reserves. Attempts have been made to establish the weta on another suitable rodent-free island.

## Hercules Beetle

*Dynastes hercules*
**Phylum:** Arthropoda
**Class:** Insecta
**Order:** Coleoptera
**Family:** Scarabaeidae
**Size:** One of the world's largest insects, reaching a length of 150–175 mm/6–7 in, half of which is taken up by a formidable pair of toothed horns resembling large claws.
**Range:** Extends from islands in the Lesser Antilles, Trinidad and Tobago into Central America (as far north as Chiapas State, Mexico), and northern South America as far south as central Bolivia.
**Habitat:** Humid neotropical forests. In the Lesser Antilles it occurs in montane areas, but this is believed to be more a matter of necessity than of choice. Larvae are found in decaying trees.
**Red Data Book category:** Vulnerable
**Threat:** In the Lesser Antilles, the chief threat arises from destruction of the native forest. On Guadeloupe, for example, the forest has been almost completely destroyed and replaced with plantations. The spraying of banana plantations with chemicals is likely to be affecting the beetles, particularly when they are feeding at night. Numbers are also being reduced by collecting. Large specimens

command high prices on the international market, and the trade is said to be widespread.

**Conservation:** Protected under French law on Guadeloupe and Martinique. The establishment of Guadeloupe National Park (175 km²/67 sq mls) and Martinique Regional Nature Park (701 km²/270 sq mls) should benefit the species. There is need to protect the beetle's habitat elsewhere in the Lesser Antilles.

## Giant Torrent Midge

*Edwardsina gigantea*
**Phylum:** Arthropoda
**Class:** Insecta
**Order:** Diptera
**Family:** Blepharoceridae
**Size:** With a wingspan of up to 25 mm /10 in, this species is one of the largest Australian representatives of the family. It resembles an outsize mosquito.
**Range:** Found only in the Snowy Mountains of New South Wales, where it is further restricted to two streams: Spencer's Creek and the Thredbo River.
**Habitat:** Confined to fast-flowing stony streams in hilly and mountainous areas. Adults live on the edges of streams, where females prey on small Diptera and males feed on nectar. The larvae occur in the torrent itself, where they avoid being swept away by attaching themselves to stones by their ventral suckers.
**Red Data Book category:** Endangered
**Threat:** Torrent midges are highly vulnerable to water pollution, changes in stream level and siltation. The construction of the Snowy Mountains hydroelectric scheme was responsible for eliminating this insect from several streams. The two streams in which it survives receive sewage. Any further lowering of water quality would seriously threaten its survival. A related species, the Tasmanian torrent midge, *E. tasmaniensis*, is similarly endangered in Tasmania.
**Conservation:** Both streams in which the species is known to occur lie within Kosciusko National Park (6,469 km² /2,497 sq mls). Water quality in the Thredbo River is being monitored with the aim of preventing any further deterioration in quality.

## Queen Alexandra's Birdwing

*Ornithoptera alexandrae*
**Phylum:** Arthropoda

**Class:** Insecta
**Order:** Lepidoptera
**Family:** Papilionidae
**Size:** The largest living butterfly. Females have a wingspan of up to 25 cm/10 in, and a head and body length averaging 7.5 cm/3 in. Males are smaller, with a wingspan of about 18 cm/7 in.
**Range:** A very limited distribution on or near the Popondetta Plain in the Northern Province of Papua New Guinea. A separate population occurs at high altitude only a short distance from the larger lowland population.
**Habitat:** Primary and secondary lowland rainforest up to a height of 400 m /1,300 ft, and secondary hill forest from 500–800 m/1,650–2,600 ft in the higher locality, where it lives in association with its larval food plant, the vine *Aristolochia schlechteri*. The leaves of this vine are often in the upper canopy, 40 m/130 ft above the ground, which explains why the larvae are seldom seen.
**Red Data Book category:** Endangered
**Threat:** Large tracts of forest known to have been previously occupied by this butterfly have disappeared from the Popondetta region as the result of development of cocoa and rubber plantations and the expansion of the oil palm industry. The 1951 eruption of Mount Lamington is reported to have destroyed 250 km²/100 sq mls of prime habitat. There is also a world-wide demand for birdwings from collectors. Properly managed, this butterfly could provide the local people with a lucrative source of income.

**Conservation:** Seven species of birdwing, including Queen Alexandra's, have been given legal protection. The law has, moreover, been stringently enforced, resulting in fines being imposed and expatriates deported. A Wildlife Management Area of 11,000 ha/27,000 acres has been created north of Popondetta. Thousands of cuttings of the vine *A. schlechteri* have been artificially propagated and planted to form a reserve and study area for this birdwing.

## Monarch Butterfly

*Danaus plexippus*
**Phylum:** Arthropoda
**Class:** Insecta
**Order:** Lepidoptera
**Family:** Danaidae
**Size:** A large butterfly, with a wingspan of 89–102 mm/3.5–4 in.

**Range:** The greater part of the North American population of monarch butterflies spends the summer breeding season in the USA and Canada, between the Pacific and the Rocky Mountains, before migrating to its wintering grounds in coastal California and Mexico.

**Habitat:** California monarchs winter in groves of trees near the coast, typically Monterey pine, Monterey cypress, and the introduced blue gum. Mexican monarchs winter in forests at high elevations up to 3,000 m / 10,000 ft or more, where they congregate in dense roosts. Aggregations are numbered in millions: one Mexican site was estimated to contain more than 14 million. The Californian colonies are smaller, seldom exceeding 90–100,000 individuals. The locations of roosting colonies vary from year to year according to changing weather conditions. In late February or March, the colonies begin to disperse, in readiness for the return migration. During the course of their migrations monarchs require temporary stopping places where they can safely obtain nectar. Larval hostplants – milkweeds – are a requirement of their summer locations.

**Threat:** The monarch's mass two-way migration is a unique phenomenon. Although the species itself is not endangered, the migration may be under threat, mainly as the result of felling of roosting trees. Virtually all the California sites are at risk from coastal development, while in Mexico the roosts are a major tourist attraction, visited by thousands of people, who inevitably create disturbance and cause fires.

**Conservation:** The IUCN/SSC Lepidoptera Specialist Group has made conservation of the monarch its first priority. A Mexican presidential decree, issued in 1980, provided a firm base on which conservation measures could be based, and has led to the creation of a number of special reserves. In California, all public land agencies have been urged to protect roosting sites.

## Wallace's Giant Bee

*Chalicodoma pluto*
**Phylum:** Arthropoda
**Class:** Insecta
**Order:** Hymenoptera
**Family:** Megachilidae
**Size:** The world's largest living bee. Females are about 39 mm / 1.5 in long; males are smaller, about 24 mm / 1 in.

**Range:** Confined to the Molucca Islands of Indonesia, where it is known to occur on the islands of Bacan, Soasiu and Halmahera. After the type specimen was collected in 1858, the giant bee remained unknown for more than a century, and was presumed extinct until its rediscovery in 1981.

**Habitat:** Lowland primary and secondary forest below 200 m / 650 ft, where it lives in association with the termite *Microcerotermes* – the only member of its family to do so. The female gathers resin from dipterocarp trees which she mixes with wood and uses to line tunnels in inhabited termite nests. Although the arboreal termite nests are abundant, the bee seems to be rare.

**Red Data Book category:** Insufficiently known

**Threat:** The lowland forest is under pressure from logging and shifting agriculture, but the extent to which the giant bee is threatened is not known.

**Conservation:** No specific measures have been taken to protect the giant bee. A comprehensive survey is needed to assess its range, status and ecological requirements, and to determine the measures necessary for its protection.

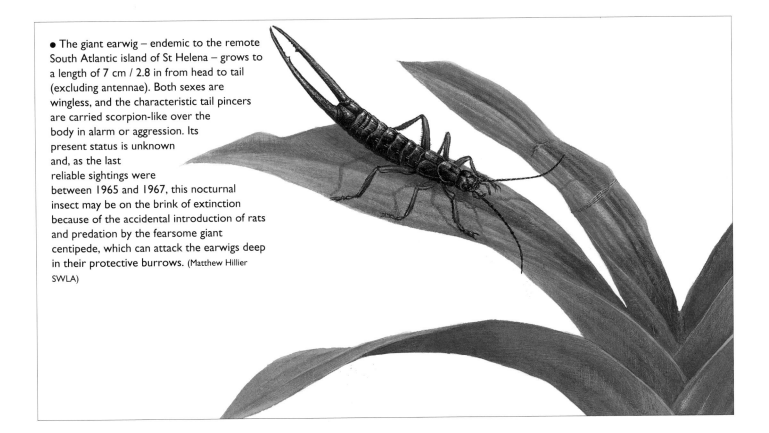

● The giant earwig – endemic to the remote South Atlantic island of St Helena – grows to a length of 7 cm / 2.8 in from head to tail (excluding antennae). Both sexes are wingless, and the characteristic tail pincers are carried scorpion-like over the body in alarm or aggression. Its present status is unknown and, as the last reliable sightings were between 1965 and 1967, this nocturnal insect may be on the brink of extinction because of the accidental introduction of rats and predation by the fearsome giant centipede, which can attack the earwigs deep in their protective burrows. (Matthew Hillier SWLA)

● Queen Alexandra's birdwing. The largest and one of the most beautiful of butterflies, it has become a target for collectors. (WWF / W. von Schmieder)

● Monarch butterflies winter in dense roosting colonies, sometimes numbering millions of individuals. (WWF / Michael Wirtz)

# RARE PLANTS

In the quest for sustainable development, plants are one of the most important yet one of the most neglected of resources. They provide our food, the timber for our homes and many of our medicines. Until the industrial era and the discovery of fossil fuels, they provided all our energy as well. Indeed, by converting the energy of the sun into carbohydrates through photosynthesis, plants make possible all animal and human life.

Yet we are losing plants faster than ever before. Indeed, in the last quarter of the 20th century, humankind is causing the most severe spasm of extinction in the 400 million-year evolution of the plant kingdom. Whereas the animal kingdom suffered a series of mass extinctions at times in the geological past, it seems that the plants did not. Extinctions certainly occurred but never, scientists believe, at the intensity at which they are happening today. Each extinct plant is a permanent loss to humanity and to the planet. Like a work of art, once a species is lost, it cannot be recreated.

## Numbers, Groups and Distributions

We see different plants all around us – in the home, in gardens, in nature – but people are often surprised at the great diversity of the plant kingdom. Botanists divide plants into a number of natural groups. The most important are the higher plants which dominate the world's flora and vegetation. The higher plants consist of three groups: the angiosperms or flowering plants, the gymnosperms (mainly the conifers and the cycads), and the ferns and their allies, such as the clubmosses and horsetails.

There are estimated to be about 250,000 species of flowering plants in the world. They are divided into about 200 plant families, such as Compositae (daisy family), Gramineae (grass family) and Leguminosae (pea family). Some families are very numerous – for example the orchids contain about 18,000 species – while other families have just a single species.

The gymnosperms are trees (or sometimes shrubs) that lack flowers and whose seeds are borne on cones. They include the 550 species of conifers, 100 species of cycads and a few other small groups. The conifers occur worldwide, but are most abundant in the temperate realms, forming the forests that dominate much of Scandinavia, north Russia, Canada and the western United States. They are the softwoods of commerce and are widely grown for timber.

The ferns lack the seeds of the flowering plants and gymnosperms, reproducing by spores. Fern species vary from tiny, delicate filmy ferns to tropical tree ferns more than 15 m/50 ft tall. There are probably about 12,000 species of ferns, with the greatest concentration in the tropics, especially in tropical rainforests.

Overall, the higher plants occur on most of the land surface of the world, being absent only from deserts, high mountain peaks and the polar regions. However, the distribution of species is very uneven, as shown in the boxed table. Roughly two-thirds of all higher plant species grow in the tropics, half of them in Latin America.

When comparisons are made from one country to another, the differences are even greater. Britain, for example, has about 1,550 species of plants, while Ireland has 892 and Iceland only 340. Yet Costa Rica has over 10,000, Cuba nearly 6,000 and Cameroon 8,000. And the larger tropical countries have even more: among the countries with the most plant species are Indonesia with 20,000, India with 15,000 and Zaïre with 11,000. The figures for Latin American countries are even greater: Colombia has 35,000 and Brazil as many as a staggering 55,000 different plants.

## The Role of Botanists

Sadly, the distribution of botanists does not match the distribution of plants. Most botanists live and work in temperate countries like Britain and North America, while most of the plants are in the tropics. As a result there is a great imbalance in our knowledge of plants from one part of the world to another.

Of all countries, the flora of Britain is probably the most studied and best known. Following 300 years or so of effort and aided by well-written field guides to identify the plants, hundreds of good amateur botanists comb the countryside to record the exact distribution of each species.

In many tropical countries, on the other hand, there is only a handful of botanists trying to study tens of thousands of different plants. Many plants have not been seen or given names. Often, once a plant has been named and described, it has not been seen again for many years; next to nothing is known about the distributions of immense numbers of tropical plants. In some regions, especially in South America, expeditions to remote areas frequently find new species never before seen by botanists. But their discovery is a race against time as plant habitats continue to decline.

Yet just as politicians are at last recognizing our concern for the future of the plant kingdom, so the funding and support for botanical institutions is drying up. Nowadays in schools and universities, botany is subsumed under biology and teaching concentrates more on genetics and microbiology than on how to classify and identify plants. The museums which are the homes of important collections of dried plants and the basis for the classification of plants are no longer scientifically fashionable, and are finding it hard to retain their research teams. Yet it is not possible to conserve the plant kingdom effectively without knowing which species exist in the world and where each of them grows.

## The Lower Plants

Even less is known about the lower plants, which are more often trodden on and overlooked than studied and conserved. The main groups here are:

**The mosses and liverworts** These little plants occur all over the world, but are most diverse and abundant in countries where the climate is cool and consistently wet, such as the Atlantic coast of Europe. Worldwide there are about 8,000 species of mosses and 6,000 of liverworts.

**Lichens** Lichens are dual organisms consisting of a fungus in association with an alga or a bacterium. They tend to be most diverse in similar areas to mosses and liverworts, typically growing on stones and on the trunks of old trees. The temperate rainforests of the southern hemisphere, especially those in Chile, New Zealand and Tasmania, are particularly important for lichens, as are some mountain regions. Botanists estimate there are around 17,000–20,000 lichen species, of which 50–70% have been discovered so far.

## TABLE OF DISTRIBUTION OF HIGHER PLANTS BY CONTINENTS

| | |
|---|---|
| Latin America | 85,000 |
| Tropical & Subtropical Africa | 40,000–45,000 |
| North Africa | 10,000 |
| Tropical Africa | 21,000 |
| Southern Africa | 21,000 |
| Tropical & Subtropical Asia | 50,000 |
| India | 15,000 |
| Malesia | 30,000 |
| China | 25,000 |
| North America | 17,000 |
| Europe | 12,500 |

Source: *Global Biodiversity: Status of the Earth's Living Resources, World Conservation Monitoring Centre, 1992.*

**Algae** Among the many different kinds of algae or seaweeds, some live in the sea and others in freshwater. The green algae (approx. 1,040 species) occur in both freshwater and the sea; the brown algae (approx. 1,500 species) are the typical seaweeds of the coastline and include the giant kelp, which is reputed to grow up to 200 m / 650 ft long; the red algae (over 2,500 species) are small and almost all marine. And these do not include the microscopic algae such as the diatoms, which are the basis for life in the seas.

**Fungi** With the bacteria, the fungi have a vital role as the decomposers of nature, returning nutrients to the soil. The fungi are even less well known than other lower plants; the species which produce mushrooms and toadstools form only a small proportion of the total number. Some experts now estimate that there may be over a million different species, of which only a small proportion is known.

## Threats to the Plant Kingdom

Plants today are in danger as never before. In virtually every country of the world, the space available for wild plants – their habitat – is declining as more and more land is taken up for the needs of human life or is wasted by exploitation. Of all the threats to plants, loss of habitat is the most serious.

Probably the largest numbers of threatened plants are in the tropical rainforests, which are home to as many as half the world's plants yet only cover a small proportion of the land surface of the earth. Estimates on the loss of rainforests vary, but all agree that they continue to be lost at an alarming rate. In some countries, such as the Philippines, the logging industry has destroyed virtually all the forests – only fragments survive. In other countries, like Gabon in Africa and Venezuela in South America, the pressures for destruction have been much less and large areas of undisturbed forest remain. In others, like Colombia, enlightened governments have ensured that large parts of their forest still survive more or less intact.

The tragedy is that so much of the destruction is for short-term greed rather than for sustainable development. The Japanese logging companies are ripping out the heart of forests in Indonesia and Malaysia, and using much of the wood

---

### THE TOROMIRO
#### SOPHORA TOROMIRO

In unravelling the stories of the famous Easter Island statues, explorer Thor Heyerdahl believed that large logs were used to move the statues into position. Easter Island was also famous for its wooden carvings. But there are no trees on the island!

In 1917 a Swedish botanist called Skottsberg had found one solitary tree on the slopes of the volcano crater. It was a species unique to Easter Island. It must have been abundant in the past as it was used for making canoes, the frames of houses – and for moving the statues.

The tree subsequently died and the species was described as extinct in the 1978 *IUCN Plant Red Data Book*. But a letter from the Goteborg Botanic Garden in Sweden soon corrected this assertion. On one of his trips, Heyerdahl had collected seeds from the last tree and given them to the botanic garden. By the late 1970s, Swedish botanists had succeeded in growing three small trees of the Toromiro.

Now aware of its importance, botanists tried to reintroduce the plant to its home. The first attempt failed, but another was made in 1988. The publicity also led to the discovery of other Toromiros in cultivation – one in Bonn, possibly another in Christchurch, New Zealand.

It may take time and it won't be easy, but one day the Toromiro may once again grace the slopes of the Rana Kao volcano on Easter Island.

The story shows how important it is to have collaboration between botanists of all nations in the struggle to make sure species are not lost for ever. No one predicted that Easter Island's only tree would turn up in a botanic garden on the opposite side of the world.

---

for consumables like chopsticks or shuttering in the building industry. In large parts of South America, forest is burnt and converted into cattle pasture, yet in most areas the soil is unsuitable for permanent grassland and after a few years the land has to be abandoned, by which time the soil has deteriorated and the forest cannot regenerate again. Such senseless destruction is an appalling indictment of our time, but until the differences between north and south – the developed and developing countries – are redressed it is unlikely that any lasting solution will be found to protect the forests that do remain.

Plants are also being lost on islands, especially on those of the tropics and subtropics. About one in six of all plants grow on oceanic islands but one in three of all known threatened plants are island endemics – plants confined to islands. In contrast to the damage to tropical rainforests, the damage to most island floras is not recent. Usually the harm was done in the era of European exploration and colonization. Sailors placed sheep and goats on remote islands to provide supplies of fresh meat, and so that should they or their fellows be shipwrecked there, they would not starve before help came.

This was immensely damaging to island floras, most of which have evolved in the absence of grazing animals and so had few defences. Whereas goats, sheep, pigs and even rabbits can be controlled, the problem of introduced invasive plants is much harder to resolve. Enthusiastic gardeners often brought to islands the plants they used to grow at home, and some of these plants out-competed the native vegetation, in effect going wild.

Plants are also being lost in parts of the arid zone. Somalia, for example, is a particularly rich centre of endemism for succulent plants, yet it has been ravaged by drought, over-grazing and war. It is doubtful whether much of the endemic flora survives.

Although the main threat to plants is loss of habitat, collecting is also a severe threat to some groups. Literally millions of bulbs are uprooted from Turkey each year for sale to gardeners in Europe. Snowdrops are thought to be particularly at risk. Cacti in the United States and Mexico are dug up at dead of night and sold to wealthy home-owners wanting instant gardens. Rare orchids are removed from remote localities and sold at barely believable prices to orchid enthusiasts.

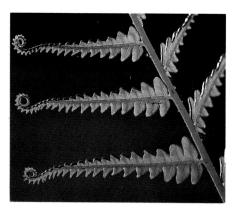

● A wood fern of the genus *Thelypteris*, photographed at the Audubon Cork Screw Swamp Sanctuary in Florida. (WWF / Fritz Polking)

## Counting the Cost

Of the higher plants, it is estimated that about 60,000 species - one in four of the global total - are in some degree of danger. So far the World Conservation Monitoring Centre has recorded about 22,000 individual plant species as threatened. The discrepancy arises because in many tropical countries, where large numbers of plants are undoubtedly under threat, it is not possible to know which individual species are in danger, simply because the floras are so large and so little known that botanists have not yet been able to determine the status of individual species.

In the late 1960s IUCN started a *Red Data Book* of threatened plants, but it soon became apparent that the number of plants that would have to be included would make the approach of a page of information on each an impractical one. Instead, IUCN prepared a *Red Data Book* in 1978 that outlined the status of 250 threatened species, as examples of the countries affected, the types of threat, and so on. Although it gave a far from complete picture, it was of great value in proving that plants were as threatened as animals - a point that is obvious nowadays but was not much appreciated in the 1960s and 1970s.

IUCN's main thrust, however, in the 1970s was to encourage each country to prepare its own *Red Data Book* of threatened plants. This policy has been very successful and today virtually all developed countries have published national *Red Data Books* on plants. The World Conservation Monitoring Centre

---

## CAFÉ MARRON
### *RAMOSMANIA HETEROPHYLLA*

On the remote island of Rodriguez, east of Madagascar, lives the rarest plant in the world – a plant so rare that only one specimen of it survives. Known as the café marron, it is a small tree, its shiny, laurel-like deep green leaves contrasting with large, starry white flowers.

Its home, Rodriguez, was once covered with forest filled with plants and animals found nowhere else in the world. But man has cut down the forests and introduced the botanists' old enemies of goats, sheep and cattle. By the 1970s, little forest remained and most of Rodriguez's unique plants were in critical danger.

The café marron was thought extinct until, in 1980, a bright schoolboy found one old tree that had somehow survived. Eminent botanists came from far and wide to see it for themselves.

Suddenly, large parts of the tree started to disappear. Local people claimed that the tree was a cure for venereal disease and hangovers. What seems more likely is that the islanders saw all the fuss over the plant and thought it must have some special property of great value.

The conservationists erected a fence around the plant – but to no avail. In 1984 they erected a second fence around the first, but this did not help either. And then, just before Christmas 1985, mealy bugs attacked the two remaining stems. This seemed like the last straw!

In desperation, local WWF botanist Wendy Strahm rang Kew Gardens and a rescue plan was hatched. Orders were given to wash the plant with soapy water – using a pesticide would be too risky. In April 1986, Strahm gingerly took one cutting from the plant, which, thanks to

● The café marron tree on the island of Rodriguez – the rarest plant in the world. (WWF / Ole Hamann)

help from British Airways, reached Kew within 24 hours.

From then on, the café marron's luck turned. The skilful gardeners at Kew rooted the cutting and now have several healthy plants. Back on Rodriguez, yet another fence – the third – was erected over the other two, but this time with a roof and a strongly padlocked door. A watchman now guards the site.

In Rodriguez, too, the situation is improving. Thanks to EC aid, fences now keep the introduced animals out of the best plant areas. Most of the endemic plants are now cultivated and the government has promised to set up a small botanic garden on the island.

---

maintains a large computer database on all the species so listed, determining which are threatened on a world scale. Some species in national *Red Data Books* are threatened in the country concerned but are common in other countries. An assessment of all national *Red Data Books* is given in IUCN's book *Plants in Danger: What do we Know?* that covers the

state of knowledge for every country and island group in the world.

So far it appears that very large numbers of plants have been greatly reduced in distribution and abundance. Thousands of species are known to have become confined to tiny sites and are in danger of being lost altogether. More worrying, a reduction in range and abundance inevitably means a reduction in genetic diversity and with it the ability to withstand change.

Yet in countries where the flora is well known, definite extinctions so far have been mercifully light. At present only a few hundred species are known to be extinct, mainly from the developed world and from islands. This could be far worse and implies that it is not yet too late to save most of the plant kingdom. It is remarkable how persistent plants can be: they have been known to survive for many years either as solitary individuals or as tiny clumps. Unlike animals, plants can persist in unfavourable circumstances, perhaps by not flowering and thus being hard to spot, or even as seeds in the soil waiting to germinate.

Yet these figures on extinct species completely ignore the situation in most of the tropics. We do not know which species are extinct but, on the evidence we have, extinctions are likely to have been many times more numerous than imagined. Botanist Al Gentry cites an example of a ridge on the Andes in Ecuador; when he first visited the ridge he found it contained many species – maybe

• An unusual fungus, *'Anthurus archeri'*, in the Wild Rivers National Park, an area of temperate rainforest in Tasmania. (Auscape International / Jean-Paul Ferrero)

• Although the orchid family is very numerous – with around 18,000 species – it contains many rare species, which frequently become the victims of unscrupulous collectors. (Auscape International / Hans and Judy Beste)

as many as a hundred – that had not been seen before and did not grow on neighbouring ridges or elsewhere; several years later he returned to find all the forest had gone. This pattern is being repeated all over the tropics, with disastrous consequences for plants.

## Climatic Change

Also, the figures given above do not take account of an even more insidious threat to plants that is now apparent. This is the spectre of a global change in the climate, specifically a warming in temperature ('global warming') due to emissions of carbon dioxide from the burning of fossil fuels, CFCs from aerosols and other gases. The experts on this issue, the Inter-Governmental Panel on Climate Change, have predicted that in the absence of any change in human behaviour the mean global temperature is likely to rise by 1–2° C by 2030. This might not sound much, but it would have dramatic effects on rainfall patterns and on vegetation. And even if we cut emissions now, much of the temperature change will still happen since the lag phase is very long. Indeed, scientists believe the temperature has already increased by 0.3–0.6° C.

To adapt to a change in climate, vegetation zones will need to move to stay in a zone of stable climate. Yet it seems likely that species will not be able to move fast enough. Studies have shown that trees with light seeds dispersed by the wind are unlikely to spread more than 2 km/1 ml a year, whereas many rainforest trees, with heavy seeds, hardly spread at all. And even if the vegetation had the potential to shift with all its species, human impact has not left space for the species to shift to. Most of our nature reserves and national parks are becoming islands of nature in a sea of agricultural land, and so could not be moved hundreds of kilometres even if such a notion was biologically possible.

## The Threat to Plant Genes

The loss of any species of wild plant is a tragedy, but the loss of some plants could have disastrous consequences. These are the plants that are of importance to people, whether as sources of timber, food, medicines or other products.

Particularly important are the wild species that are related to the ancestors of modern crop plants. Examples are wild wheats in the Middle East, wild

coffee in Ethiopia and wild tomatoes in South America. Their importance lies in the part they play in plant breeding.

The major food crops of the world are mostly highly bred varieties – termed cultivars. Because they are very uniform, with most of nature's variety removed, and because they are grown on a very large scale, they are susceptible to pests and diseases. Plant breeders try to make the cultivars as resistant to pests and diseases as possible but, all too often, the resistance soon breaks down as new strains of pests and diseases evolve. So plant breeders are engaged in a race against time, constantly having to produce new, disease-resistant cultivars.

For the most part, they find the characteristics of disease-resistance in the wild relatives of the crops, rather than other varieties of the crop itself. Thus the wild relatives of crops are vital in ensuring world food supplies. The loss of some of these plants could be critical.

Since plant breeders are looking for individual characteristics in the wild plants, it is not sufficient just to conserve one example of each species. In the early 1970s, for instance, the International Rice

Research Institute in the Philippines was searching for rice plants resistant to the grassy stunt virus which was decimating rice crops in Asia. They screened thousands of rice cultivars but found resistance in none of them. They then screened hundreds of samples of wild species of rice. They found just one variant that was resistant – a population of *Oryza nivara* collected in 1963 from a water-logged field in Uttar Pradesh, India. It was not the species as a whole that was resistant, nor just one population, but only three plants from the 30 or so plants they grew from the seed in their collection. The gene for disease resistance to grassy stunt virus is now found in every high-yielding rice cultivar in Asia.

## Medicinal Plants

Wild plants are also vital in medicine. The World Health Organization estimates that 80% of the people of the world still depend on traditional medicine for their primary health care needs, and most of the drugs and cures employed come from plants. In some parts of the world the plants are cultivated; for example, Indonesian housewives grow a collec-

---

# DOWNY WOUNDWORT
## *STACHYS GERMANICA*

Britain may no longer be a leader in conservation, as in the 1950s and 1960s, but it does have the largest and most knowledgeable group of field naturalists in the world. The story of the downy woundwort shows how important this can be to conservation.

The downy woundwort grows beside ancient drove roads over oolitic limestone. It used to grow in several counties, but today can only be found in four or five small sites, all in Oxfordshire.

It likes disturbed ground and its original habitat was probably woodland edges and clearings. The cattle and sheep which walked the drove roads for hundreds of years must have created an ideal alternative habitat. But as the roads ceased to be used, they became overgrown and the downy woundwort retreated.

One day, in 1984, amateur botanist Jo Dunn was out recording wildlife near her home. She saw a plant she did not recognize, and realized that it was a new sighting of the rare downy woundwort.

Its site was where a farmer had cut back a hedge two years before. Since it was the first time the hedge had been cut for 35–40 years, the seed must have been dormant in the soil for all those years.

With help from the farmer, Jo Dunn cared for the little colony, keeping the old man's beard, *Clematis vitalba*, from smothering it, protecting it from rabbits and pheasants, and even weeding it. As a local newspaper put it, she became a fairy godmother to the downy woundwort.

Thanks to her example, conservation groups are planning to cut back small areas of scrub in other sites where the downy woundwort once grew to see if it will regenerate from dormant seed in the soil. This is only possible because the exact locations where downy woundwort grew 30 years ago are known. The long tradition of plant recording in Britain may make it possible to restore the downy woundwort — and many other threatened species — to their ancient sites.

tion of medicinal plants in pots around their homes as a home pharmacy, but many medicinal plants are still collected from the wild with drastic consequences. The range of plants used is very large – traditional Chinese medicine uses about 5,000 different plants, including the famous ginseng root.

Plants are important too in Western, scientific medicine. In the past virtually all drugs came from plants, and although most are now synthesized, many of them would never have been discovered without the knowledge first gained from plants. Aspirin, for example, the world's most widely used drug, is today synthesized but was first discovered in willow bark. Scientists have estimated that about 120 key drugs in Western medicine still come from plants. Foxgloves provide digoxin which gives relief to millions of people with heart problems; mandrake, henbane and thornapples provide scopolomine, which enables people to sleep deeply and breathe easily during operations; the bark of the cinchona tree *Cinchona ledgeriana* provides quinidine and quinine, still the best preventions against malaria. And so the list continues.

Other groups of plants of great importance for genetic conservation are timber trees, forage plants (grasses in particular), spices and plants that produce essential oils. There is also an important group of plants, the so-called under-exploited crops – plants with great potential as crops but which have not yet been widely used. So far humans have only used a tiny fraction of the plant kingdom for major food crops, and there are many others that could be important sources of foods. It would be tragic to lose them and their essential genetic diversity before they can be put to use.

## Saving the Plant Kingdom

In 1984, IUCN and WWF launched a Plants Conservation Programme, and as a result we now know far more about how to save plants. In fact the programme proved that plants are far easier to save than animals. They don't move, they often persist for long periods of time, and they can be propagated from a few seeds or cuttings taken from the wild without harming the wild population. The models provided by the IUCN–WWF Plants Conservation Programme now need to be scaled up to cover not just a few places but the whole world.

## GOULIMY'S RED TULIP
### TULIPA GOULIMYI

Greece in the springtime is a mecca for botanists. Among the plants they most want to see are the wild tulips. All the Greek wild tulips are rare, but one, Goulimy's red tulip, is in particular danger.

This beautiful plant, with orange-red flowers and wavy leaves, grows in a handful of places in Lakonia (the remote southeastern part of the Peloponnese) and on nearby islands. Even there, it is apparently rare.

Its habitat is sandy fields or stony and gravelly places, where it tends to be threatened by the many changes to farming in the region – the growing use of herbicides, deep ploughing, and the ever-increasing erection of plastic greenhouses. Another threat is the local children's habit of digging up and eating the sweet-tasting bulbs.

Goulimy's red tulip is just one of the nearly 600 endemic plants in Greece. Mediterranean countries are particularly rich in endemics, and there are similar numbers in Spain, Italy, Morocco and Turkey. The same is true of other regions with a Mediterranean climate, notably Western Australia, coastal Chile, California and the Cape of South Africa – all countries that are high priorities for global plant conservation.

## Rescuing Threatened Species

In the last ten years, there have been many attempts to rescue critically endangered species, and as a result hundreds, possibly thousands, of plant extinctions have been avoided. Most of the best results have occurred in temperate countries and on islands.

A good example is the work of the local Botanic Garden on the island of Gran Canaria. It is one of the Canary Islands, which have nearly 600 endemic plants. Staff from the Garden combed Gran Canaria, plotting the localities of all the rare and endemic species they found. They subsequently prepared a *Red Data Book* listing 146 species that are threatened, and identifying the native species that may be of value to mankind, such as wild olives, wild date-palms, wild cabbages and forage grasses.

Based on this information they were able to prepare a plan for the creation of nature reserves on Gran Canaria. The plan, now accepted by the Island Council, envisages over half the island being protected in some way. This is made possible by many people having left the hills, where they practised inefficient and environmentally harmful agriculture, and moved to the coast, where most of the crops are now grown and where tourist hotels provides many jobs. Steadily the Island Council is acquiring small reserves for each of the endemic plants, on the advice of the Botanic Garden staff.

Simply setting up a nature reserve is often not enough. For example the Flor de Mayo Leñosa *Senecio hadrosomus* was reduced to a mere 10 specimens on a single damp cliff in the mountains. Staff from the Botanic Garden perilously collected some seed from these plants, and were able to grow on hundreds of plants in the Garden. These have now been planted in the wild, boosting the wild population by a factor of ten.

In some cases research was needed to find out how to conserve a rare species. When a plant is very rare, removing the external threat is often not enough to save it – positive action may be needed to help its population increase to a safe level. Research is the first step in the process, to find out why the plant does not reproduce and increase. Does it flower enough, for instance, or, if the flowers are pollinated, do the seeds set? Do those seeds germinate or do they get eaten first by predators? Answers to questions like these can indicate how the plant can be saved.

One vital technique is micropropagation, where very large numbers of plants can be made from a single shoot or leaf. The 'microgardener' manipulates the tiny tips of the shoot under a microscope and then grows on the resulting plantlets in test-tubes. This approach is essential where the population of the plant is so low that taking any seed at all could be harmful, or where the plant does not seem to produce viable seed – as with Flor de Mayo Leñosa.

In other cases, dramatic management is needed to restore the habitat for rare plants. Gran Canaria provides a good example of this. The Canary Islands (Madeira also) are famous for their laurel forests, forests that are dense, dark and moist inside. Many of the island's endemic plants grow in the shade of evergreen laurel trees. On Gran Canaria, less than 1% of the laurel forests survive, the rest having been felled for timber and grazing. The best remaining site is at Los Tilos de Moya, a popular picnic site. Because of its great biological importance, the Island Council bravely decided to close it and to invite the Botanic Garden to restore its vegetation. Over the last ten years the Garden has replanted many of the main tree species and the area is beginning to look like a forest again. This in itself is ensuring the survival of some previously endangered endemic plants.

With the knowledge gained, the Garden persuaded the Island Council to buy a large farm in rolling hills nearby. It was the largest surviving country estate on Gran Canaria and had been farmed in the traditional way. The intention is to restore the laurel forest there. The council plans eventually to increase the native forest to about 20% of its original extent in Gran Canaria as a whole.

Besides safeguarding the endemic plants, this will also have other benefits. As well as providing villagers with a source of wood and other products, the restored forest will be an attraction for the many tourists who visit Gran Canaria each year. Above all, it will increase the supply of freshwater to the towns.

In the past, streams in Gran Canaria ran all year round but, as the laurel forests were removed, the springs dried up in the hot summer months. Water supply is a critical factor in Gran Canaria which depends very heavily on its tourist industry. At present the island relies on a desalination plant, powered by oil at considerable expense. A large area of laurel forest, on land of low value for

● Plants endemic to islands have proved particularly vulnerable to introduced grazing and browsing animals. This subspecies of *Pachypodium baroni* is found only in a very limited area in northern Madagascar, and is classified as vulnerable. (WWF / Olivier Langrand)

● Tapping rubber trees in the forests of Brazil is an example of the sustainable harvesting of wild plants and their products. (WWF / Edward Parker)

● A forest killed by acid rain in Poland. Acid rain strips the leaves of their protective covering of wax and, besides the direct damage that the acid causes to their cells, the leaves are left unprotected from desiccation and from fungal and bacterial infections. (WWF / Wlodzimierz Lapinski)

farming, would greatly reduce the need for the desalination plant.

Programmes on saving plants can also be used for educational purposes. Gran Canaria's Botanic Garden involves school-children in the work. For example, children were invited to grow plants of the dragon's-blood tree *Dracaena draco* for later replanting in the wild. They soon learnt that the plants grew very slowly and could easily be killed – a practical lesson on the vulnerability of nature.

## Conserving Plants

Worldwide, very encouraging progress is being made in conserving individual threatened plants, particularly in the creation of nature reserves for rare species. In the United States, for example, the Nature Conservancy, a private conservation group, has been acquiring nature reserves at the rate of about one a day for 20 years or more, many of them chosen because of the presence of a rare or threatened plant – a most impressive achievement.

So far only two countries – Czecho-slovakia and Poland – can claim that all the plants in their *Red Data Books* are represented in nature reserves, national parks or other types of protected area. Of course this alone does not guarantee successful conservation. Following close behind is New Zealand, where about 70% of the threatened species are in permanent protected areas. Indeed, New Zealand may be the more impressive example, since in Eastern Europe, the Communist regimes created large networks of nature reserves and the like on paper but did not always implement them on the ground. Today, too, former land-owners are claiming their land back, on the grounds that the Communists nation-alized it illegally.

Two remarkable cases are Australia and South Africa, both countries of great importance for plants. Almost exactly half of Australia's threatened plant species are in conservation reserves. When one learns that 3,635 plant species are listed as threatened, the size of Australia's achievement becomes apparent. In South Africa, some 74% of all the region's plants – a staggering total of 23,300 species – are reported to be in nature reserves. Although this is only an estimate, not a count as in the case of Australia, it does show that the nature reserve system protects a high proportion of the flora and, by implication, of threatened species also.

These figures are very encouraging. They imply that where there have been long-running programmes to identify and conserve threatened plants, success is possible. Yet there is still a long way to go, even in the exemplary countries. And there are still many countries with far lower proportions of their threatened plants in nature reserves. Indeed, some countries cannot point to having protected a single threatened plant.

Of course, simply declaring a nature reserve around the site of a rare plant does not guarantee its survival. There are even cases where species have disappeared from the reserves set up to safeguard them! One of the best such examples comes from Britain. In 1932, botanists purchased a minute site of 290 m²/346 sq yds in western England, to protect one of two localities in Britain for the adders-tongue spearwort *Ranunculus ophioglossifolius*. It was simply a damp patch with a few trees in the middle of a grass field. In 1933, when the site was full of the plant, they erected a stout fence around it. Devoid of the grazing and trampling by cattle that had maintained its habitat for centuries, the plant rapidly declined. In some years, there were even plants outside the reserve but none inside! By 1962, the botanists decided something had to be done. They first cut back the vegetation over a small part of the reserve. The plant quickly reappeared, germinating from the reserve of seed that had remained in the soil since the 1930s. Today, it is abundant each year, due to careful and regular management of the vegetation. This is perhaps an extreme example, but it does indicate that just declaring a nature reserve is often not enough.

## Protecting Centres of Plant Diversity

In much of the tropics, the species are too numerous and the botanists too few to identify individual threatened plants. Here the best approach is to identify the areas that are richest in plants and conserve them.

This approach is helped by the fact that the distribution of plants over the earth is very uneven. There are large areas of the world where plants are uniform and few in number: the north temperate forests and arctic regions, for example, and many dry areas of Africa. In others, such as many mountains, rainforests and tropical islands, there is a wealth of unique plants. By identifying these places and concentrating their efforts on them, botanists can conserve a large proportion of the plant kingdom.

With funding from British Aid and the European Community, WWF and IUCN are at present identifying about 300 sites around the world which, if conserved, would between them 'catch' the great majority of species in the plant kingdom. Already WWF have projects to help set up and implement large national parks in about half the 300 sites, and the hope is that once the list of sites is agreed it will become a pillar of international efforts to conserve biodiversity. These are sites of such importance for plants that their conservation needs the help and support not just of the nations in which they occur but of all countries.

Most of the areas are specific sites. Some are defined geographically, such as a mountain or mountain range. Good examples are Mount Cameroon and Mount Nimba in West Africa, Mount Kinabalu in Borneo and the Darien Mountains in Costa Rica. Others are simply the only areas left of a certain vegetation type. The Taï Forest National Park in Ivory Coast, for example, is the only large area of West African lowland rainforest to have survived. Virtually all the rest has been felled. Saving it will secure the survival of many of the characteristic species of that type of forest.

Among other chosen areas are rainforests along the eastern coast of Brazil. Distinct from the Amazon forest and separated from them, these are believed to be the most endangered rainforests in the world. Only small fragments still remain, most having been replaced by agriculture. Here the approach is not to

conserve any one site but rather to build a network of reserves containing all the best fragments of forest that survive.

Most of the sites are in the tropics, simply because the tropics are home to about two-thirds of the world's plants. At least half the sites are of tropical rainforest, where it is felt that large protected areas are the best way of conserving their plant diversity. In some cases, it may be possible for local inhabitants to act as forest guardians. For example, in the state of Acre, Brazil, the rubber-tappers who have tapped the wild rubber trees in the rainforest for a hundred years or more without destroying it have been given rights to manage a large area of plant-rich tropical rainforest. Collecting rubber and many other plant products is quite compatible with conserving rare plants, and ensures that local people benefit from the forest. Indeed, it has been calculated that if the forest is not felled but maintained and used in a sustainable way, the economic return is much greater.

## Cultivating the Plants

For plants required for human needs, cultivation rather than collection from nature is the answer. This is particularly true of medicinal plants. The great developments in health care and the revival of traditional medicine in many countries is leading to decimation of some medicinal plants in nature, with dire consequences for health care in the future. The first country to set up a systematic programme to counter this trend is Sri Lanka, where WWF is helping the Ministry of Indigenous Medicine with a conservation programme. This ministry now gives householders medicinal plants to grow on their patios and in their gardens, and has set up several nurseries in different parts of the island to cultivate the plants needed by hospitals and doctors. One nursery is even in the grounds of a large general hospital, where patients can wander round the garden seeing the plants used to cure them. The ministry is also planning laws to control the collecting of medicinal plants from the wild and is setting up a network of nature reserves of its own – entirely for medicinal plants. Working with the World Health Organization, IUCN and WWF are promoting this fascinating model to other countries around the world.

Large-scale cultivation is also the answer for plants threatened by collecting for gardens. By growing bulbs, cacti and orchids on a large scale, cultivation can take the pressure off vulnerable wild stocks from collectors and greatly reduce the price of the plants to hobbyists.

Trade between countries in many thousands of rare plants and their look-alikes is monitored and to some extent controlled by the Convention on International Trade in Endangered Species (CITES), but it is very difficult to distinguish small illegal shipments of wild-collected rarities from the millions of legitimate shipments of cultivated plants. Nevertheless, there have been a number of recent successes and many hobbyists, too, are insisting on buying only cultivated stocks, not wild-collected ones.

## Ex situ Conservation

The strange phrase *ex situ* simply means 'off-site' but, with its opposite of *in situ*, has become very prevalent in conservation circles. *Ex situ* conservation is a valuable back-up to *in situ* conservation, especially if the plant's original habitat has been destroyed. It is, however, rather more of an engineer's than a biologist's approach to conservation. Conserved in seed banks or gardens, plants no longer continue to evolve as they would in nature and, although easily accessible for use by man, they cannot have any truly long-term future.

The best means of *ex situ* conservation is in seed banks. A seed bank is simply a collection of dried seed stored in a deep freeze. The drying and freezing greatly increase the viability of the seed – the length of time it can be stored before being successfully germinated. In some plants, notably cereals, it has been predicted that longevity of the seeds could be extended as much as a thousand years, though periods of a hundred years or so seem more likely. The great advantage of the seed bank is that very large numbers of seeds can be stored in a small space – and each seed is genetically different from every other seed. One seed bank can store a great deal of diversity, both within and between species.

Just putting the precious seeds into the 'bank' is not enough. The techniques of seed banking were developed for a few crop species, and as species vary greatly in their reaction to seed banking some research may be needed

for each of the wild species deposited in the 'bank'. The other main task is to test the viability – the proportion of seeds that germinate – of the samples on a regular basis, say every year or so. When the viability drops so that less than half the sample germinates, it is time either to collect a new batch of seed from nature or, if that is no longer possible, to germinate all the seeds of that collection and harvest a new crop of seeds. This 'growing on' has to be done under very careful conditions to ensure the plants are not cross-pollinated by other species and that as many as possible of the seeds grown reach harvest stage. Otherwise one is unconsciously selecting for forms of the plant that do best in cultivation rather than in the wild.

About a fifth of the world's plants do not produce seeds at all, or produce seeds that do not retain their viability in storage. This is particularly true for trees such as oaks, which have large heavy seeds. For them the best form of *ex situ* conservation is the field gene bank, which is simply a plot of land in which the species are grown in rows, with as many individuals of each species as possible. The snag to this approach is that it takes a lot of space and labour, and the plants are vulnerable to disease epidemics.

Cultivating a specimen or two of an endangered plant in a greenhouse or outdoor bed is much less use for conservation, principally because so small a proportion of the genetic diversity of the species is represented. Also labels can get muddled, plants can be lost – or even stolen! – and seedlings are often of hybrids from one species to another. But, sadly, most of the endangered plants so far conserved in botanic gardens are grown in this way. Bodies like Botanic Gardens Conservation International, founded by IUCN in 1987, are trying to rectify this situation and to encourage the development of seed banks.

But even so, *ex situ* conservation is best seen as a temporary measure, allowing plants to be bulked up and re-established in the wild, and as a complement to *in situ* conservation rather than an alternative. It is no long-term solution.

## Towards a Sustainable Use of Plants

The solutions listed above are valuable and important but are only short-term answers. Nor do they address the important issue of the decline of our common plants – the primroses, wild daffodils and buttercups that once graced our countryside, but are now seen so much less.

We have to bring our lives into balance with nature and find new ways of living that are both satisfying and sustainable. One of the most fundamental needs is to redress the imbalance between north and south, between rich and poor countries. At present citizens all over the developing world have little alternative to destroying for short-term needs the resources they will need in the future. Nomads in the Sahel have to cut down valuable and ancient trees to collect their firewood, so increasing the spread of the desert. Landless peasants in Central and South America have to colonize inclement areas of jungle to scratch a living for their families from the soil. The results are neither productive nor sustainable for very long.

At present, many Third World governments see no option but to sell parts of their forests to pay the foreign debts built up by previous governments, often for nefarious and ill-thought-out industrialization schemes that all too often have not worked. They are forced to replace natural forest with plantation crops like coffee, tea and rubber, which they can sell for foreign exchange. At present, the repayment of debt from south to north exceeds aid flow from north to south. There will not be a permanent solution to saving the world's tropical plants until this issue is resolved.

One encouraging sign is the degree to which developed countries are reconsidering their aid policies, on the one hand tying them to good government and, on the other, considering to a far greater extent the environmental aspects of the projects they finance. Many aid agencies are now prepared to fund projects for establishing and managing national parks in rainforests, and to support ecologically sustainable development in the surrounding areas. However, the amount allocated for this purpose is only a small proportion of total aid.

At the Rio Earth Summit, in 1992, governments agreed a Biodiversity Convention – an international treaty between all nations that defines obligations on conservation of plants and animals. In return for developing countries undertaking to conserve their biological diversity – their plants and animals – the developed countries will contribute to the costs of them doing so through an international fund. Already the World Bank and the United Nations Development Programme have set up a fund of US$1.5 billion – the Global Environment Fund or GEF – as grants for environmental improvement, but there is still a long way to go.

## Biosphere Reserves

A new approach to protected areas is also needed. In the past national parks were seen as places set aside, removed from the vital process of a nation's socio-economic development. Nowadays the opposite is seen as more appropriate: national parks make a vital contribution to development – they attract tourists, prevent soil erosion, help moderate the climate, protect vital water supplies and conserve resources such as crop relatives and medicinal plants.

A particularly valuable approach is that of the biosphere reserve, promoted by UNESCO under its Man and the Biosphere Programme. A biosphere reserve is a large area that has three functions: conservation, monitoring and serving as a model of sustainable local development. A large protected area at the core fulfils the conservation function, protecting a valuable area for plants and animals. This core area is also used as a benchmark for global monitoring, in comparison with other altered areas. And the area around the strictly protected part is used to demonstrate how to achieve sustainable development for the people living in that region without destroying the environment. Governments are understandably keen to develop such reserves, combining as they do conservation and development.

In developed countries, major changes are needed in farming methods. Over large parts of Europe, notably eastern England, northern France, Denmark and Holland, the traditional landscape rich in wild plants has given way to a type of prairie agriculture, made possible by massive inputs of artificial fertilizers, herbicides and pesticides. Few, if any, wild plants survive. With the European Community continuing to produce food surpluses, farming should be seen more as a service industry than a production industry. Farmers should continue to receive the current subsidies, but should receive them for both maintaining the rural landscape and producing food, not just for the amount of food they produce. Britain and other EC countries are now designating Environmentally Sensitive Areas where farmers are paid to farm in traditional, non-harmful ways. This approach needs to be extended over more of the countryside.

New and more sophisticated forms of regional planning are also needed to conserve plant habitats. In many developed countries, land covered by natural or semi-natural vegetation is such a small proportion of the total land area that landowners with such assets should have to accept obligations on what they can do. There are many novel schemes of this nature, especially in Italy, Germany and Switzerland; but perhaps the most revolutionary is in Denmark. If in Denmark you own land on which there is a salt marsh, a peat bog or certain other types of valuable wild habitats, you can only alter it or destroy it with permission from the local authority. No compensation is due if permission is refused. In the European Community as a whole, the Flora, Fauna and Habitats Directive, adopted by the member states in 1991, obliges each government to conserve its threatened habitats and species to acceptable standards. This new and little-known piece of EC law is absolutely vital to the plants of Europe.

We also need a massive programme to rapidly complete the inventory of the plant kingdom started by Linnaeus 300 years ago. Computers and modern means of travel and communication mean the links between botanists from one country to another are stronger than ever. We need to reverse the absurd decline in botany as a science and train as many as 1,500 plant taxonomists to work in the tropics. Otherwise plants will be lost before we discover them or know anything about them.

# DIRECTORY OF SPECIES

Letters in brackets show the category of threat given in the *Red Data Book*:

| | | | |
|---|---|---|---|
| Extinct | (Ex) | Indeterminate | (I) |
| Endangered | (E) | Insufficiently known | (K) |
| Vulnerable | (V) | Threatened | (T) |
| Rare | (R) | Commercially threatened | (CT) |

## MAMMALS

**Monotremata** / **Monotremes**

| | | |
|---|---|---|
| *Tachyglossus aculeatus* | Echidna; spiny anteater | |
| *Zaglossus bruijni* | Long-beaked echidna | (V) |
| *Ornithorhynchus anatinus* | Duck-billed platypus | |

**Marsupialia** / **Marsupials**

| | | |
|---|---|---|
| *Caluromysiops irrupta* | Black-shouldered opossum | |
| *Phascogale* spp. | Phascogale; marsupial 'mice' | |
| *Antechinomys* spp. | Jerboa marsupials | |
| *Dasyurops maculatus* | Spotted-tailed quoll | |
| *Sarcophilus harrisi* | Tasmanian devil | |
| *Thylacinus cynocephalus* | Thylacine | (Ex) |
| *Myrmecobius fasciatus* | Numbat; banded anteater | (E) |
| *Notoryctes typhlops* | Marsupial mole | |
| *Phalanger maculatus* | Spotted cuscus | |
| *Petaurus australis* | Yellow-bellied glider | |
| *Phascolarctos cinereus* | Koala | |
| *Hemibelideus lemuroides* | Ring-tailed possum | |
| *Vombatus ursinus* | Wombat | |
| *Lagorchestes* spp. | Hare wallabies | |
| *Petrogale penicillata* | Brush-tailed rock wallaby | |
| *Macropus robustus* | Wallaroo or euro | |
| *Macropus rufus* | Red or plains kangaroo | |
| *Dendrolagus lumholtzi* | Lumholtz's tree kangaroo | |
| *Dendrolagus bennettianus* | Bennett's tree kangaroo | |
| *Bettongia lesueur* | Burrowing bettong | (R) |
| *Bettongia penicillata* | Brush-tailed bettong | (E) |
| *Hypsiprymnodon moschatus* | Musky rat-kangaroo | |

**Insectivora** / **Insectivores**

| | | |
|---|---|---|
| *Limnogale mergulus* | Aquatic tenrec | (I) |
| *Eremitalpa granti* | Grant's golden mole | (R) |
| *Anourosorex squamipes* | Sichuan burrowing shrew | |
| *Chimmarogale platycephala* | Himalayan water shrew | |
| *Nectogale elegans* | Sichuan water shrew | |
| *Uropsilus soricipes* | Mole shrew | |
| *Desmana moschata* | Russian desman | (V) |
| *Galemys pyrenaicus* | Pyrenean desman | (V) |
| *Talpa micrura* | Short-tailed mole | |

**Chiroptera** / **Bats**

| | | |
|---|---|---|
| *Pteropus niger* | Mauritian flying fox | (V) |
| *Pteropus rodricensis* | Rodriguez flying fox | (E) |
| *Pteropus seychellensis aldabrensis* | Seychelles flying fox | (V) |
| *Pteropus subniger* | Lesser Mascarene flying fox | (Ex) |
| *Myzopoda aurita* | Sucker-footed bat | (V) |
| *Lasiurus cinereus semotus* | Hawaiian hoary bat | (I) |

**Primates** / **Primates**

| | | |
|---|---|---|
| *Propithecus verreauxi* | Verreaux's sifaka | (V) |
| *Indri indri* | Indri | (E) |
| *Daubentonia madagascariensis* | Aye-aye | (E) |
| *Nycticebus pygmaeus* | Pygmy loris | (V) |
| *Arctocebus calabarensis* | Angwantibo | (K) |
| *Perodicticus potto* | Potto | |
| *Galago crassicaudatus* | Thick-tailed bushbaby | |
| *Galago thomasi* | Thomas's bushbaby | (K) |
| *Tarsius syrichta* | Philippine tarsier | (E) |
| *Alouatta caraya* | Black howler monkey | |
| *Saimiri oerstedi* | Central American squirrel monkey | (E) |
| *Ateles geoffroyi* | Geoffroy's spider monkey | (V) |
| *Brachyteles arachnoides* | Woolly spider monkey | (E) |
| *Lagothrix lagotricha* | Woolly monkey | (V) |
| *Leontopithecus chrysomelas* | Golden-headed lion tamarin | (E) |
| *Saguinus imperator* | Emperor tamarin | (I) |
| *Macaca fascicularis* | Crab-eating macaque | |
| *Papio hamadryas* | Hamadryas baboon | (R) |
| *Papio cynocephalus* | Yellow baboon | |
| *Papio ursinus* | Chacma baboon | |
| *Theropithecus gelada* | Gelada baboon | (R) |
| *Cercopithecus mitis* | Blue monkey | |
| *Rhinopithecus avunculus* | Tonkin snub-nosed monkey | (E) |
| *Rhinopithecus bieti* | Yunnan snub-nosed monkey | (E) |
| *Rhinopithecus brelichi* | Guizhou snub-nosed monkey | (E) |
| *Rhinopithecus roxellana* | Golden snub-nosed monkey | (V) |
| *Nasalis larvatus* | Proboscis monkey | (V) |
| *Colobus polykomos* | Black-and-white colobus | |
| *Procolobus badius* | Red colobus | |

| | | |
|---|---|---|
| *Hylobates pileatus* | Pileated gibbon | (E) |
| *Pongo pygmaeus* | Orang-utan | (E) |
| *Pan paniscus* | Pygmy chimpanzee | (V) |
| *Pan troglodytes* | Chimpanzee | (V) |
| *Pan troglodytes verus* | West African chimpanzee | (E) |
| *Gorilla gorilla* | Gorilla | (V) |
| *Gorilla gorilla berengei* | Mountain gorilla | (E) |
| *Gorilla gorilla graueri* | Eastern lowland gorilla | (E) |

| **Edentata** | **Edentates** | |
|---|---|---|
| *Myrmecophaga tridactyla* | Giant anteater | (V) |
| *Tamandua tetradactyla* | Tamandua; lesser anteater | |
| *Cyclopes didactylus* | Dwarf or silky anteater | |
| *Bradypus infuscatus* | Three-toed sloth | |
| *Choloepus didactylus* | Two-toed sloth | |
| *Choloepus hoffmanni* | Hoffmann's sloth | |
| *Priodontes giganteus* | Giant armadillo | (V) |
| *Dasypus novemcinctus* | Nine-banded armadillo | |
| *Chlamyphorus truncatus* | Pink fairy armadillo | (K) |

| **Lagomorpha** | **Lagomorphs** | |
|---|---|---|
| *Ochotona princeps* | American pika | |
| *Caprolagus hispidus* | Hispid hare | (E) |
| *Lepus oiostolus* | Woolly hare | |
| *Oryctolagus cuniculus* | European rabbit | |

| **Rodentia** | **Rodents** | |
|---|---|---|
| *Sciurus niger avicennia* | Mangrove fox squirrel | |
| *Marmota marmota* | Alpine marmot | |
| *Marmota caligata* | Hoary marmot | |
| *Marmota bobak* | Himalayan or bobak marmot | |
| *Cynomys parvidens* | Utah prairie dog | (V) |
| *Dipodomys* spp. | Kangaroo rats | |
| *Castor fiber* | Beaver | |
| *Oryzomys bauri* | Santa Fé rice rat | |
| *Nesoryzomys narboroughi* | Fernandina rice rat | |
| *Megaoryzomys curoio* | Galápagos giant rat | |
| *Hypogeomys antimena* | Malagasy giant rat | |
| *Ondatra zibethicus* | Muskrat | |
| *Neofiber alleni* | Round-tailed muskrat | |
| *Tachyoryctes* spp. | East African mole rat | |
| *Rattus rattus* | Black rat | |
| *Rattus norvegicus* | Brown rat | |
| *Rattus lutreolis* | Swamp rat | |
| *Mastacomys fuscus* | Broad-toothed rat | |
| *Cavia porcellus* | Guineapig | |
| *Hydrochoerus hydrochaeris* | Capybara | |
| *Cuniculus paca* | Paca | |
| *Lagostomus maximus* | Viscacha | |
| *Chinchilla brevicaudata* | Short-tailed chinchilla | (I) |
| *Chinchilla lanigera* | Long-tailed chinchilla | (I) |
| *Myocastor coypus* | Coypu | |

| **Cetacea** | **Cetaceans** | |
|---|---|---|
| *Platanista gangetica* | Ganges river dolphin | (V) |
| *Platanista minor* | Indus river dolphin | (E) |
| *Lipotes vexillifer* | Yangtze river dolphin | (E) |
| *Inia geoffrensis* | Amazon river dolphin | (V) |
| *Neophocaena phocaenoides* | Finless porpoise | |
| *Sousa sinensis* | Indopacific hump-backed dolphin | |
| *Tursiops truncatus* | Bottle-nose dolphin | |
| *Orcaella brevirostris* | Irrawaddy dolphin | (K) |

| | | |
|---|---|---|
| *Orcinus orca* | Killer whale | |
| *Balaena mysticetus* | Bowhead whale | (V) |
| *Eubalaena australis* | Southern right whale | (V) |
| *Eubalaena glacialis* | Northern right whale | (E) |
| *Balaenoptera acutorostrata* | Minke whale | |
| *Balaenoptera borealis* | Sei whale | |
| *Balaenoptera musculus* | Blue whale | (E) |
| *Balaenoptera physalus* | Fin whale | (V) |
| *Megaptera novaeangliae* | Humpback whale | (V) |

| **Carnivora** | **Carnivores** | |
|---|---|---|
| *Canis lupus* | Grey wolf | (V) |
| *Canis latrans* | Coyote | |
| *Canis simensis* | Ethiopian wolf | (E) |
| *Canis dingo* | Dingo | |
| *Vulpes vulpes* | Red fox | |
| *Nyctereutes procyonoides* | Raccoon dog | |
| *Dusicyon griseus* | Grey zorro | (V) |
| *Dusicyon vetulus* | Hoary zorro | (K) |
| *Chrysocyon brachyurus* | Maned wolf | (V) |
| *Speothus venaticus* | Bush dog | (V) |
| *Cuon alpinus* | Asiatic wild dog; dhole | (V) |
| *Lycaon pictus* | African wild dog | (E) |
| *Tremarctos ornatus* | Spectacled bear | (V) |
| *Ursus thibetanus* | Asiatic black bear | (V) |
| *Ursus americanus* | American black bear | |
| *Ursus arctos* | Brown bear | |
| *Ailurus fulgens* | Red panda | (K) |
| *Ailuropoda melanoleuca* | Giant panda | (E) |
| *Mustela erminea* | Stoat | |
| *Mustela nigripes* | Black-footed ferret | (E) |
| *Mustela vison* | American mink | |
| *Martes americana* | American marten | |
| *Gulo gulo* | Wolverine | (V) |
| *Lutra lutra lutra* | European otter | (V) |
| *Lutra provocax* | Southern river otter | (V) |
| *Lutra perspicillata* | Smooth-coated otter | (K) |
| *Pteronura brasiliensis* | Giant otter | (V) |
| *Aonyx cinerea* | Small-clawed otter | (K) |
| *Osbornictis piscivora* | African water civet | |
| *Herpestes auropunctatus* | Small Indian mongoose | |
| *Cryptoprocta ferox* | Fossa | (K) |
| *Proteles cristatus* | Aardwolf | |
| *Hyaena brunnea* | Brown hyena | (V) |
| *Felis lynx* | Northern lynx | |
| *Felis lynx isabellina* | Tibetan lynx | |
| *Felis pardina* | Pardel lynx | (E) |
| *Felis manul* | Steppe cat | |
| *Felis viverrina* | Fishing cat | |
| *Felis pardalis* | Ocelot | (V) |
| *Felis wiedii* | Margay | (V) |
| *Felis tigrina* | Little spotted cat | (V) |
| *Felis jacobita* | Andean cat | (R) |
| *Felis yagouarundi* | Jaguarundi | (I) |
| *Felis concolor coryi* | Florida cougar | (E) |
| *Neofelis nebulosa* | Clouded leopard | (V) |
| *Panthera tigris* | Tiger | (E) |
| *Panthera tigris virgata* | Caspian tiger | (Ex) |
| *Panthera pardus* | Leopard | (T) |
| *Panthera onca* | Jaguar | (V) |

| | | |
|---|---|---|
| *Panthera uncia* | Snow leopard | (E) |
| *Acinonyx jubatus venaticus* | Asiatic cheetah | (E) |

### Pinnipedia — Pinnipeds – seals

| | | |
|---|---|---|
| *Arctocephalus pusillus* | South African fur seal | |
| *Arctocephalus forsteri* | New Zealand fur seal | |
| *Arctocephalus philippii* | Juan Fernandez fur seal | (V) |
| *Zalophus californianus* | Californian sealion | |
| *Phoca vitulina* | Common seal | |
| *Lobodon carcinophagus* | Crabeater seal | |
| *Ommatophoca rossi* | Ross seal | |
| *Hydrurga leptonyx* | Leopard seal | |
| *Leptonychotes weddelli* | Weddell seal | |
| *Mirounga leonina* | Southern elephant seal | |

### Tubulidentata — Aardvarks

| | | |
|---|---|---|
| *Orycteropus afer* | Aardvark; antbear | |

### Proboscidea — Elephants

| | | |
|---|---|---|
| *Elephas maximus* | Indian elephant | (E) |
| *Loxodonta africana* | African elephant | (V) |

### Hyracoidea — Hyraxes

| | | |
|---|---|---|
| *Procavia capensis* | Rock hyrax | |
| *Dendrohyrax arboreus* | Tree hyrax | |
| *Dendrohyrax validus* | Eastern tree dassie | (K) |

### Sirenia — Sirenians

| | | |
|---|---|---|
| *Dugong dugon* | Dugong | (V) |
| *Hydrodamalis gigas* | Steller's sea cow | (Ex) |
| *Trichechus manatus* | West Indian manatee | (V) |
| *Trichechus inunguis* | Amazonian manatee | (V) |
| *Trichechus senegalensis* | West African manatee | (V) |

### Perissodactyla — Odd-toed ungulates

| | | |
|---|---|---|
| *Equus przewalskii* | Przewalski's horse | (Ex?) |
| *Equus hemionus* | Asiatic wild ass | (E) |
| *Equus hemionus kiang* | Kiang; Tibetan wild ass | |
| *Equus hemionus kulan* | Kulan; Turkmenian wild ass | |
| *Equus zebra zebra* | Cape mountain zebra | (E) |
| *Equus zebra hartmannae* | Hartmann's mountain zebra | (V) |
| *Equus burchelli* | Plains zebra | |
| *Equus grevyi* | Grevy's zebra | (E) |
| *Tapirus indicus* | Malayan tapir | (E) |
| *Tapirus terrestris* | South American tapir | |
| *Tapirus pinchaque* | Mountain or Andean tapir | (V) |
| *Tapirus bairdii* | Central American tapir | (V) |
| *Rhinoceros unicornis* | Great Indian rhinoceros | (E) |
| *Rhinoceros sondaicus* | Javan rhinoceros | (E) |
| *Didermocerus sumatrensis* | Sumatran rhinoceros | (E) |
| *Ceratotherium simum cottoni* | Northern square-lipped rhino | (E) |
| *Diceros bicornis* | Black rhinoceros | (E) |

### Artiodactyla — Even-toed ungulates

| | | |
|---|---|---|
| *Potamochoerus porcus* | Bushpig | |
| *Sus scrofa* | Wild boar | |
| *Sus salvanius* | Pygmy hog | (E) |
| *Hippopotamus amphibius* | Hippopotamus | |
| *Lama glama* | Llama | |
| *Lama guanicoe* | Guanaco | |
| *Lama pacos* | Alpaca | |
| *Vicugna vicugna* | Vicuña | (V) |
| *Camelus bactrianus* | Wild Bactrian camel | (V) |
| *Moschus chrysogaster* | Himalayan musk deer | |

| | | |
|---|---|---|
| *Axis axis* | Chital or axis deer | |
| *Axis porcinus* | Hog deer | |
| *Cervus unicolor* | Sambar | |
| *Cervus timorensis* | Javan deer | |
| *Cervus duvauceli* | Swamp deer or barasingha | (E) |
| *Cervus albirostris* | Thorold's deer | (V) |
| *Cervus elaphus* | Red deer; elk (USA) | |
| *Cervus elaphus bactrianus* | Bactrian deer | (E) |
| *Cervus elaphus hanglu* | Hangul; Kashmir stag | (E) |
| *Cervus elaphus wallichi* | Shou | (E) |
| *Elaphurus davidianus* | Père David's deer | (E) |
| *Odocoileus hemionus* | Mule deer | |
| *Odocoileus virginianus* | White-tailed deer | |
| *Odocoileus virginianus clavium* | Florida Key deer | (R) |
| *Hippocamelus antisensis* | North Andean huemul | (V) |
| *Hippocamelus bisulcus* | South Andean huemul | (E) |
| *Blastocerus dichotomus* | Marsh deer | (V) |
| *Ozotoceros bezoarticus celer* | Argentinian pampas deer | (E) |
| *Pudu mephistopheles* | Northern pudu | (I) |
| *Alces alces* | Moose | |
| *Hydropotes inermis* | Chinese water-deer | (R) |
| *Capreolus capreolus* | Roe deer | |
| *Okapia johnstoni* | Okapi | |
| *Giraffa camelopardalis* | Giraffe | |
| *Antilocapra americana* | Pronghorn | (E) |
| *Tragelaphus buxtoni* | Mountain nyala | (E) |
| *Tragelaphus spekei* | Sitatunga | |
| *Tragelaphus scriptus* | Bushbuck | |
| *Tragelaphus imberbis* | Lesser kudu | |
| *Tragelaphus strepsiceros* | Greater kudu | |
| *Taurotragus oryx* | Eland | |
| *Taurotragus eurycerus* | Bongo | |
| *Bubalus bubalis* | Wild Asiatic buffalo | (E) |
| *Bos gaurus* | Gaur | (V) |
| *Bos sauveli* | Kouprey | (E) |
| *Bos grunniens mutus* | Wild yak | (E) |
| *Syncerus caffer* | African buffalo | |
| *Bison bonasus* | European bison | (V) |
| *Bison bison* | American bison | |
| *Cephalophus jentinki* | Jentink's duiker | (E) |
| *Cephalophus spadix* | Abbott's duiker | (V) |
| *Kobus ellipsiprymnus* | Waterbuck | |
| *Kobus kob leucotis* | White-eared kob | |
| *Kobus vardoni* | Puku | |
| *Kobus leche* | Lechwe | (V) |
| *Kobus leche kafuensis* | Kafue lechwe | |
| *Kobus leche smithemani* | Black lechwe | |
| *Kobus megaceros* | Nile lechwe | |
| *Redunca arundinum* | Reedbuck | |
| *Redunca fulvorufula* | Mountain reedbuck | |
| *Hippotragus equinus* | Roan antelope | |
| *Hippotragus niger* | Sable antelope | |
| *Hippotragus niger variani* | Giant sable antelope | (E) |
| *Oryx gazella* | Oryx; gemsbok | |
| *Oryx leucoryx* | Arabian oryx | (E) |
| *Oryx dammah* | Scimitar-horned oryx | (E) |
| *Addax nasomaculatus* | Addax | (E) |
| *Damaliscus lunatus tiang* | Tiang | |
| *Damaliscus dorcas dorcas* | Bontebok | (R) |
| *Damaliscus hunteri* | Hunter's antelope | (V) |

| | | |
|---|---|---|
| *Alcelaphus buselaphus swaynei* | Swayne's hartebeest | (E) |
| *Alcelaphus lichtensteini* | Lichtenstein's hartebeest | |
| *Connochaetes taurinus* | Wildebeest | |
| *Raphicerus campestris* | Steinbok | |
| *Madoqua kirki damarensis* | Damara dikdik | |
| *Aepyceros melampus* | Impala | |
| *Aepyceros melampus petersi* | Black-faced impala | (E) |
| *Ammodorcas clarkei* | Dibatag | (V) |
| *Litocranius walleri* | Gerenuk | |
| *Gazella cuvieri* | Cuvier's gazelle | (E) |
| *Gazella dama* | Dama gazelle | (E) |
| *Gazella dorcas* | Dorcas gazelle | (V) |
| *Gazella gazella* | Mountain gazelle | (V) |
| *Gazella leptocerus* | Slender-horned gazelle | (E) |
| *Gazella rufifrons* | Red-fronted gazelle | (V) |
| *Gazella soemmerringi* | Soemmerring's gazelle | (V) |
| *Gazella spekei* | Speke's gazelle | (V) |
| *Gazella thomsoni* | Thomson's gazelle | |
| *Gazella subgutturosa* | Goitred gazelle | |
| *Antidorcas marsupialis* | Springbok | |
| *Procapra picticaudata* | Tibetan gazelle | |
| *Procapra gutturosa* | Mongolian gazelle | |
| *Pantholops hodgsoni* | Tibetan antelope | |
| *Saiga tatarica* | Saiga antelope | |
| *Nemorhaedus goral* | Goral | |
| *Capricornis sumatraensis* | Serow | |
| *Oreamnos americanus* | Rocky mountain goat | |
| *Rupicapra rupicapra* | Chamois | |
| *Budorcas taxicolor bedfordi* | Golden takin | (R) |
| *Ovibos moschatus* | Musk ox | |
| *Hemitragus jemlahicus* | Himalayan tahr | |
| *Capra ibex* | Ibex | |
| *Capra pyrenaica pyrenaica* | Pyrenean ibex | (E) |
| *Capra walie* | Walia ibex | (E) |
| *Pseudois nayaur* | Bharal; blue sheep | |
| *Ammotragus lervia* | Barbary sheep | (V) |
| *Ovis ammon* | Argali | |
| *Ovis ammon polii* | Marco Polo sheep | |
| *Ovis canadensis* | Bighorn sheep | |

# BIRDS

| | | |
|---|---|---|
| **Struthionidae** | **Ostriches** | |
| *Struthio camelus* | Ostrich | |
| **Rheidae** | **Rheas** | |
| *Rhea americana* | Rhea | |
| **Casuariidae** | **Cassowaries** | |
| *Casuarius casuarius* | Cassowary | |
| **Dromaiidae** | **Emus** | |
| *Dromaius novaehollandiae* | Emu | |
| **Apterygidae** | **Kiwis** | |
| *Apteryx australis australis* | South Island brown kiwi | |
| *Apteryx owenii* | Little spotted kiwi | (V) |
| *Apteryx haasti* | Great spotted kiwi | |
| **Spheniscidae** | **Penguins** | |
| *Aptenodytes forsteri* | Emperor penguin | |

| | | |
|---|---|---|
| *Eudyptes crestatus* | Rockhopper penguin | |
| *Eudyptes pachyrhynchus* | Crested penguin | |
| *Spheniscus demersus* | Jackass penguin | (K) |
| *Spheniscus mendiculus* | Galápagos penguin | |
| **Podicipedidae** | **Grebes** | |
| *Tachybaptus pelzelni* | Madagascar little grebe | (K) |
| *Podiceps australis* | Crested grebe | |
| *Podiceps cristatus* | Great crested grebe | |
| **Diomedeidae** | **Albatrosses** | |
| *Diomedia exulans* | Wandering albatross | |
| *Diomedia irrorata* | Waved albatross | |
| *Diomedia chlororhynchos* | Yellow-nosed albatross | |
| **Procellariidae** | **Petrels** | |
| *Pterodroma macroptera* | Long-winged petrel | |
| *Pterodroma aterrima* | Mascarene black petrel | (I) |
| *Pterodroma arminjoniana* | Round Island petrel | |
| *Pterodroma baraui* | Barau's petrel | |
| *Pterodroma phaeopygia* | Dark-rumped petrel | (R) |
| *Pterodroma madeira* | Freira | (E) |
| *Procellaria gravis* | Great shearwater | |
| *Puffinus pacificus* | Wedge-tailed shearwater | |
| **Pelecanidae** | **Pelicans** | |
| *Pelecanus occidentalis* | Brown pelican | |
| **Sulidae** | **Gannets; Boobies** | |
| *Sula nebouxii* | Blue-footed booby | |
| *Sula capensis* | Cape gannet | |
| *Sula dactylatra* | Masked booby | |
| *Sula sula* | Red-footed booby | |
| *Phaëton rubricauda* | Red-tailed tropic bird | |
| *Phaëton lepturus* | White-tailed tropic bird | |
| **Phalacrocoracidae** | **Cormorants** | |
| *Phalocrocorax capensis* | Cape cormorant | |
| *Nannopterum harrisi* | Galápagos flightless cormorant | (R) |
| **Fregatidae** | **Frigatebirds** | |
| *Fregata minor* | Greater frigate bird | |
| *Fregata ariel* | Lesser frigate bird | |
| **Ardeidae** | **Herons; bitterns** | |
| *Ixobrychus minutus* | Little bittern | |
| *Cochlearius cochlearius* | Boat-billed heron | |
| *Egretta vinaceigula* | Slaty egret | |
| *Egretta sacra* | Reef heron | |
| **Balaenicipitidae** | **Shoebills** | |
| *Balaeniceps rex* | Shoebill | (K) |
| **Ciconiidae** | **Storks** | |
| *Mycteria americana* | Wood ibis | |
| **Threskiornithidae** | **Ibises; spoonbills** | |
| *Pseudibis davisoni* | White-shouldered ibis | (V) |
| *Pseudibis gigantea* | Giant ibis | (E) |
| *Geronticus eremita* | Northern bald ibis | (E) |
| *Platalea leucorodia* | Spoonbill | |
| *Platalea ajaja* | Roseate spoonbill | |
| **Phoenicopteridae** | **Flamingos** | |
| *Phoenicopterus ruber* | Greater flamingo | |
| *Phoenicopterus ruber chilensis* | Chilean flamingo | |

| | | |
|---|---|---|
| *Phoenicopterus minor* | Lesser flamingo | |
| *Phoenicoparrus andinus* | Andean flamingo | (K) |
| *Phoenicoparrus jamesi* | Puna flamingo | (K) |

**Anhimidae** — **Screamers**
*Chauna torquata* — Crested screamer (*kamichi*)

**Anatidae** — **Swans; geese; ducks**

| | | |
|---|---|---|
| *Dendrocygna autumnalis* | Red-winged whistling duck | |
| *Cygnus cygnus* | Whooper swan | |
| *Anser fabalis* | Bean goose | |
| *Anser indicus* | Bar-headed goose | |
| *Anser brachyrhynchus* | Pink-footed goose | |
| *Branta sandvicensis* | Néné or Hawaiian goose | (R) |
| *Branta bernicla* | Brent goose | |
| *Tadorna ferruginea* | Ruddy shelduck | |
| *Tadorna variegata* | Paradise shelduck | |
| *Tadorna tadorna* | Shelduck | |
| *Plectropterus gambensis* | Spurwinged goose | |
| *Cairina scutulata* | White-winged duck | (V) |
| *Lophonetta specularioides* | Crested duck | |
| *Merganetta armata* | Torrent duck | |
| *Nettapus auritus* | Pygmy goose | |
| *Anas flavirostris* | Yellow-billed teal | |
| *Anas clypeata* | Shoveler | |
| *Anas puna* | Puna teal | |
| *Anas aucklandica* | Brown teal | (R) |
| *Rhodonessa caryophyllacea* | Pink-headed duck | (Ex) |
| *Hymenolaimus malacorhynchos* | Blue (or mountain) duck | |
| *Netta rufina* | Red-crested pochard | |
| *Aythya nyroca* | Ferruginous duck | |
| *Somateria mollissima* | Eider | |
| *Melanitta nigra* | Scoter | |
| *Mergus merganser* | Goosander | |

**Cathartidae** — **New World vultures; condors**
*Vultur gryphus* — Andean condor

**Accipitridae** — **Eagles; hawks**

| | | |
|---|---|---|
| *Pandion haliaetus* | Osprey | |
| *Rostrhamus sociabilis* | Everglades kite | |
| *Haliaeetus leucogaster* | White-bellied sea eagle | |
| *Haliaeetus leucoryphus* | Pallas's fish eagle | (R) |
| *Haliaeetus leucocephalus* | Bald eagle | |
| *Haliaeetus albicilla* | White-tailed sea eagle | (R) |
| *Haliaeetus pelagicus* | Steller's sea eagle | (R) |
| *Buteo solitarius* | Hawaiian hawk or i'o | (R) |
| *Morphnus guianensis* | Crested eagle | (R) |
| *Harpia harpyja* | Harpy eagle | (R) |
| *Aquila rapax nipalensis* | Steppe eagle | |
| *Aquila chrysaetos* | Golden eagle | |
| *Aquila audax* | Wedge-tailed eagle | |
| *Aquila verreauxi* | Verreaux's eagle | |
| *Stephanoaetus coronatus* | Crowned eagle | |
| *Polemaetus bellicosus* | Martial eagle | |

**Sagittariidae** — **Secretary bird**
*Sagittarius serpentarius* — Secretary bird

**Falconidae** — **Falcons**

| | | |
|---|---|---|
| *Falco deiroleucus* | Orange-breasted falcon | (K) |
| *Falco naumanni* | Lesser kestrel | (R) |
| *Falco punctatus* | Mauritius kestrel | (E) |

**Cracidae** — **Curassows; guans**
*Penelope superciliaris* — Rusty-margined guan

**Phasianidae** — **Quail; partridges; pheasants**

| | | |
|---|---|---|
| *Lagopus mutus* | Rock ptarmigan | |
| *Lagopus leucurus* | White-tailed ptarmigan | |
| *Lyrurus tetrix* | Black grouse | |
| *Tetrao urogallus* | Capercaillie | |
| *Lerwa lerwa* | Snow partridge | |
| *Tetraogallus himalayensis* | Tibetan snowcock | |
| *Francolinus gularis* | Swamp francolin | (V) |
| *Ophrysia superciliosa* | Himalayan mountain quail | (Ex) |
| *Ithaginis cruentus* | Blood pheasant | |
| *Tragopan melanocephalus* | Western tragopan | (V) |
| *Tragopan satyrus* | Satyr tragopan | |
| *Tragopan blythi* | Blyth's tragopan | (I) |
| *Lophophorus impejanus* | Himalayan monal | |
| *Gallus gallus* | Red jungle fowl | |
| *Catreus wallichi* | Cheer | (V) |
| *Afropavo congensis* | Congo peacock | (K) |
| *Numida meleagris* | Wattle-nosed guineafowl | |

**Opisthocomidae** — **Hoatzin**
*Opisthocomus hoazin* — Hoatzin

**Gruidae** — **Cranes**

| | | |
|---|---|---|
| *Grus nigricollis* | Black-necked crane | (R) |
| *Grus japonicus* | Japanese crane | (V) |
| *Grus antigone sharpii* | Eastern sarus crane | |
| *Grus leucogeranus* | Siberian crane | (R) |
| *Bugeranus carunculatus* | Wattled crane | (K) |

**Rallidae** — **Rails**

| | | |
|---|---|---|
| *Gallirallus australis* | South Island weka, or wood hen | |
| *Dryolimnas cuvieri* | White-throated rail | |
| *Atlantisia rogersi* | Inaccessible Island rail | (R) |
| *Gallinula nesiotis nesiotis* | Tristan 'island hen' | (Ex) |
| *Porphyrula martinica* | American purple gallinule | |
| *Notornis mantelli* | Takahé | |
| *Fulica gigantea* | Giant coot | |
| *Fulica cornuta* | Horned coot | (R) |

**Heliornithidae** — **Finfoot (sungrebe)**
*Heliornis fulica* — Finfoot

**Otididae** — **Bustards**

| | | |
|---|---|---|
| *Tetrax tetrax* | Little bustard | (R) |
| *Otis tarda* | Great bustard | (R) |
| *Chlamydotis undulata* | Houbara bustard | (V) |
| *Afrotis atra* | Black korhaan | |
| *Houbaropsis bengalensis* | Bengal florican | (E) |

**Haematopodidae** — **Oystercatchers**
*Haematopus ostralegus* — Oystercatcher

**Ibidorhynchidae** — **Ibisbill**
*Ibidoryncha struthersii* — Ibisbill

**Recurvirostridae** — **Avocets; stilts**
*Recurvirostra avosetta* — Avocet

**Charadriidae** — **Plovers**

| | | |
|---|---|---|
| *Vanellus gregarius* | Sociable plover | |
| *Charadrius mongolus* | Mongolian sandplover | |
| *Charadrius asiaticus* | Caspian plover | |

**Scolopacidae** — Snipe; sandpipers
Limosa lapponica — Bar-headed godwit
Numenius tenuirostris — Slender-billed curlew (K)
Numenius arquata — Curlew
Tringa totanus — Redshank
Tringa glareola — Wood sandpiper
Calidris temmincki — Temmincks's stint

**Stercorariidae** — Skuas
Stercorarius parasiticus — Arctic skua

**Laridae** — Gulls; terns
Larus saundersi — Saunders' gull
Sterna paradisaea — Arctic tern
Sterna dougalli — Roseate tern
Anoüs stolidus pileatus — Aldabra noddy
Anoüs albus — White noddy or fairy tern

**Alcidae** — Auks
Brachyramphus marmoratus — Marbled murrelet

**Raphidae** — Dodo
Raphus cucullatus — Dodo (Ex)
Raphus solitarius — Réunion solitaire (Ex)
Pezophaps solitaria — Rodriguez solitaire (Ex)

**Columbidae** — Pigeons
Ectopistes migratorius — Passenger pigeon (Ex)
Nesoenas mayeri — Mauritius pink pigeon (E)

**Psittacidae** — Parrots
Psephotus varius — Mulga parrot
Neophema bourki — Bourke's grass parakeet
Neophema chrysogaster — Orange-bellied parrot (R)
Melopsittacus undulatus — Budgerigar
Pezoporus wallicus — Ground parrot (E)
Geopsittacus occidentalis — Night parrot (I)
Poicephalus gulielmi — Red-headed parrot
Psittacula eques — Mauritius parakeet (E)
Anodorhynchus hyacinthinus — Hyacinth macaw (K)
Nestor notabilis — Kea
Nestor meridionalis — Kaka
Strigops habroptilus — Kakapo (E)

**Musophagidae** — Turacos
Tauraco bannermani — Bannerman's turaco (E)
Tauraco hartlaubi — Hartlaub's turaco

**Strigidae** — Owls
Scotopelia peli — African fishing owl
Sceloglaux albifacies — Laughing owl; whekau (Ex)

**Apodidae** — Swifts
Collocalia francica — Mauritius cave swiftlet

**Trogonidae** — Trogons
Pharomacrus mocinno — Resplendent quetzal (V)
Harpactes erythrocephalus — Red-headed trogon

**Alcedinidae** — Kingfishers
Ceryle lugubris — Crested kingfisher
Ceryle rudis — Pied kingfisher
Dacelo novaeguineae — Kookaburra

**Coraciidae** — Rollers; ground rollers
Uratelornis chimaera — Long-tailed ground roller (R)

**Leptosomatidae** — Cuckoo-rollers
Leptosomus discolor — Cuckoo-roller

**Picidae** — Woodpeckers
Picoides tridactylus — Northern three-toed woodpecker
Colaptes rupicola — Andean rock flicker
Dryocopus martius — Black woodpecker
Campephilus principalis — Ivory-billed woodpecker (E)

**Xenicidae** — New Zealand wrens
Xenicus longipes longipes — New Zealand bushwren (Ex)

**Philepittidae** — Asitys
Philepitta castanea — Velvet asity
Philepitta schlegeli — Schlegel's asity
Neodrepanis coruscans — Wattled false sunbird
Neodrepanis hypoxanthus — Small-billed false sunbird (I)

**Alaudidae** — Larks
Melanocorypha maxima — Long-billed calandra lark
Calandrella rufescens — Lesser short-toed lark

**Campephagidae** — Cuckoo-shrikes
Coracina typica — Mauritius cuckoo-shrike (V)
Coracina newtoni — Réunion cuckoo-shrike (V)

**Pycnonotidae** — Bulbuls
Hypsipetes olivaceus — Mauritius bulbul (or 'merle') (V)

**Vangidae** — Vanga shrikes
Hypositta corallirostris — Coral-billed nuthatch

**Troglodytidae** — Wrens
Stipiturus ruficeps — Rufous-crowned emu-wren

**Prunellidae** — Accentors
Prunella collaris — Alpine accentor
Prunella rubeculoides — Robin accentor

**Muscicapidae** — Thrushes; babblers; warblers
Phoenicurus schisticeps — White-throated redstart
Sialia currucoides — Mountain bluebird
Oenanthe deserti — Desert wheatear
Nesocichla eremita — Tristan thrush or 'starchy'
Phaeornis obscurus — Hawaiian thrush or omao
Turdus torquatus — Ring ouzel
Babax waddelli — Giant babax
Babax koslowi — Kozlov's babax
Picathartes gymnocephalus — White-necked picathartes (V)
Picathartes oreas — Grey-necked picathartes (V)
Acrocephalus seychellensis — Seychelles brush warbler (R)
Acrocephalus rodericanus — Rodriguez brush warbler (E)
Nesillas aldabranus — Aldabra warbler (E)
Platysteira laticincta — Banded wattle-eye (E)
Crateroscelis gutturalis — Australian fern wren
Turnagra capensis capensis — Piopio (Ex)

**Paridae** — Tits; chickadees
Parus gambeli — Mountain chickadee

**Certhiidae** — Treecreepers
Tichodroma muraria — Wallcreeper

**Nectariniidae** — Sunbirds
Nectarinia johnstoni — Scarlet-tufted malachite sunbird
Aethopyga nipalensis — Nepal sunbird

**Zosteropidae** — White-eyes
Zosterops chloronothus — Mauritius olive white eye (V)

**Meliphagidae** — Honeyeaters; sugarbirds
*Meliphaga frenata* — Bridled honeyeater
*Meliphaga penicillata* — White-plumed honeyeater
*Anthornis melanura* — Bellbird

**Emberizidae** — Buntings; grosbeaks
*Emberiza koslowi* — Koslov's bunting
*Rowettia goughensis* — Gough Island bunting (R)
*Nesospiza acunhae* — Tristan bunting (R)
*Nesospiza wilkinsi* — Tristan grosbeak (R)
*Cactospiza pallida* — Galápagos woodpecker finch
*Geospiza difficilis* — Sharp-billed ground finch

**Drepanididae** — Honeycreepers
*Hemignathus munroi* — Akiapola'au (V)
*Loxops coccineus* — Akepa (R)
*Psittirostra psittacea* — O'u (E)

**Fringillidae** — Finches
*Carpodacus puniceus* — Rose-breasted rose finch
*Carpodacus roborowski* — Tibetan rose finch

**Ploceidae** — Weavers; sparrows
*Montifringilla nivalis* — Snow finch
*Montifringilla taczanowskii* — Taczanowski's snow finch
*Foudia rubra* — Mauritius fody (E)
*Foudia flavicans* — Rodriguez fody (E)

**Sturnidae** — Starlings; oxpeckers
*Cinnyricinclus femoralis* — Abbott's starling (K)

**Callaeidae** — Wattlebird
*Callaeas cinerea* — Kokako (V)

**Ptilonorhynchidae** — Bowerbirds
*Scenopeetes dentirostris* — Tooth-billed bowerbird

**Corvidae** — Crows
*Nucifraga caryocatactes* — Nutcracker
*Pyrrhocorax graculus* — Alpine chough
*Corvus albicollis* — African white-necked raven

# REPTILES

Order **Testudines** — Turtles; tortoises
Family **Cheloniidae**
*Caretta caretta* — Loggerhead turtle (V)
*Chelonia mydas* — Green turtle (E)
*Eretmochelys imbricata* — Hawksbill turtle (E)

Family **Dermochelyidae**
*Dermochelys coriacea* — Leatherback (E)

Family **Testudinidae**
*Aldabrachelys elephantina* — Aldabran giant tortoise (R)
*Geochelone nigra* — Galápagos giant tortoise (V)
*Geochelone yniphora* — Angonoka (E)

Order **Rhynchocephalia** — Tuatara
Family **Sphenodontidae**
*Sphenodon punctatus* — Tuatara (R)

Order **Sauria** — Lizards
Family **Agamidae**
*Aporosaura anchietae* — Namib or snouted lizard
*Chlamydosaurus kingii* — Frilled lizard

*Moloch horridus* — Moloch or spiny desert lizard

Family **Chamaeleontidae**
*Chamaeleo oustaleti* — Oustalet's chameleon
*Brookesia spp.* — Stump-tailed chameleon

Family **Geckkonidae**
*Nactus serpensinsula* — Serpent Island gecko (R)
*Phelsuma edwardnewtonii* — Rodriguez day-gecko (E)
*Phelsuma guentheri* — Round Island day-gecko (E)
*Palmatogecko rangei* — Web-footed gecko
*Uroplatus fimbriatus* — Leaf-tailed gecko

Family **Iguanidae**
*Amblyrhynchus cristatus* — Galápagos marine iguana (R)
*Conolophus pallidus* — Barrington land iguana (R)
*Conolophus subcristatus* — Galápagos land iguana (V)

Family **Scincidae**
*Leiolopsima telfairii* — Round Island skink
*Pseudomeia palfreyman* — Pedra Branca skink

Order **Serpentes** — Snakes
Family **Boidae**
*Bolyeria multicarinata* — Round Island boa (E)
*Casarea dussumieri* — Round Is. keel-scaled boa (E)

Order **Crocodylia** — Crocodilians
Family **Alligatoridae**
*Alligator mississippiensis* — American alligator
*Alligator sinensis* — Chinese alligator (E)
*Caiman crocodilus* — Common caiman
*Caiman latirostris* — Broad-snouted caiman (E)
*Caiman yacare* — Yacare
*Melanosuchus niger* — Black caiman (E)

Family **Crocodylidae**
*Crocodylus acutus* — American crocodile (E)
*Crocodylus intermedius* — Orinoco crocodile (E)
*Crocodylus mindorensis* — Philippine crocodile (E)
*Crocodylus moreletii* — Morelet's crocodile (E)
*Crocodylus niloticus* — Nile crocodile (V)
*Crocodylus palustris* — Mugger crocodile (V)
*Crocodylus porosus* — Saltwater crocodile (V)
*Crocodylus rhombifer* — Cuban crocodile (E)
*Crocodylus siamensis* — Siamese crocodile (E)
*Tomistoma schlegelii* — Tomistoma (E)

Family **Gavialidae**
*Gavialis gangeticus* — Gharial (E)

# AMPHIBIANS

Order **Anura** — Frogs; toads
Family **Myobatrachidae**
*Rheobatrachus silus* — Conondale mouth-brooding frog (E)

Family **Bufonidae**
*Bufo marinus* — Cane toad
*Bufo periglenes* — Golden toad (E)

Family **Hylidae**
*Litoria burrowsi* — Tasmanian tree frog

# FISH

Order *Lepidosireniformes*
Family **Lepidosirenidae**
*Lepidosiren paradoxa* — South American lungfish

Order *Acipeniseriformes*
Family **Acipenseridae**
*Acipenser sturio* — Common sturgeon (E)
*Acipenser ruthenus* — Sterlet

Order *Clupeiformes*
Family **Clupeidae**
*Clupea harengus* — Herring
*Sardinia pilchardus* — Sardine
*Engraulis encrasicholus* — Anchovy

Order *Pleuronectiformes*
Family **Pleuronectidae**
*Pleuronectes platessa* — Plaice
*Hippoglossus hippoglossus* — Halibut
Family **Soleidae**
*Solea solea* — Common Sole

Order *Cypriniformes*
Family **Cyprinidae**
*Cyprinus carpio* — Carp
*Rutilus rutilus* — Roach

Order *Siluriformes*
Family **Pangasiidae**
*Pangasianodon gigas* — Mekong giant catfish (V)
Family **Clariidae**
*Clarias batrachus* — Walking catfish

Order *Salmoniformes*
Family **Galaxiidae**
*Galaxias fontanus* — Swan galaxias (E)
*Galaxias johnstoni* — Clarence galaxias (E)
*Galaxias parvus* — Small Pedder galaxias (V)
*Galaxias pedderensis* — Pedder galaxias (V)
*Galaxias tanycephalus* — Saddled galaxias (V)
Family **Salmonidae**
*Onchorhynchus nerka* — Sockeye
*Onchorhynchus ishikawai* — Satsukimasu salmon
*Salmo salar* — Atlantic salmon
*Salmo trutta* — Brown trout
*Salmo gairdneri* — Rainbow trout

Order *Perciformes*
Family **Cichlidae**
*Oreochromis alcalicus grahami* — Lake Magadi tilapia (R)
*Tilapia* spp. — Tilapia
*Haplochromis pyrrhocephalus* — Lake Victoria cichlid (E)
Family **Centropomidae**
*Lates niloticus* — Nile perch

# GASTROPODS

Order *Mesogastropoda*
Family **Cymatiidae**

*Charonia tritonis* — Triton's trumpet (R)

# BIVALVES

Order *Veneroida*
Family **Tridacnidae**
*Tridacna gigas* — Giant clam (V)

# ANNELID WORMS

Order *Haplotaxida*
Family **Megascolecidae**
*Megascolides australis* — Giant Gippsland earthworm (V)

# SPIDERS

Order *Aranae*
Family **Theraphosidae**
*Brachypelma smithi* — Red-kneed tarantula (K)

# CRUSTACEANS

Order *Euphausiacea*
Family **Euphausiidae**
*Euphausia superba* — Krill

Order *Anaspidacea*
Family **Anaspidae**
*Allanaspides helonomus* — Tasmanian anaspid crustacean (V)
*Allanaspides hickmani* — Tasmanian anaspid crustacean (V)

Order *Decapoda*
Family **Parastacidae**
*Parastacoides tasmanicus* — Burrowing freshwater crayfish
Family **Coenobitidae**
*Birgus latro* — Coconut or robber crab (R)

# INSECTS

Order *Odonata* — *Dragonflies; damselflies*
Family **Epiophlebiidae**
*Epiophlebia laidlawi* — Relict Himalayan dragonfly (V)
Family **Gomphidae**
*Isomma hieroglyphicum* — Nosy Bé dragonfly
Family **Neopetaliidae**
*Archipetalia auriculata* — Archaic Tasmanian dragonfly (I)
Order *Orthoptera* — *Grasshoppers; crickets*
Family **Stenopelmatidae**

| | | | |
|---|---|---|---|
| *Deinacrida rugosa* | Stephens Island weta | (V) | |

**Order Coleoptera**
**Family Scarabaeidae**

| | | |
|---|---|---|
| *Dynastes hercules* | Hercules beetle | (V) |

**Order Diptera**
**Family Blepharoceridae**

| | | |
|---|---|---|
| *Edwardsina gigantea* | Giant torrent midge | (E) |

**Order Lepidoptera**
**Family Papilionidae**

| | | |
|---|---|---|
| *Ornithoptera alexandrae* | Queen Alexandra's birdwing | (E) |
| *Ornithoptera euphorion* | Cairns birdwing | |
| *Papilio sjoestedti* | Kilimanjaro swallowtail | |
| *Papilio aristodemus ponceanus* | Schaus' swallowtail | (E) |

**Family Danaidae**

| | | |
|---|---|---|
| *Danaus plexippus* | Monarch butterfly | (TP) |

**Family Pyralidae**

| | |
|---|---|
| *Cactoblastis cactorum* | Cactus moth |

**Family Saturniidae**

| | |
|---|---|
| *Argema mittrei* | Emperor moth |
| *Coscinocera hercules* | Hercules moth |

**Order Hymenoptera**
**Family Formicidae**

| | |
|---|---|
| *Iridomyrmex humilis* | Argentine ant |

**Family Megachilidae**

| | | |
|---|---|---|
| *Chalicodoma pluto* | Wallace's giant bee | (K) |

# STARFISH

**Order Spinulosida**
**Family Acanthasteridae**

| | |
|---|---|
| *Acanthaster planci* | Crown-of-thorns starfish |

---

# PLANTS

## CONIFEROPHYTA — CONIFERS

**Pinaceae**

| | |
|---|---|
| *Abies lasiocarpa* | Alpine fir |
| *Larix decidua* | Larch |
| *Pinus cembra* | Arolla pine |
| *Pinus contorta* | Lodgepole pine |
| *Pinus montana* | Mountain pine |
| *Pinus radiata* | Monterey pine |
| *Pinus sylvestris* | Scots pine |
| *Picea abies* | Norway spruce |
| *Picea engelmannii* | Engelmann's spruce |
| *Picea glauca* | White spruce |
| *Picea sitchensis* | Sitka spruce |
| *Pseudotsuga menziesii* | Douglas fir |
| *Tsuga heterophylla* | Western hemlock |

**Araucariaceae**

| | |
|---|---|
| *Araucaria auracana* | Monkey puzzle tree |
| *Araucaria cunninghamii* | Hoop pine |

**Cuppressaceae**

| | |
|---|---|
| *Juniperus procera* | African cedar |
| *Chamaecyparis nootkatensis* | Yellow cypress |
| *Athrotaxis cupressoides* | Smooth Tasmanian cedar |
| *Athrotaxis laxifolia* | Summit cedar |
| *Athrotaxis selaginoides* | King Billy pine |
| *Fitzroya cupressoides* | Patagonian cypress; alerce |
| *Cryptomeria japonica* | Japanese red cedar |
| *Sequoia sempervirens* | Coast redwood |
| *Sequoiadendron giganteum* | Sierra redwood |

**Podocarpaceae**

| | |
|---|---|
| *Podocarpus dacrydioides* | Kahikatea |
| *Podocarpus ferrugineus* | Miro |
| *Podocarpus milanjianus* | Podocarp |
| *Podocarpus nivalis* | Mountain totara |
| *Dacrydium cupressinum* | Rimu; red pine |
| *Dacrydium franklinii* | Tasmanian huon pine |
| *Phyllocladus alpinus* | Alpine toatoa |

## GNETOPHYTA — GNETOPHYTES

**Gnetoceae**

| | |
|---|---|
| *Welwitschia mirabilis* | Welwitschia |

## MAGNOLIOPHYTA — FLOWERING PLANTS

**Order Laurales**
**Family Lauraceae**

**Laurels**

| | |
|---|---|
| *Laurus* spp. | Laurel |
| *Ocotea usambarensis* | Camphorwood |

**Order Ranunculales**
**Family Ranunculaceae**

**Buttercups; anemones**

| | |
|---|---|
| *Pulsatilla vernalis* | Anemone |
| *Ranunculus lyalli* | Mountain buttercup |
| *Ranunculus ophioglossifolius* | Adders-tongue spearwort |

**Order Fagales**
**Family Fagaceae**

**Beeches; oaks; sweet chestnuts**

| | |
|---|---|
| *Nothofagus antarctica* | Southern beech |
| *Nothofagus cliffortioides* | Mountain beech |
| *Nothofagus cunninghamii* | Myrtle beech |
| *Nothofagus fusca* | Red beech |
| *Nothofagus menziesii* | Silver beech |
| *Nothofagus solandri* | Black beech |

**Order Betulales**
**Family Betulaceae**

**Birches; alders; hazels**

| | |
|---|---|
| *Alnus viridis* | Green alder |
| *Betula glandulosa* | Dwarf birch |
| *Betula pubescens* | White birch |

Order **Caryophyllales**
Family **Chenopodiaceae** | Sugar beet; beetroot, etc.
*Arthrophytum aphyllum* | Black saxaul
*Arthrophytum persicum* | White saxaul
*Atriplex* spp. | Saltbush
*Kochia* spp. | Bluebush

Order **Cactales**
Family **Cactaceae** | Cacti
*Opuntia echios* | Prickly pear

Order **Salicales**
Family **Salicaceae** | Aspens; poplars; willows
*Populus balsamifera* | Poplar
*Populus tremuloides* | Trembling aspen

Order **Ericales**
Family **Ericaceae** | Heaths; rhododendrons; bilberry

*Erica arborea* | Tree heath
*Philippia* spp. | Tree heath

Family **Epacridaceae**
*Archeria traversii* | Red mountain heath
*Dracophyllum* sp. | Grass tree

Order **Ebenales**
Family **Ebenaceae** | Ebonies
*Diospyros reticulata* | Mascarene ebony
*Diospyros terselavia* | Black ebony

Order **Malvales**
Family **Bombacaceae** | Baobab, balsa
*Adansonia* spp. | Baobab

Order **Rosales**
Family **Rosaceae** | Roses; hawthorn, etc.
*Sorbus aucuparia* | Mountain ash; rowan

Order **Fabales**
Family **Caesalpiniaceae** | Caesalpinias
*Caesalpinia echinata* | Braza tree
Family **Leguminosae**
*Acacia tortilis* | Umbrella thorn
*Cercis siliquastrum* | Judas tree
*Dalbergia nigra* | Brazilian rosewood
*Sophora toromiro* | Pagoda tree

Order **Myrtales**
Family **Myrtaceae** | Myrtles; eucalyptuses; cloves

*Eucalyptus camaldulensis* | River red gum
*Eucalyptus obliqua* | Messmate
*Eucalyptus radiata* | Peppermint
*Eucalyptus regnans* | Mountain ash
*Eugenia caryophyllata* | Clove tree
*Metrosideros imbellata* | Southern rata

Family **Combretaceae** | Terminalia; combretum, etc.
*Terminalia prunoides* | Wild prune

Family **Thymelaeaceae**
*Gonstylus bancanas* | Ramin

Order **Rutales**
Family **Anacardiaceae** | Cashew; mango, etc.
*Schinus* sp. | Brazilian pepper

Family **Meliaceae** | Mahoganies
*Entandophragma utile* | Utile

Family **Rutaceae** | Citrus fruits
*Zanthoxylum paniculatum* | Endemic Rodriguez angiosperm

Order **Cornales**
Family **Cornaceae** | Dogwoods
*Cornus nuttallii* | Pacific dogwood

Order **Rhamnales**
Family **Rhamnaceae** | Buckthorns
*Phylica arborea* | Island tree
*Phylica nitida*

Order **Oleales**
Family **Oleaceae** | Olive; ash
*Olea africana* | Wild olive
*Olea europaea* | Common olive

Order **Gentianales**
Family **Rubiaceae** | Gardenias, coffee, quinine
*Cinchona ledgeriana* | Cinchona tree
*Ramosmania heterophylla* | Café marron

Family **Apecynaceae**
*Pachypodium namaquanum* | Halfmen or Elephant's trunk

Order **Scrophulariales**
Family **Gesneriaceae** | African violets
*Saintpaulia ionantha* | African violet

Order **Lamiales**
Family **Lamiaceae** | Mints
*Stachys germanica* | Downy woundwort

Order **Campanulales**
Family **Campanulaceae** | Bellflowers
*Wahlenbergia pygmaea* | Maori bluebell

Family **Lobeliaceae** | Lobelias
*Lobelia rhynchopetalum* | Giant lobelia

Order **Asterales**
Family **Asteraceae** | Asters
*Helichrysum* spp. | Everlastings
*Leontopodium alpinum* | Edelweiss
*Leucogenes grandiceps* | South Island edelweiss
*Senecio hadrosomus* | Flor de Mayo leñosa
*Senecio* spp. | Giant groundsel

Order **Najadales**
Family **Aponogetonaceae**
*Aponogeton fenestralis* | Madagascan lace plant

Order **Liliales**
Family **Agavaceae** | Sisal, hemp, dragon trees
*Dracaena draco* | Dragon's blood tree
*Yucca brevifolia* | Joshua tree

Family **Pontederiaceae** | Water hyacinth
*Eichhornia crassipes* | Water hyacinth

Family **Xanthorrhoeaceae** | Grass trees
*Xanthorrhoea* spp. | Australian grass tree

Order **Zingiberales**
Family **Zingiberaceae** | Gingers
*Ravenela madagascariensis* | Traveller's palm

Order **Cyperales**
Family **Cyperaceae**
*Carex* spp.
*Cyperus giganteus*

**Reeds, sedges**
Sedges
Giant papyrus-like sedge

Order **Bromeliales**
Family **Bromeliaceae**
*Puya raimondii*

**Bromeliads**
Puya tree

Order **Poales**
Family **Poaceae (Gramineae)**

**Grasses, cereals, bamboo, reeds**

*Andropogon gerardi* — Blue stem grass
*Andropogon tectorum* — Tall blue stem
*Andropyron spicatum* — Wheat grass
*Aristida* spp. — Needle grasses
*Arundinaria alpina* — Mountain bamboo
*Astrebla* spp. — Perennial tussock grasses
*Cenchrus ciliaris* — African foxtail
*Chloris roxburghiana* — Horse tail
*Chrysopogon aucheri* — Golden beard grass
*Cladium jamaicensis* — Sawgrass
*Cymbopogon* spp. — Lemon grass
*Danthonia* spp. — Snow-tussock grasses
*Digitaria* spp. — Finger grass
*Echinochloa* spp. — Antelope grass
*Eragrostis* spp. — Love grass
*Festuca scabrella* — Fescue

*Hyparrhenia* spp. — Thatching grass
*Loudetia* spp. — Russet grass
*Oryza nivara* — Wild rice
*Panicum maximum* — Guinea grass
*Pennisetum purpureum* — Elephant grass
*Phragmites australis* — Reed
*Setaria* spp. — African timothy grass
*Spartina arundinacea* — Tussock grass
*Themeda triandra* — Red oat grass
*Stipa speciosa* — Feather grass
*Triodia* spp. — Porcupine grass
*Zea diploperennis* — Perennial teosinte

Order **Arecales**
Family **Arecaceae**

**Palms**

*Artocarpus altilis* — Breadfruit tree
*Dictyosperma album* — Round Island hurricane palm
*Eugeissona utilis* — Sago palm
*Hyophorbe lagenicaulis* — Bottle palm
*Latania loddigesii* — Fan palm
*Licuala ramsayi* — Fan palm
*Livistona mariae* — Central Australian cabbage palm
*Lodoicea maldivica* — Coco de Mer palm
*Nypa fruticans* — Nypa palm

Order **Arales**
Family **Araceae**
*Pistia stratiotes* — Nile cabbage

# GLOSSARY

**acclimatization** The adjustment of an organism to new living conditions.

**adaptive radiation** The process whereby animals spread out from their centre of origin to exploit (and adapt to) the ecological opportunities.

**aestivate** To spend the hottest time of year in a torpid state.

**alluvial fan** A fan-shaped alluvial deposit formed by a stream where its velocity is abruptly decreased, as at the mouth of a ravine or at the foot of a mountain.

**alluvium** The sedimentary matter formed by flowing water and deposited in recent times, especially in large river valleys.

**altiplano** A discontinuous series of plateaux and basins in the northern Andes, of varying size and elevation, separated by mountain ranges and deep canyons.

**anadromous** Relating to fish or other marine organisms that live in salt water but migrate to fresh water to spawn.

**angiosperm** A plant with its seeds enclosed in an ovary.

**aquifer** A layer of rock that holds water and allows water to percolate through it.

**arboreal** Adapted for life and movement in trees.

**artesian water** A water-bearing stratum lying beneath an impermeable layer which, when tapped, rises by hydrostatic pressure.

**barkhan** Crescent-shaped sand dune with horns pointing downwind, found especially in the deserts of Turkmenistan.

**biological control** The use of predatory or parasitic organisms to control plant or animal pests.

**biodiversity** The variety of living organisms in all their forms and combinations.

**biomass** The total weight of living organisms supported by a given area.

**biome** A major biotic community of plants and animals occupying a wide geographical region or climatic zone.

**biosphere** The thin mantle of atmosphere surrounding the earth that sustains life.

**biota** The animal and plant life of a region or period.

**biotic community** The totality of plants, animals and micro-organisms in a given area of land or water, characterized by interrelationships with each other and with the physical environment.

**biotope** A geographical unit of habitat occupied by a species or community.

**bird of prey** A bird that habitually hunts and kills other animals for food. A raptor.

**bryophyte** A group of plants comprising the true mosses and liverworts.

**buffer zone** An adjunct to a national park, reserve or other protected area.

**caatinga** Low grade, dry, semi-deciduous thorny forest in northeastern Brazil.

**campo** (pl. *campos*) Sub-humid wooded grassland.

**catchment area** The area draining into a lake, river or other body of water.

**capillary action** The elevation or depression of liquids in the soil due to surface tension.

**carrying capacity** The biomass that can be sustained in a given area.

**chaco** Part of the Gran Chaco region in central South America, covering about 260,000 km²/100,000 sq mls.

**cirque** A natural amphitheatre.

**classification** The grouping of organisms into categories: the subject matter of taxonomy.

**clear-felling** The removal of an entire stand of trees.

**climax** The culminating stage in the ecological succession or evolution of a plant/animal community that has attained stability (equilibrium) and is self-perpetuating.

**cordillera** One in a sequence of broadly parallel mountain ranges.

**cover** Vegetation or other natural feature used by an animal to avoid detection by its enemies.

**crepuscular** Neither wholly diurnal nor nocturnal.

**cryptic coloration** Coloration designed for concealment.

**deciduous** Shedding leaves annually.

**delta** An alluvial plain at the mouth of a river where the river divides into a series of channels before entering the sea or lake.

**diurnal** Active by day.

**divide** The border between adjacent catchment areas.

**ecological balance** The dynamic stability in an ecosystem due to the totality of interacting processes and components within it.

**ecological niche** The place of a species in an ecosystem.

**ecological separation** The system whereby specialist feeders (e.g. the large herbivores) coexist without competing for food. This is achieved by each species having a preference for a particular kind of grass or a particular stage of growth.

**ecology** The study of plants, animals and other organisms in relation to their environment and to one another.

**ecosystem** An interdependent system of living organisms of all kinds that together with non-living components make up a particular environment.

**edaphic** Relating to soil or topography rather than climate.

**elevated atoll** An atoll raised above sea level by geologic activity.

**endemic** A species or other classified group of plants or animals peculiar to a particular locality.

**ephemeral** A plant or other organism that is short-lived.

**epiphyte** A non-parasitic plant growing on trees and possessing aerial roots having the ability to absorb moisture from the atmosphere.

**equinox** The time when the sun crosses the plane of the earth's Equator, making night and day all over the world of equal length. The vernal equinox occurs about March 21st, and the autumnal equinox about September 22nd.

**eutrophication** The process whereby effluents entering lakes, rivers and other wetlands generate large quantities of nitrogen, phosphates and other plant nutrients, leading to accelerated growth of algae. Their death and decomposition exhausts the dissolved oxygen, causing the water to become biologically dead.

**evolutionary convergence** The development of superficially similar characteristics and habits by totally unrelated species living under comparable ecological conditions, but in isolation from each other.

**fire-climax** The established vegetation resulting from frequent burning.

**firn** The imperfectly consolidated granular snow found on glaciers.

**flash flood** A sudden flood of water down a normally dry river bed, caused by a rainstorm some distance upstream.

**floodplain** A plain cut by a river that is subject to seasonal or periodic flooding.

**flora** The plants of a particular region or period; a work systematically listing and describing such plants.

**fluvial** Pertaining to or produced by a river.

**flyway** The route followed by migratory birds during the course of their regular migrations from their summer breeding areas to their wintering grounds.

**food chain** A sequence of organisms feeding on one another, the larger preying on the smaller.

**fumerole** A vent in or near a volcano from which vapour is emitted.

**gallery forest** A long, narrow strip of forest bordering one or both banks or a river.

**game animal** An animal hunted and taken for sporting purposes.

**game management** The manipulation of wild populations or their habitat with the aim of maintaining ecological equilibrium.

**geothermal** Pertaining to the internal heat of the earth.

**genetic diversity** The variety and frequency of different genes and/or genetic stocks.

**grazing capacity** The optimum number of animals that can be supported on a particular pasture or range.

**grazing succession** A sequence of grazing by different species, each of which tends to make the habitat suitable for the next.

**groundwater** Water that has accumulated beneath the surface of the soil above the first impermeable layer.

**groundwater forest** Forest located in a relatively arid area watered by seepage.

**habitat** The natural habitation of an animal or plant.

**halophyte** A plant that grows in salty or alkaline soils.

**hanging valley** A tributary valley in a mountainous region, which joins the main valley by a sudden sharp descent caused by glacial erosion.

**herbivore** An animal that feeds on plants.

**hunting reserve** An area reserved for game animals where hunting is allowed under controlled conditions.

**hydrology** The science dealing with water both on and under the surface of the land.

**igapó** Amazonian floodplain forest either seasonally or near-permanently inundated by clear water poor in nutrients.

**immobilization** Rendering an animal temporarily immobile with the purpose of catching, marking, treating, or transporting it.

**irruption** A spontaneous migratory movement occasioned by an animal population rapidly increasing in numbers to the extent of outgrowing its food supply, thereby creating conditions that will no longer support life.

**jheel** Hindi name for a marsh, the size of which may vary from year to year, according to rainfall.

**lacustrine** Pertaining to a lake.

**leach** The process of dissolving and washing away soluble matter by the action of a percolating liquid: e.g. as with water percolating through soil and in the process removing nutrients.

**levee** A natural embankment along a river or canal formed by deposition of silt during flooding.

**littoral** Pertaining to the shore of a lake, sea or ocean.

**llano** (pl. **llanos**) An extensive and largely treeless grass plain.

**Malesia** Region consisting of Malaysia, Brunei, Indonesia, the Philippines and Papua New Guinea.

**massif** A mountain range or block of mountainous country containing one or more summits or dominant heights.

**migration** The seasonal movement, often over long distances, of animals (chiefly birds but including insects, fish and mammals) from one habitat to another.

**miombo** Deciduous woodland in the African sub-humid zone in which the tree genus *Brachystegia* is dominant. (From the Kinyamwezi name for the common tree *Brachystegia boehmii*, 'muyombo', plural 'miyombo').

**monocotyledon** A flowering plant characterized chiefly by a single seed leaf (cotyledon) and by an endogenous (growing from within) mode of growth.

**monoculture** The intensive or protracted culture of a single species of plant or animal.

**monogamous** Having only a single mate.

**monotypic** A genus comprising only one species, or a species not divisible into subspecies.

**muskeg** A type of bog in the Canadian Arctic formed in hollows or depressions by the accumulation of water and growth of sphagnum mosses.

**nivation** The disintegration of rocks around a patch of snow, brought about by alternate freezing and thawing.

**nocturnal** Active by night.

**nomadism** The movement of a tribe or other group of people from place to place in quest of pasturage and water for their livestock.

**nomenclature** The scientific naming of species and other scientifically recognized categories of animals and other organisms in which they are classified.

**non-renewable natural resources** Finite natural resources that become depleted or exhausted when over-exploited.

**nyika** An arid region of largely *Commiphera* bush in the coastal hinterland of Somalia, Kenya and Tanzania.

**Palaearctic** The zoogeographical region comprising Europe, Asia north of the Himalayas, and Africa north of the Sahara.

**palaeobotany** The branch of palaeontology relating to fossil plants.

**palaeontology** The science of forms of life existing in former geological periods, as represented by fossil plants and animals.

**palustrine** Pertaining to a marsh.

**páramo (pl. páramos)** Alpine meadow above the tree line in the northern Andes, consisting of grasses and herbaceous plants.

**pelagic** Pertaining to the seas or oceans, far from land.

**peneplain** A tract of land reduced almost to a plain by erosion.

**Pleistocene** A geological epoch lasting from 1.6 million to 10,000 years ago.

**polder** A tract of low-lying land, reclaimed from the sea, and protected by dykes.

**polygynous** Characterized by a single male having more than one female partner.

**population** The total number of individuals of a particular species inhabiting a given region.

**precipitation** Atmospheric condensation appearing in the form of mist or fog or falling as rain, snow or hail.

**predation** The habitual preying of one species of animal upon another for food.

**predator** An animal whose way of life is based upon killing other animals for food.

**primary forest** Undisturbed natural forest.

**pronk** To make stiff-legged leaps high into the air.

**prosimian** A suborder of primates (the highest order of vertebrates, which includes monkeys, apes and man) that accommodates the tree shrews, lemurs, lorises, tarsiers, bushbabies, and their relatives.

**protection forest** Forest maintained or planted for the purpose of protecting the land (e.g. against erosion) or for hydrological regulation.

**puna** High, cold, arid plateauland in the Peruvian Andes, the vegetation typically consisting of rough grasses and a variety of herbaceous plants, shrubs and cacti.

**rain shadow** The leeward side of a mountain range receiving little or no rain owing to the prevailing rain-bearing winds having already deposited it on the weather side.

**range** The distributional area in which a plant or animal occurs.

**raptor** Bird of prey.

**reedbrake** A place overgrown with reeds; a reed thicket.

**relict** Remnant pockets of plants or animals that are all that remain of once much larger populations.

**renewable natural resources** Natural resources that perpetuate themselves, provided the rate at which they are used does not exceed their capacity for regeneration.

**riparian** Of, or pertaining to, the banks of a river or body of water.

**run-off** Rain which flows off the surface of the land.

**salinization** The process by which soluble salts accumulate in or on the soil.

**schleropyll** Any of a number of plants with thickened and tough leaves.

**secondary forest** Natural regeneration that follows the elimination of the primary forest.

**siltation** The deposition of fine-grained sediments from standing or slow-flowing water.

**slough** A tract of soft muddy ground. In the USA and Canada: a marshy or reed-covered pool, pond or inlet.

*solonchak* A salty area, such as the dry bed of a salt-lick, where the water table is high.

**solstice** Either of the two occasions in the year when the sun is at its greatest distance from the celestial equator and apparently does not move either north or south; about June 21st (summer solstice) and December 22nd (winter solstice).

**speciation** The process by which new species are formed.

**strict nature reserve** A reserve from which any human activity is rigidly excluded except for strictly controlled scientific studies.

**succession** The natural replacement of one type of vegetation by another.

**sustainability** A characteristic of a process or state that can be maintained indefinitely.

**sustainable use** The use of an organism, ecosystem or other renewable natural resource within its capacity for renewal.

**sustainable yield** The number of animals or the amount of plant material that may be periodically removed from a population without affecting the total supply.

**symbiosis** The living together of two different organisms to their mutual advantage.

**systematics** The systematic description and orderly arrangement of living organisms.

*takyr* An expanse of bare clay, sometimes quite extensive, having a hardened surface polished smooth by wind-blown sand, and covered with water in spring, which later dries out, when it cracks and forms into polygons, and for a short while thereafter supports a temporary cover of lichens and blue-green algae.

**taxonomy** The science of classification (often used synonymously with systematics).

*toich* Seasonally inundated grassland in the Upper Nile region of southern Sudan.

**trophic** Relating to diet. The trophic pyramid illustrates the general quantitative pattern of food chains: the relatively large quantity (biomass) of plant material (the base of the pyramid) eaten by a smaller biomass of herbivorous animals (middle level of pyramid) and the smallest biomass of the carnivores (or parasites) which subsist upon them (top of pyramid). Some food chains have a fourth or fifth level where carnivores are eaten or parasitized.

*tugai* Thickets of reeds, trees, shrubs and creepers, sometimes so dense as to be virtually impenetrable, associated with saline groundwater on the seasonally inundated floodplains of some Central Asian rivers.

**ungulate** A hoofed mammal.

*várzea* Amazonian floodplain, seasonally inundated by rivers carrying sediments relatively rich in nutrients, and sustaining forest adapted to this hydrological regime.

**vegetation** The plant life of a particular region considered as a whole.

**viviparous** Giving birth to living young.

**watershed** The ridge or crest line separating two drainage areas.

**wilderness** An area of undeveloped land in pristine condition, protected and managed to retain its natural character.

**wintering ground** An area in which animals spend the winter.

**xerophyte** A plant adapted to living under arid conditions.

# BIBLIOGRAPHY

**ALBIGNAC, Roland** (1987) Status of the Aye-aye in Madagascar. *Primate Conservation* 8:44-45.

**ALLAN, J.A.** (1984) Oases. In: *Sahara Desert* (ed. J.L.Cloudsley-Thompson). Pergamon Press, Oxford.

**ALLEN, Glover M.** (1938-40) The Mammals of China and Mongolia. *Natural History of Central China*, vol. 11. American Museum of Natural History, New York.

**ALLEN, Glover M.** (1942) Extinct and Vanishing Mammals of the Western Hemisphere with the Marine Species of all the Oceans. *Spec. Publ. Amer. Comm. Int. Wildlife Protection*, No. 11.

**ANADU, P.A.** (1987) Prospects for Conservation of Forest Primates in Nigeria. *Primate Conservation* 8:154-157.

**AVELING, Rosalind and Conrad** (1987) Report from the Zaire Gorilla Conservation Project. *Primate Conservation* 8:162-164.

**AZEEZ, Gundula** (1991) The Aral Sea: its Environment and Deterioration. Mimeograph.

**BACON, Edward** (ed.) (1963) Vanished Civilizations: Forgotten Peoples of the Ancient World. Thames & Hudson, London.

**BAKER, Sir Samuel W.** (1886) Ismailia: a Narrative of the Expedition to Central Africa for the Suppression of the Slave Trade. MacMillan and Co, London.

**BALLY, Peter R.D.** (1964) Recent Floristic and Faunistic Changes in the Somali Republic, with particular reference to Specialised Desert Forms. *IUCN Bull.*, N.S. 1(11):6-7.

**BANNIKOV, A.G.** (1960) The Ecology of *Saiga tatarica L.* In: *Eurasia, its Distribution and Management*. IUCN 8th Technical Meeting, Warsaw.

**BARRETT, Suzanne W.** (1978) The Amazon National Park – Preliminary Management Plan. IUCN/WWF, Gland, Switzerland.

**BEAMISH, Tony** (1970) Aldabra Alone. George Allen & Unwin Ltd., London.

**BENNETT, Elizabeth L.,** *et al.* (1987) Current Status of Primates in Sarawak. *Primate Conservation* 8:184-186.

**BENNETT, Elizabeth L.** (1988) Proboscis Monkeys and their Swamp Forests in Sarawak. *Oryx* 22(2):69-74.

**BERGAMINI, David** (1964) The Land and Wildlife of Australia. *Life Nature Library*. Time Inc., New York.

**BIGALKE, R.C.** (1961) Some Observations on the Ecology of the Etosha Game Park, South West Africa. *Ann. Cape Prov. Mus.* 149-67.

**BLOWER, J.** (1968) The Wildlife of Ethiopia. *Oryx* 9:276-285.

**BOCK, K.R.** (1979) Tourism and Kenya's Coral Reef. *Swara* 2(5):28-34.

**BOLTON, M.** (1972) Report on a Wildlife Survey of South-eastern Ethiopia. *Walia*, 4:19-31.

**BOLTON, M.** (1973) Hartebeests in Ethiopia. *Oryx* 12:99-108.

**BOLTON, M.** (1973) Notes on the Current Status and Distribution of some Large Mammals in Ethiopia (excluding Eritrea). *Mammalia* 37:562-586.

**BOURLIÈRE, F.** (1963) The Wild Ungulates of Africa: Ecological Characteristics and Economic Implications. In: *Conservation of Nature and Natural Resources in Modern African States*. IUCN Publ. N.S. No. 1:102-105.

**BOURLIÈRE, François** (1964) The Land and Wildlife of Eurasia. *Life Nature Library*. Time Inc., New York.

**BOURLIÈRE, F.** (ed) (1983) Tropical Savannas. Vol. 13, Ecosystems of the World. Elsevier, Amsterdam.

**BRADEN, Kathleen** (1986) Wildlife Reserves in the USSR. *Oryx* 20(3):165-169.

**BREMNER, Charles** (1990) Intruder in Nature's Wonderland. *The Times Review*, 14th April.

**BROWN, Leslie** (1965) Africa: A Natural History. Hamish Hamilton, London.

**BROWN, L.H.** (1969) Observations on the Status, Habitat and Behaviour of the Mountain Nyala, *Tragelaphus buxtoni*, in Ethiopia. *Mammalia* 33:545-597.

**BROWN, Leslie H.** (1971) The Biology of Pastoral Man as a Factor in Conservation. *Biological Conservation* 3(2):93-100.

**BROWN, Leslie** (1971) East African Mountains and Lakes. East African Publishing House, Nairobi.

**BROWN, Leslie H.** (1973) Conservation for Survival: Ethiopia's Choice. Haile Selassie 1 University, Addis Ababa.

**BULLOCK, David, and NORTH, Steven** (1984) Round Island in 1982. *Oryx* 18:36-41.

**BURTON, R.W.** (1952) A History of Shikar in India. *J. Bombay Nat. Hist. Soc.* 50(4):847-848.

**CARLQUIST, Sherwin** (1965) Island Life: a Natural History of the Islands of the World. American Museum of Natural History, The Natural History Press, New York.

**CARP, Eric** (ed.) (1980) Directory of Wetlands of International Importance in the Western Palaearctic. IUCN, Gland, Switzerland and UNEP, Nairobi.

**CARROLL, Richard W.** (1986) Status of the Lowland Gorilla and other Wildlife in the Dzanga-Sangha Region of Southwestern Central African Republic. *Primate Conservation* 7:38-41.

**CARRUTHERS, D.** (1913) Unknown Mongolia: A Record of Travel and Exploration in North-West Mongolia and Dzungaria. 2 vols. Hutchinson, London.

**CARRUTHERS, D.** (1949) Beyond the Caspian: a Naturalist in Central Asia. Oliver and Boyd, Edinburgh & London.

**CAUGHLEY, Graeme** (1970) *Cervus elaphus* in Southern Tibet. *J. Mammal.* 51(3):611-614.

**CAVE, F.O., and CRUICKSHANK, A.** (1940) A Note on Game Migration in the South Eastern Sudan. *Sudan Notes and Records* 23:341-344.

**CHAUVET, M.** (1972) The Forest of Madagascar. In: *Biogeography and Ecology in Madagascar* (R. Battistini and G. Richard-Vindard, eds). *Monographiae Biologicae* 21, W. Junk, the Hague, pp. 191-200.

**CHEKE, Anthony** (1987) The Legacy of the Dodo – Conservation in Mauritius. *Oryx* 21:29-36.

**CLARK, M.R., and DINGWELL, P.R.** (1985) Conservation of Islands in the Southern Ocean: a Review of the Protected Areas of Insulantarctica. IUCN, Gland, Switzerland, and Cambridge, UK.

**CLOUDSLEY-THOMPSON, J.L.** (1964) Life in Deserts. G.T. Foulis, London.

**CLOUDSLEY-THOMPSON, J.L.** (ed.) (1984) Sahara Desert, Pergamon Press, Oxford.

**COBB, Stephen** (1981) Wildlife in Southern Sudan. *Swara* 4(5):28-31.

**COLE, Monica M.** (1986) The Savannas: Biogeography and Geobotany. Academic Press/London, New York.

**COLLAR, N.J., and STUART, S.N.** (1988) Key Forests for Threatened Birds in Africa. ICBP Monograph No. 3. Cambridge.

**COTT, Hugh B.** (1961) Scientific Results of an Inquiry into the Ecology and Economic Status of the Nile Crocodile (*Crocodilus niloticus*) in Uganda and Northern Rhodesia. *Trans. Zool. Soc. Lond.* 29(4):211-356.

**COUPLAND, R.T.** (ed) (1979) Grassland Ecosystems of the World: Analysis of Grasslands and their Uses. IBP 18, Cambridge.

**CRAVEN, Patricia** (1988) Save our Flora! *African Wildlife* 42(2):107-110.

**CRONWRIGHT-SCHREINER, S.C.** (1925) The Migratory Springbucks of South Africa (The Trekbokke). Fisher Unwin, London.

**CURL, David** (1986) The Rarest Tortoise on Earth. *Oryx* 20(1):35-39.

**CURZON, G.N.** (1896) The Pamirs and the Source of the Oxus. *Geogr. J.* 8(1):15-54; 8(2):97-119; 8(3):239-260.

**DARLING, F. Fraser** (1961) African Wild Life as a Protein Resource. *Span* 4(3):100-103.

**DARLING, F. Fraser** (1970) Wilderness and Plenty: the Reith Lectures, 1969. BBC, London.

**DAVIES, Glyn** (1986) The Orang-utan in Sabah. *Oryx* 20(1):40-45.

**DAVIS, Stephen D.,** *et al* (1986) Plants in Danger. What do we know? IUCN, Gland, Switzerland, and Cambridge, UK.

**DEFLER, Thomas R.** (1986) The Giant River Otter in El Tuparro National Park, Colombia. *Oryx* 20(2):87-88.

**DIXON, Alexandra, and JONES, David** (eds.) (1988) Conservation and Biology of Desert Antelopes. Christopher Helm, London.

**DORST, Jean** (1967) South America and Central America: A Natural History. Hamish Hamilton, London.

**DORST, Jean** (1974) Parks and Reserves on Islands. In: *Proc. 2nd World Conf. on National Parks*, 267-276, IUCN, Morges, Switzerland.

**DRAZ, O.** (1985) The Hema System of Range Reserves in the Arabian Peninsula. In: *Culture and Conservation: the Human Dimension in Environmental Planning.* (McNeely, J.A., and Pitt, D., eds), IUCN, Gland, Switzerland.

**DUGAN, Patrick J.** (ed.) (1990) Wetland Conservation: a Review of Current Issues and Required Action. IUCN, Gland, Switzerland.

**DUNCAN, Patrick** (ed.) (1992) Zebras, Asses and Horses: Global Survey and Action Plan for the Conservation of Wild Equids: 1992-1997. IUCN/SSC Equid Specialist Group, Gland, Switzerland.

**EAST, R.** (ed.) (1988) Antelopes: Global Survey and Regional Action Plans. Part 1: East and Northeast Africa. IUCN/SSC Antelope Specialist Group, Gland, Switzerland.

**EAST, R.** (ed.) (1989) Antelopes: Global Survey and Regional Action Plans. Part 2: Southern and South-central Africa. IUCN/SSC Antelope Specialist Group, Gland, Switzerland.

**EAST, R.** (ed.) (1990) Antelopes: Global Survey and Regional Action Plans. Part 3: West and Central Africa. IUCN/SSC Antelope Specialist Group, Gland, Switzerland.

**EDROMA, E.L.** (1980) Road to extermination in Uganda. *Oryx* 15:451-452.

**EISENBERG, J.F., and GOULD, Edwin** (1970) The Tenrecs: A Study in Mammalian Behavior and Evolution. Smithsonian Contributions to Zoology, No. 27. Smithsonian Institution Press, Washington, DC.

**ELLIOTT, H.F.I.** (1953) The Fauna of Tristan da Cunha. *Oryx* 2(1):41-53.

**ELTON, Charles S.** (1958) The Ecology of Invasions by Animals and Plants. Methuen, London.

**EUDEY, A.A.** (ed.) (1987) Action Plan for Asian Primate Conservation: 1987-91. IUCN/SSC Primate Specialist Group, Gland, Switzerland.

**F.A.O./MACKINNON, J.R.** (1981/82) National Conservation Plan for Indonesia. 8 vols.

**FINLAYSON, C.M.** (ed) (1992) A Strategy and Action Plan to Conserve the Wetlands of the Lower Volga. Developed at an International Workshop held in the Russian city of Astrakhan during October 1991. International Waterfowl and Wetlands Research Bureau, Slimbridge, UK.

**FISHER, James, SIMON, Noel, and VINCENT, Jack** (1969) The Red Book: Wildlife in Danger. Collins, London.

**FITZSIMONS, V.** (1963) The Namib Desert. *African Wild Life* 17(3):215-227.

**FLINT, J.H.** (1967) Conservation Problems on Tristan da Cunha. *Oryx* 9:28-32.

**FOSTER-TURLEY, Pat, MACDONALD, Sheila, and MASON, Chris** (1990) Otters: an Action Plan for their Conservation. IUCN/SSC Otter Specialist Group, Gland, Switzerland.

**FOX, Joseph L.** (1989) A Review of the Status and Ecology of the Snow Leopard (*Panthera uncia*). International Snow Leopard Trust.

**FRAMPTON, George T., Jr.** (1988) Conservation Begins at Home. In: *For the Conservation of Earth* (Vance Martin, ed.). *Proc. 4th World Wilderness Congress.*

**FRASER, M.W.** (1989) The Inaccessible Island Rail: smallest flightless bird in the world. *African Wildlife* 43(1):14-19.

**FRASER, M.W.** (1990) The Birds of Inaccessible Island. Part 1: Seabirds. *African Wildlife* 44(6):347-353.

**FRASER, M.W.** (1991) Birds of Inaccessible Island. Part 2: Landbirds. *African Wildlife* 45(1):20-23.

**FRENCH, N.R.** (ed) (1979) Perspectives in Grassland Ecology. Vol. 32, *Ecological Studies.* Springer Verlag, New York.

**FRIES, Carl** (1959) The Fate of Arcadia: Land Use and Human History in the Mediterranean Region. *Proc. IUCN Seventh Technical Meeting, Athens.*

**GANZHORN, Jörg U., and RABESOA, Joseph** (1986) Sightings of Aye-ayes in the Eastern Rainforest of Madagascar. *Primate Conservation* 7:45.

**GASS, I.G.** (1963) The Royal Society's Expedition to Tristan da Cunha, 1962. *Geogr. J.* 129(3):283-289.

**GAYMER, R.** (1968) The Indian Ocean Giant Tortoise *Testudo gigantea* on Aldabra. *J. Zool.* 154:341-363.

**GINSBERG, J.R., and MACDONALD, D.W.** (1990) Foxes, Wolves, Jackals, and Dogs: an Action Plan for the Conservation of Canids. IUCN/SSC Canid Specialist Group, Gland, Switzerland.

**GLOVER, T.R.** (1945) Springs of Hellas and other Essays. Cambridge University Press.

**GODFREY, Laurie** (1986) *Hapalemur simus*: Endangered Lemur once Widespread. *Primate Conservation* 7:92-96.

**GOODALL, D.W.,** *et al* (eds) (1979) Arid-land Ecosystems: Structure, Functioning and Management, vol 1. Cambridge University Press.

**GREENWOOD, P.H.** (1986) The Nile Perch in Lake Victoria. *Oryx* 20(4):249.

**GRIMWOOD, Ian** (1988) 'Operation Oryx': the start of it all. In: *Conservation and Biology of Desert Antelopes* (eds A. Dixon and D. Jones). Christopher Helm, London.

**GROOMBRIDGE, Brian** (ed.) (1982) The IUCN Amphibia–Reptilia Red Data Book. Part 1: Testudines, Crocodylia, Rhynchocephalia. IUCN, Gland, Switzerland, and Cambridge, UK.

**GROVES, Colin P.** (1988) A Catalogue of the genus *Gazella*. In: *Conservation and Biology of Desert Antelopes* (eds A. Dixon and D. Jones). Christopher Helm, London.

**HACHISUKA, Masauji** (1953) The Dodo and Kindred Birds, or the Extinct Birds of the Mascarene Islands. H.F. & G. Witherby, London.

**HARCOURT, A.H.** (1981) Can Uganda's Gorillas Survive? – a Survey of the Bwindi Forest Reserve. *Biological Conservation* 19:269-282.

**HARCOURT, C., and THORNBACK, J.** (1990) Lemurs of Madagascar and the Comoros. The IUCN Red Data Book. IUCN, Gland, Switzerland, and Cambridge, UK.

**HARPER, Francis** (1945) Extinct and Vanishing Mammals of the Old World. *Spec. Publ. Amer. Comm. Int. Wildlife Protection* No. 12.

**HART, John A., and HART, Teresa B.** (1986) The Ituri Forest of Zaïre: Primate Diversity and Prospects for Conservation. *Primate Conservation* 7:42-43.

**HEPTNER, V.G., NASIMOVIC, A.A., and BANNIKOV, A.G.** (1966) Die Säugetiere der Sowjetunion, Band 1: Paarhufer und Unpaarhufer. Gustav Fischer Verlag, Jena.

**HILLMAN, J.C.** (1986) Conservation in Bale Mountains National Park, Ethiopia. *Oryx* 20(2):89-94.

**HILLMAN-SMITH, Kes**, *et al* (1986) A Last Chance to Save the Northern White Rhino. *Oryx* 20(1):20-26.

**HILLS, E.S.** (ed) (1966) Arid Lands: a Geographical Appraisal. Methuen, London.

**HOLDGATE, M.W.** (1965) The Fauna of the Tristan da Cunha Islands. *Phil. Trans. R. Soc.* Ser. B. 249:361-402.

**HOFFMANN, L.** (1968) Project MAR: its Principles, Objectives and Special Significance for the Near and Middle East Region. *IUCN Publ.* N.S. No.12:36-40.

**HUTCHISON, Robert A.** (ed.) (1991) Fighting for Survival: Insecurity, People and the Environment in the Horn of Africa, based on study by Spooner, B.C., and Walsh, N.; IUCN, Gland, Switzerland.

**ISAKOV, Y.A.,** (1968) The Status of Waterfowl Populations Breeding in the USSR and Wintering in S.W. Asia and Africa. *IUCN Publ.*, N.S. No. 12:175-186.

**ISAKOV, Y.A., and SHEVAREVA, T.P.** (1968) Interrelationship of Waterfowl Breeding and Wintering Areas in the Central Palearctic. *IUCN Publ.*, N.S. No. 12:165-174.

**IUCN/UNEP** (1987) The IUCN Directory of Afrotropical Protected Areas. IUCN, Gland, Switzerland, and Cambridge, UK.

**IUCN** (1990) 1990 IUCN Red List of Threatened Animals. IUCN, Gland, Switzerland, and Cambridge, UK.

**IUCN** (1990) 1990 United Nations List of National Parks and Protected Areas. IUCN, Gland, Switzerland, and Cambridge, UK.

**IUCN** (1991) A Strategy for Antarctic Conservation. IUCN, Gland, Switzerland, and Cambridge, UK.

**IUCN** (1991) IUCN Directory of Protected Areas in Oceania. Prepared by the World Conservation Monitoring Centre. IUCN, Gland, Switzerland, and Cambridge, UK.

**IUCN** (1991) The Lowland Grasslands of Central and Eastern Europe. Information Press, Oxford.

**IUCN** (1992) Protected Areas of the World: a review of national systems. Vol. 1: Indomalaya, Oceania, Australia and Antarctic. IUCN, Gland, Switzerland, and Cambridge, UK.

**IUCN** (1992) Protected Areas of the World: a review of national systems. Vol. 2: Palaearctic. IUCN, Gland, Switzerland, and Cambridge, UK.

**IUCN** (1992) Protected Areas of the World: a review of national systems. Vol. 3: Afrotropical. IUCN, Gland, Switzerland, and Cambridge, UK.

**JAEGER, Edmund C.** (1957) The North American Deserts. Stanford University Press, California.

**JOLLY, Alison, OBERLÉ, Philippe, and ALBIGNAC, Roland** (eds) (1984) Madagascar. Key Environments Series. Pergamon Press, Oxford.

**JONES, D.M.** (1991) The Society and Conservation in Africa. *Lifewatch*, Spring Issue: 19-21.

**KASSAS, Mohamed** (1974) National Parks in Arid Regions. *Proc. 2nd World Conf. on National Parks*, 199-208. IUCN, Morges, Switzerland.

**KAYANJA, F.I.B., and DOUGLAS-HAMILTON, I.** (1983) Impact of the Unexpected: a Case History of the Uganda National Parks. *Swara* 6(3): 8-14.

**KEAST, Allen** (1966) Australia and the Pacific Islands: A Natural History. Hamish Hamilton, London.

**KENDALL, Beryl** (1977) A Threat to Wildlife in the Sudan? *Africana* 6(7):17-18, 33.

**KHAN, M.A.R., and AHSAN, M.F.** (1986) The Status of Primates in Bangladesh and a Description of their Forest Habitats. *Primate Conservation* 7:102-109.

**KLINOWSKA, Margaret** (1991) Dolphins, Porpoises and Whales of the World: *The IUCN Red Data Book*. IUCN, Gland, Switzerland, and Cambridge, UK.

**LAMPREY, Hugh F.** (1974) Management of Flora and Fauna in National Parks. In: *Proc. 2nd World Conf. on Nat. Parks*, IUCN, Morges, Switzerland. pp. 237-249.

**LAMPREY, H.F., KRUUK, H., & NORTON-GRIFFITHS, M.** (1971) Research in the Serengeti. *Nature* 230:497-499.

**LEE, P.C., THORNBACK, J., and BENNETT, E.L.** (1988) Threatened Primates of Africa. *The IUCN Red Data Book*. IUCN, Gland, Switzerland, and Cambridge, UK.

**LEGUAT, François** (1708) Voyages et aventures de François Leguat et ses compagnons en deux isles désertes des Indes Orientales (1650-1689). Amsterdam & London.

**LEITCH, William C.** (1990) South America's National Parks: a Visitor's Guide. The Mountaineers, Seattle.

**LEOPOLD, A. Starker** (1959) Wildlife of Mexico: the Game birds and Mammals. University of California Press, Berkeley.

**LEOPOLD, A. Starker** (1969) The Desert. *Life Nature Library*, Time Inc., New York.

**LEOPOLD, A. Starker, and DARLING, F. Fraser** (1953) Wildlife in Alaska: an Ecological Reconnaissance. The Ronald Press Company, New York.

**LIU ZHENHE** *et al* (1987) Field Report on the Hainan Gibbon. *Primate Conservation* 8:49-50.

**LOUW, G.N., and SEELY, M.K.** (1982) Ecology of Desert Organisms. Longman.

**LUCAS, Gren, and SYNGE, Hugh** (eds) (1978) The IUCN Plant Red Data Book. IUCN, Morges, Switzerland.

**LUNA, Mariella Leo** (1987) Primate Conservation in Peru: a Case Study of the Yellow-tailed Woolly Monkey. *Primate Conservation* 8:122-123.

**LYTH, R.E.** (1947) The Migration of Game in the Boma Area. *Sudan Notes and Records* 28:191-192.

**MACKINNON, John, and MACKINNON, Kathy** (1986) Review of the Protected Area System in the Afrotropical Realm. UNEP-IUCN, 2 vols.

**MACKINNON, John, and MACKINNON, Kathy** (1987) Conservation Status of the Primates of the Indo-Chinese Subregion. *Primate Conservation* 8:187-195.

**MACKINNON, Kathy** (1987) Conservation Status of Primates in Malesia, with special reference to Indonesia. *Primate Conservation* 8:175-183.

**MACKINNON, John R., and STUART, Simon N.** (1989) The Kouprey: an Action Plan for its Conservation. IUCN, Gland, Switzerland.

**MACLEAN, Fitzroy** (1974) To the Back of Beyond: An Illustrated Companion to Central Asia and Mongolia. Jonathan Cape, London.

**MALLINSON, Jeremy J.C.** (1987) International Efforts to Secure a Viable Population of the Golden-headed Lion Tamarin. *Primate Conservation* 8:124-125.

**MALTBY, Edward** (1986) Waterlogged Wealth: Why Waste the World's Wet Places? Earthscan, London.

**MARGALEF, Ramon** (ed.) (1985) Western Mediterranean. *Key Environments Series*, Pergamon Press, Oxford.

**MARKHAM, Clements R.** (1967) The Incas of Peru. Editorial Grafica Pacific Press S.A., Lima, Peru.

**MAY, John** (1989) The Greenpeace Book of Antarctica: a New View of the Seventh Continent. Dorling Kindersley, London.

**McGINNIES, William G., GOLDMAN, Bram J., and PAYLORE, Patricia** (1968) Deserts of the World. University of Arizona Press, Tucson.

**MEIER, Bernhard, and RUMPLER, Yves** (1987) Preliminary Survey of *Hapalemur simus* and of a New Species of *Hapalemur* in Eastern Betsileo, Madagascar. *Primate Conservation* 8:40-43.

**MEINERTZHAGEN, R.** (1912) On the Birds of Mauritius. *Ibis* 6:52-108.

**MERRON, Glenn S.** (1989) The Annual Okavango Catfish Run. *African Wildlife* 43(6):302-305.

**MILLER, Norman N.** (1982) Wildlife - Wild Death. *Swara* 5(3):8-18.

**MILLS, M.G.L.** (1990) Kalahari Hyaenas: Comparative Behavioural Ecology of Two Species. Unwin Hyman, London.

**MIRIMANIAN, Kh. P.** (1972) Mountain National Parks and Nature Reserves. *Proc. 2nd World Conf. on National Parks*. IUCN, Morges, Switzerland. 209-219.

**MITCHELL, A.H.** (1984) Primate Conservation and the Gunung Leuser National Park, Indonesia. IUCN/SSC Primate Specialist Group Newsletter No. 4:40-41.

**MITTERMEIER, Russel A.**, *et al* (1987) Current Distribution of the Murique in the Atlantic Forest Region of Eastern Brazil. *Primate Conservation* 8:143-149.

**MITTERMEIER, R.A.** (1984) The World's Endangered Primates: an introduction and a case study - the monkeys of Brazil's Atlantic Forests. In: *Primates and the Tropical Forest*, R.A. Mittermeier and M.J. Plotkin (eds), World Wildlife Fund/L.S.B. Leakey Foundation, Washington, DC, pp. 11-12.

**MITTERMEIER, R.A.**, *et al* (1982) Conservation of Primates in the Atlantic Forest Region of Eastern Brazil. *Int. Zoo Yb.* 2-17.

**MITTERMEIER, R.A.**, *et al* (1990) Conservation in the Pantanal of Brazil. *Oryx* 24(2):103-112.

**MONDOR, C., and KUN, S.** (1982) The Long Struggle to Protect Canada's Vanishing Prairie. *Ambio* 2:143-145.

**MOORE, I.** (1966) Grass and Grasslands. Collins, London.

**MOREAU, R.E.** (1966) The Bird Faunas of Africa and its islands. Academic Press, London & New York.

**MORRIS, Desmond** (1965) The Mammals: a Guide to the Living Species. Hodder & Stoughton, London.

**MOUTOU, François** (1984) Wildlife on Réunion. *Oryx* 18(3):160-162.

**NEWBY, J.E.** (1984) Large Mammals. In: *Sahara Desert* (ed. J.L. Cloudsley-Thompson). *Key Environments Series*. Pergamon Press, Oxford.

**NEWBY, John** (1988) Aridland Wildlife in Decline: the Case of the Scimitar-horned Oryx. In: *Conservation and Biology of Desert Antelopes* (ed. A. Dixon and D. Jones). Christopher Helm, London.

**NEWBY, John E.** (1992) Parks for People - a Case Study from the Aïr Mountains of Niger. *Oryx* 26(1):19-28.

**NEWMAN, J.C. and R.W.** (1969) Land Use and Present Conditions. In: *Arid Lands of Australia*, R.O. Slatyer and R.A. Perry, (eds) 105-132. Australian National University Press, Canberra.

**NEWMARK, William D.** (ed) (1991) The Conservation of Mount Kilimanjaro. IUCN, Gland, Switzerland, and Cambridge.

**NICHOLSON, E.M.** (1970) The Environmental Revolution. Hodder & Stoughton, London.

**NICHOLSON, E. Max** (1974) What is Wrong with the National Park Movement? In: *Proc. 2nd World Conf. on National Parks*, 32-38. IUCN, Morges, Switzerland.

**OATES, J.F.** (1985) Action Plan for African Primate Conservation: 1986-90. IUCN/SSC, Gland, Switzerland.

**O'CONNOR, Sheila**, *et al* (1986) Conservation Program for the Andohahela Reserve, Madagascar. *Primate Conservation* 7:48-52.

**OLMOS, Fábio** (1992) Serra da Capivara National Park and the Conservation of North-eastern Brazil's Caatinga. *Oryx* 26(3):142-146.

**OWEN-SMITH, Garth** (1986) The Kaokoveld, South West Africa/Namibia's Threatened Wilderness. *African Wildlife* 40(3):104-115.

**PAIN, Stephanie** (1990) Last Days of the Old Night Bird. *New Scientist* 126(1721):37-41.

**PEARSALL, W.H.** (1950) Mountains and Moorlands. New Naturalist Series. Collins, London.

**PERRIN, William F.**, *et al* (1989) Biology and Conservation of the River Dolphins. Occasional Papers of the IUCN Species Survival Commission (SSC), No. 3. IUCN, Gland, Switzerland.

**PERRIN, W.F.** (1989) Dolphins, Porpoises and Whales: an Action Plan for the Conservation of Biological Diversity: 1988-1992. 2nd Edition. IUCN, Gland, Switzerland.

**PERRY, R.** (ed.) (1984) Galapagos. *Key Environments Series*. Pergamon Press, Oxford.

**PETROV, M.P.** (1976) Deserts of the World. Halsted Press, New York.

**PETTET, A.** (1984) Migratory Birds. In: *Sahara Desert* (ed. J.L. Cloudsley-Thompson). *Key Environments Series*. Pergamon Press, Oxford.

**PETTER, J.-J., and SIMON, N.M.** (1967) Summary Report of the 1967 IUCN Mission to Madagascar. Cyclostyled.

**PFEFFER, Pierre** (1968) Asia: A Natural History. Hamish Hamilton, London.

**PODUSCHKA, Walter, and RICHARD, Bernard** (1986) The Pyrenean Desman - an Endangered Insectivore. *Oryx* 20(4):230-232.

**POLLOCK, J.** (1975) Field Observations on *Indri indri*: a preliminary report. In: *Lemur Biology* (I. Tattersall and R.W. Sussman, eds). Plenum Press, New York, pp 287-311.

**POLLOCK, Jonathan** (1986) A Note on the Ecology and Behavior of *Hapalemur griseus*. *Primate Conservation* 7:97-101.

**POPOV, G.B., WOOD, T.G., and HARRIS, M.J.** (1984) Insect Pests of the Sahara. In: *Sahara Desert* (ed. J.L. Cloudsley-Thompson). *Key Environments Series*. Pergamon Press, Oxford.

**POVILITIS, Anthony** (1983) The Huemul in Chile: National Symbol in Jeopardy? *Oryx* 17(1):34-40.

**PREJEVALSKY, N.** (1879) From Kulja, across the Tian Shan to Lob-Nor. Sampson Low, Marston, Searle and Rivington, London.

**RAXWORTHY, Chris J.** (1986) The Lemurs of Zahamena Reserve. *Primate Conservation* 7:46-48.

**REDFORD, Kent H.** (1985) Emas National Park and the Plight of the Brazilian Cerrados. *Oryx* 19(4):210-214.

**RYDER, Oliver A.** (1988) Przewalski's Horse - Putting the Wild Horse Back in the Wild. *Oryx* 22(3):154-157.

**SABOUREAU, Pierre** (1963) The Natural Resources of Madagascar. *IUCN Publ.*, N.S. No. 12:235-320.

**SANTOS, Ilmar B.**, *et al* (1987) The Distribution and Conservation Status of Primates in Southern Bahia, Brazil. *Primate Conservation* 8:126-142.

**SCHÄFER, Ernst** (1937) Zur Kenntniss des Kiang (*Equus kiang* Moorcroft). *Zool. Gart.* 9(3-4):122-139.

**SCHALLER, George B.** (1972) The Serengeti Lion: a Study of Predator-Prey Relationships. University of Chicago Press, Chicago & London.

**SCHALLER, George B.** (1973) Serengeti: a Kingdom of Predators. Collins, London.

**SCHALLER, George B.** (1991) The New Chang Tang Wildlife Reserve in Northwestern Tibet. Mimeographed.

**SCHALLER, George B.**, *et al* (1985) The Giant Pandas of Wolong. University of Chicago Press, Chicago.

**SCHALLER, George B.**, *et al* (1988) The Snow Leopard in Xinjiang, China. *Oryx* 22:197-204.

**SCHALLER, George B.**, *et al* (1990) Javan Rhinoceros in Vietnam. *Oryx* 24:77-80.

**SCHALLER, George B., Ren Junrang and Qiu Mingjiang** (1991) Observations on the Tibetan antelope (*Pantholops hodgsoni*). *Appl. Anim. Behav. Sci.*, 29:361-378.

**SCHALLER, George B., and SIMON, Noel M.** (1970) The Endangered Large Mammals of South Asia. *IUCN Publ.* N.S. No.18:11-23.

**SCHREIBER, A., WIRTH, R., RIFFEL, M., and ROMPAEY, H. van** (1989) Weasels, Civets, Mongooses, and Their Relatives: an Action Plan for the

Conservation of Mustelids and Viverrids. IUCN/SSC Mustelid and Viverrid Specialist Group, Gland, Switzerland.

SCOTT, D.A. (ed.) (1989) A Directory of Asian Wetlands. IUCN, Gland, Switzerland, and Cambridge, UK.

SHANTZ, H.L. (1954) The Place of Grasslands in the Earth's Cover of Vegetation. *Ecology*, 35, 2:143-145.

SHORTRIDGE, G.C. (1934) The Mammals of South West Africa: a Biological Account of the Forms Occurring in that Region. 2 vols. William Heinemann Ltd, London.

SIMON, Noel (1962) Between the Sunlight and the Thunder: the Wild Life of Kenya. Collins, London.

SIMON, Noel (1966) Red Data Book, vol.1, Mammalia. IUCN/SSC, Morges, Switzerland.

SIMONS, Hilary J., and LINDSAY, N.B.D. (1987) Survey Work on Ruffed Lemurs (*Varecia variegata*) and Other Primates in the Northeastern Rain Forests of Madagascar. *Primate Conservation* 8:88-91.

SIMPSON, George C. (1953) The Major Features of Evolution. Columbia University Press.

SLATYER, R.O. and PERRY, R.A. (eds) (1969) Arid Lands of Australia. Australian National University Press, Canberra.

SMITH, G. (1984) Climate. In: *Sahara Desert* (ed. J.L. Cloudsley-Thompson). *Key Environments Series*. Pergamon Press, Oxford.

SOWERBY, Arthur de Carle (1924) Approaching Desert Conditions in North China. *China J. Sci. Arts*, 2(3):199-203.

SOWERBY, Arthur de Carle (1924) Forestry in China. *China J. Sci. Arts* 2(4):299-303.

SOWERBY, Arthur de Carle (1924) Famine, Floods and Folly. *China J. Sci. Arts* 2(5):395-399.

SPRAGUE, David S. (1986) Conservation of the Monkeys and Forests of Yakushima, Japan. *Primate Conservation* 7:55-57.

STEWART, John Massey (1991) Lake Baikal: on the brink? IUCN East European Programme. Gland, Switzerland.

St. GEORGE, G. (1974) Soviet Deserts and Mountains. Time-Life Books, Amsterdam.

STODDART, D.R. (1968) The Conservation of Aldabra. *Geogr. J.* 134(4):471-486.

STODDART, D.R. (1968) Catastrophic Human Interference with Coral Atoll Ecosystems. *Geography* 53, January 1968.

STODDART, D.R., and WRIGHT, C.A. (1967) Ecology of Aldabra Atoll. *Nature* 213:1174-1177.

STONE, Peter B. (ed.) (1992) The State of the World's Mountains: a Global Report. Zed Books Ltd., London and New Jersey.

STRAHM, Wendy (1983) Rodrigues: can its Flora be Saved? *Oryx* 17:122-125.

SUSSMAN, Robert W., and RICHARD, Alison F. (1986) Lemur Conservation in Madagascar: the Status of Lemurs in the South. *Primate Conservation* 7:86-92.

TATTERSALL, I. (1982) The Primates of Madagascar. Columbia University, New York.

TEMPLE, Stanley A. (1974) Wildlife in Mauritius Today. *Oryx* 12:584-590.

TENAZA, Richard (1987) The Status of Primates and their Habitats in the Pagai Islands, Indonesia. *Primate Conservation* 8:104-110.

THOMSON, Sir A. Landsborough (ed.) (1964) A New Dictionary of Birds. Nelson, London, and McGraw-Hill, New York.

THORBJARNARSON, John (1992) Crocodiles: an Action Plan for their Conservation. IUCN, Gland, Switzerland.

THORNBACK, Jane, and JENKINS, Martin (eds) (1982) The IUCN Mammal Red Data Book. Part 1: Threatened Mammalian Taxa of the Americas and the Australasian Zoogeographic Region. IUCN, Gland, Switzerland, and Cambridge, UK.

TINLEY, K.L. (1971) Etosha and the Kaokoveld. *African Wildlife* (Suppl.) 25(1).

TRAIN, Russell E. (1972) An Idea whose Time has Come: the World Heritage Trust, a World Need and a World Opportunity. *Proc. 2nd World Conf. on National Parks*: 377-381. IUCN, Morges, Switzerland.

TUTIN, Caroline E.G., and FERNANDEZ, Michel (1987) Gabon: a Fragile Sanctuary. *Primate Conservation* 8:160-161.

UNEP/IUCN (1988) Coral Reefs of the World. Vol.1: Atlantic and Eastern Pacific; Vol. 2: Indian Ocean, Red Sea and Gulf; Vol. 3: Central and Western Pacific. UNEP Regional Seas Directories and Bibliographies. IUCN, Gland, Switzerland, and Cambridge, UK/UNEP, Nairobi, Kenya.

VAN ORSDOL, K.G. (1986) Agricultural Encroachment in Uganda's Kibale Forest. *Oryx* 20:115-117.

VAUGHAN, R.E. (1968) Mauritius and Rodriguez. In: *Conservation of Vegetation in Africa South of the Sahara* (eds O. Hedberg and I. Hedberg). *Acta Phytogeographica Suecica* 54:265-272.

VAURIE, Charles (1972) Tibet and its Birds. H.F. & G. Witherby Ltd., London.

VEDDER, Amy (1987) Report from the Gorilla Advisory Committee on the status of *Gorilla gorilla*. *Primate Conservation* 8:75-81.

VERSCHUREN, Jacques (1985) Mauritania: its Wildlife and a Coastal Park. *Oryx* 19(4):221-224.

VERSCHUREN, Jacques, HEYMANS, Jean-Claude, and DELVINGT, Willy (1989) Conservation in Benin - with the Help of the European Economic Community. *Oryx* 23(1):22-26.

VESEY-FITZGERALD, D. (1971) Fire and Animal Impact on Vegetation in Tanzania National Parks. *Proc. Annual Tall Timbers Fire Ecology Conference*.

VESEY-FITZGERALD, D. (1973) East African Grasslands. East African Publishing House, Nairobi.

VILJOEN, S. (1988) The Desert-dwelling Elephant - hardy survivor. *African Wildlife* 42(2):111-115.

VINSON, J. (1950) Round Island and Snake Island. *Proc. R. Soc. Arts Sci. Maurit.* 1:32-52.

WACE, N.M. (1969) The Discovery, Exploitation and Settlement of the Tristan da Cunha Islands. *Proc. S. Aust. Brch R. Geogr. Soc. Aust.* 70:11-40.

WALKER, B.H. (ed.) (1979) Management of Semi-arid Ecosystems. Elsevier, Amsterdam.

WARREN, A. (1984) The Problems of Desertification. In: *Sahara Desert* (ed. J.L. Cloudsley-Thompson). *Key Environments Series*, Pergamon Press, Oxford.

WASSER, Samuel K. (1987) The Values and Problems of Wildlife Conservation in Tanzania. *Primate Conservation* 8:167-168.

WIETERSHEIM, Anton von (1988) Game Farming - the Economics of Conservation. *African Wildlife* 42(2):69-75.

WELLS, Susan M., PYLE, Robert M., and COLLINS, Mark (1983) The IUCN Invertebrate Red Data Book. IUCN, Gland, Switzerland, and Cambridge, UK.

WESTERN, David, and VIGNE, Lucy (1985) The Deteriorating Status of the African Rhinos. *Oryx* 19(4):215-220.

WHALLEY, R.C.R. (1932) Southern Sudan Game and its Habits. *Sudan Notes and Records* 15:261-267.

WILLIAMSON, Douglas, and WILLIAMSON, Jane (1984) Botswana's Fences and the Depletion of Kalahari Wildlife. *Oryx* 18(4):218-222.

WILSON, Jane M. (1987) The Crocodile Caves of Ankarana, Madagascar. *Oryx* 21(1):43-47.

WILSON, Jane M., STEWART, Paul D., and FOWLER, Simon V. (1988) Ankarana - a Rediscovered Nature Reserve in Northern Madagascar. *Oryx* 22(3):163-171.

WORLD CONSERVATION MONITORING CENTRE (1992) Global Diversity: Status of the Earth's Living Resources. Chapman & Hall, London.

# INDEX

An asterisk after an entry indicates a photograph.

9 11/62